The Palgrave Handbook of Blue Heritage

Rosabelle Boswell · David O'Kane ·
Jeremy Hills
Editors

The Palgrave Handbook of Blue Heritage

palgrave
macmillan

Editors
Rosabelle Boswell
Nelson Mandela University
Port Elizabeth, South Africa

David O'Kane
Max Planck Institute for Social Anthropology
Halle, Germany

Jeremy Hills
University of the South Pacific
Suva, Fiji

ISBN 978-3-030-99346-7   ISBN 978-3-030-99347-4 (eBook)
https://doi.org/10.1007/978-3-030-99347-4

© The Editor(s) (if applicable) and The Author(s), under exclusive license to Springer Nature Switzerland AG 2022
This work is subject to copyright. All rights are solely and exclusively licensed by the Publisher, whether the whole or part of the material is concerned, specifically the rights of translation, reprinting, reuse of illustrations, recitation, broadcasting, reproduction on microfilms or in any other physical way, and transmission or information storage and retrieval, electronic adaptation, computer software, or by similar or dissimilar methodology now known or hereafter developed.
The use of general descriptive names, registered names, trademarks, service marks, etc. in this publication does not imply, even in the absence of a specific statement, that such names are exempt from the relevant protective laws and regulations and therefore free for general use.
The publisher, the authors and the editors are safe to assume that the advice and information in this book are believed to be true and accurate at the date of publication. Neither the publisher nor the authors or the editors give a warranty, expressed or implied, with respect to the material contained herein or for any errors or omissions that may have been made. The publisher remains neutral with regard to jurisdictional claims in published maps and institutional affiliations.

Cover credit: Yadid Levy/Alamy Stock Photo

This Palgrave Macmillan imprint is published by the registered company Springer Nature Switzerland AG
The registered company address is: Gewerbestrasse 11, 6330 Cham, Switzerland

To all those committed to saving the planet and to safeguarding our blue heritage.

Acknowledgements

This book developed out of a desire to foreground the contribution of the humanities and social sciences in ocean conservation and management. The book was also partly inspired by earlier research in South Africa and the southwest Indian Ocean region, and via collaboration between researchers in the UKRI funded One Ocean Hub project. The latter includes transdisciplinary research on small scale fishers in Ghana, Fiji, Namibia and South Africa. As the contributors to this book show, human interaction with the oceans and coasts is thoroughly complex. National governments need to do more to advance a more inclusive ocean management strategy and to consider the multifaceted nature of the human–ocean dynamic.

The work of researchers presented in this book also reveals the increasingly transdisciplinary field of ocean management and the intellectual flexibility required to understand diverse disciplinary frameworks and epistemologies in the ocean sciences. We are therefore very grateful to the authors in this book. Many are working in extraordinarily difficult and fluid circumstances but still managed to submit their chapters to this edited book. The added challenge of Covid-19 made their task even more challenging and because of Covid-19, some researchers did not manage to offer a chapter. In the 'field', researchers were also required to engage with communities embedded in highly unequal societies, where primary concerns were for subsistence and health care needs. For doing the difficult work of engaging research participants in local communities, sharing challenging stories, collating data and analysing it and working as part of a team, we thank the authors, the editors, the research assistants and community people across the world, who have made this book possible. Sustainable ocean use and management is an urgent and collaborative effort requiring input from scientists (social and natural), as well as those who do not describe themselves as scholars.

Contents

1	**Introduction: Blue Heritage, Human Development and the Climate Crisis** *Rosabelle Boswell*	1

Part I   Humans and Humanities in Ocean Heritage

2	**The Evolutionary Roots of Blue Heritage** *Curtis Marean*	13
3	**Islands and the Sea** *Godfrey Baldacchino*	25
4	**Oceanic Humanities for Blue Heritage** *Isabel Hofmeyr and Charne Lavery*	31
5	**World and Marine Heritages** *Rosabelle Boswell*	41

Part II   Historical, Cultural and Literary Perspectives

6	**Evolving Hegemonies of Blue Heritage: From Ancient Greece to Today** *Evanthie Michalena and Jeremy Hills*	51

x   Contents

7   Tales of Ocean, Migration and Memory in Ancestral
    Homespaces in Goa   75
    Pedro Pombo

8   Salt, Boats and Customs: Maritime Princely States
    in Western India, 1910–1932   93
    Varsha Patel

9   Navigating Environment, History, and Archaeology
    in Portsmouth Island, USA   109
    Lynn Harris

10 Multi-Spirited Waters in Lynton Burger's She Down There   141
    Confidence Joseph

11 Women Rising, Women Diving in Vanuatu and Australia   161
    Cobi Calyx

12 Speaking with the Sea: Divination and Identity in South
    Africa   181
    Dominique Santos and Rev Thebe Shale

Part III   Environmental, Legal and Political Realities

13 Selo! Oral Accounts of Seychelles' Maritime Culture   203
    Penda Choppy

14 Waking up to Wakashio: Marine and Human Disaster
    in Mauritius   225
    Rosabelle Boswell

15 Blue Heritage Among Fishermen of Mafia Island,
    Tanzania   251
    Mariam de Haan

16 Blue Heritage in the Blue Pacific   273
    *Jeremy Hills, Kevin Chand, Mimi George, Elise Huffer,
    Jens Kruger, Jale Samuwai, Katy Soapi, and Anita Smith*

17 Policy, Epistemology, and the Challenges of Inclusive
    Oceans Management in Sierra Leone   303
    David O'Kane

18	**Navigating a Sea of Laws: Small-Scale Fishing Communities and Customary Rights in Ghana and South Africa**	325
	Anthea Christoffels-DuPlessis, Bola Erinosho, Laura Major, Elisa Morgera, Jackie Sunde, and Saskia Vermeylen	
19	**Narratives of Non-Compliance in "Tuesday Non-Fishing Day" in Ghana**	357
	John Windie Ansah, Georgina Yaa Oduro, and David Wilson	
20	***Lalela uLwandle*: An Experiment in Plural Governance Discussions**	383
	Kira Erwin, Taryn Pereira, Dylan McGarry, and Neil Coppen	
21	**'Other' Social Consequences of Marine Protection in Tsitsikamma, South Africa**	411
	Jessica Leigh Thornton and Ryan Pillay	
22	**Sustaining the Underwater Cultural Heritage**	427
	Elena Perez-Alvaro	
23	**Underwater Cultural Heritage, Poverty and Corruption in Mozambique**	451
	Anezia Asse	
Afterword		477
Index		485

Notes on Contributors

Dr. John Windie Ansah is Senior Lecturer in political economy at the Department of Sociology and Anthropology of the University of Cape Coast. He was appointed a faculty member of the Institute of Global Law and Policy (IGLP), Harvard Law School, in 2015 where he lectured in the area of 'Society of Global Corporations' in a writing workshop held in Qatar. He served as a co-consultant to the International Labour Organization on projects which produced background data on employment in the agriculture and agro-processing sectors as well as the infrastructure sector in 2017 and as a co-consultant for the Ministry of Energy, Ghana, on a Charcoal Value Chain Analysis project. He has also served as a resource person to a number of workshops on Energy Justice (2017) and Sustainable Livelihoods (2018) organised by Kultur Studier, in The Netherlands. He is, currently, a co-investigator in the One Ocean Hub Project aimed at promoting sustainable livelihoods and ocean governance.

Anezia Asse is a maritime archaeologist from Mozambique, with experience in documentation and preservation of underwater shipwrecks as well as an experience in heritage management. She has a master's degree in Heritage from University of Witwatersrand under WiSER (Wits Institute for Science and Economic Research) and Oceanic Humanities for the Global South. Her research interests include maritime heritage (tangible and intangible) and community based research. She is currently a Ph.D. candidate in the

NRF Research Chair in Ocean Cultures and Heritage Programme at Nelson Mandela University.

Godfrey Baldacchino is Professor of Sociology, Department of Sociology at the University of Malta, Malta. He served as an Island Studies Teaching Fellow, UNESCO co-Chair and Canada Research Chair in Island Studies at the University of Prince Edward Island (UPEI), Canada, between 2003 and 2020. He is founding Executive Editor of *Island Studies Journal* (ISSN: 1715-2593), and since 2018 founding Executive Editor of *Small States & Territories* journal (ISSN: 2616-8006). He served as Visiting Professor of Island Tourism at the Universita' di Corsica Pascal Paoli, France (2012–2015). He was Member and Chair of the Malta Board of Cooperatives (1994–2003) and core member of the Malta-European Union Steering & Action Committee (MEUSAC). In 2008–2010, he was Vice-President of the Prince Edward Island Association for Newcomers to Canada. In 2014, he was elected President of the International Small Islands Studies Association (ISISA). In June 2015, he was elected Chair of the Scientific Board of RETI, the global excellence network of island universities. In 2021, he was appointed (thematic) Malta Ambassador for islands and small states. He served as Pro-Rector for International Development and Quality Assurance (2016–2021) during the first Rectorate of Professor Alfred Vella at the University of Malta (Source of Biography, University of Malta).

Rosabelle Boswell is Professor of anthropology and Research Chair of Ocean Cultures and Heritage funded by the Department of Science and Innovation, National Research Foundation and Nelson Mandela University, South Africa. A graduate of Vrije Universiteit, Amsterdam, her work involves transdisciplinary research on coastal and island cultures and heritage, including a focus on oceans conservation in the selected field sites of coastal South Africa, Kenya, Namibia and Mozambique. Rose has done anthropological field research in South Africa, Mauritius, Zanzibar, Madagascar and Seychelles. She is author of *Le Malaise Creole: Ethnic Identity in Mauritius* (Oxford, Berghahn, 2006), *Representing Heritage in Zanzibar and Madagascar* (Addis Ababa, Eclipse, 2008); Challenges *to Identifying and* Managing *Intangible Cultural Heritage in Mauritius, Zanzibar and Seychelles* (Dakar, CODESRIA, 2011) and *Postcolonial African Anthropologies* (co-edited with F. Nyamnjoh, Pretoria, HSRC Press, 2016). Her two poetry books are published by RPCIG, Bamenda and New York. They are entitled: *Things Left Unsaid* (2019) and *Pandemix* (2020). A forthcoming poetry book *Between Worlds* (RPCIG: Cameroon) examines issues of ocean sustainability and anthropogenic impacts on oceans and coasts.

Dr. Cobi Calyx is a researcher for Australia's Centre for Social Impact through UNSW and the Climate and Sustainability Policy Research group at Flinders University, on Kaurna country in Adelaide where she lives with family. She has a Ph.D. in science communication as well as postgraduate education and health promotion qualifications. Dr. Calyx has expertise in environmental governance, science communication, health promotion and disaster response, with experience of Australian Aid-funded projects in Asia and the Pacific as well as the UN in Geneva. Her contribution in this book relates to reflecting on Mauritian Creole ancestry among European heritage and Australian citizenship.

Kevin Chand is a program officer at the High Seas Alliance. Kevin has an LLM from Stanford Law School, a Masters' in Environmental Management from the University of Queensland, and an LLB from the University of the South Pacific. Kevin taught International Environmental Law, Oceans and Illegal, Unreported, and Unregulated (IUU) fishing policy at Stanford University. He has also held numerous roles at Stanford including Lecturer at the Program in International Relations, Design Thinking Fellow at the Law and Policy Lab at Stanford Law, and Ocean Design Fellow at the Stanford Center for Ocean Solutions. Kevin has previously worked for the IUCN on climate change mitigation efforts in Fiji related to the deployment of REDD+ projects and entered private practice as a solicitor specialising in environmental and commercial law. In private practice, his work revolved around supporting environmental conservation groups and foundations He has provided ocean policy advice to several regional organisations in the Pacific. Kevin is committed to supporting negotiations towards the Biodiversity Beyond National Jurisdiction (BBNJ) Treaty.

Dr. Penda Choppy is the Director of the Creole Language and Culture Research Institute at the University of Seychelles. She was the Director of the Creole Institute of Seychelles from 1999 to 2014 when she became its CEO. She moved to the University of Seychelles in 2016. Penda studied English at Leeds University and has completed a research degree at Masters' level at the University of Birmingham, UK, in 2018. She is currently doing a doctoral research degree with the Islands and Small States Institute, University of Malta. Her main research interests are the oral and written creole traditions of the Western Indian Ocean, in the postcolonial context.

Neil Coppen is a renowned and prolific South African storyteller, working across a variety of mediums and disciplines including theatre, journalism and film. Coppen has won several major awards for his writing, design and direction work including Naledi's, Fiesta and Kanna Awards, the Standard Bank

Young Artist Award for Drama 2011 and the 2019 Olive Schreiner Prize for Drama. Seminal to Neil's work across a range of cultures and communities in South Africa, is a social-justice, theatre-making methodology titled *Empatheatre* which Coppen co-founded. *Empatheatre* has been responsible for launching several ground-breaking theatre projects over the last decade in South Africa including *Soil & Ash* (focusing on rural communities facing pressure from coal-mining companies), *Ulwembu* (street-level Drug addiction), *The Last Country* (female migration stories), *Boxes* (Urban land justice inequalities in the city of Cape Town) and *Lalela ulwandle* (an international project supporting sustainable transformative governance of our oceans).

Mariam de Haan is currently working as the African collections curator at the British Library. She obtained her Research Masters in African Studies at Leiden University in 2020. She has researched extensively in East Africa, with a focus on Tanzanian culture, economics and politics. Her future research aims to examine manuscripts and items in Kiswahili that are held by the British Library.

Bola Erinosho is at the University of Cape Coast in Ghana where she teaches and researches in the areas of environmental and natural resources law and policy. She is a qualified barrister and solicitor. She obtained an LLM and Ph.D. in International Environmental Law from the Universities of Nottingham and Sheffield, in the UK, respectively. She is a co-investigator on the GCRF funded One Ocean Hub where she leads a work package on ocean legal pluralism.

Dr. Kira Erwin is an urban sociologist, and senior researcher, at the Urban Futures Centre at the Durban University of Technology in Durban, South Africa. Her research and publications focus largely on race, racialisation, racism and anti-racism work within the urban context. Her past projects explore narratives of home and belonging within the context of migration, gender and inclusion; as well as state delivered housing projects in the city. She is currently working on two environmental justice projects in Durban. The first, Lalela uLwandle is a collaboration with a team of researchers and civil society activists to think through how people's economic, spiritual, scientific and symbolic meanings of the sea should be part of ocean governance decisions. The second is a Zero-Waste project working with informal workers. She uses creative participatory methods and collaborates with colleagues in various creative fields to produce forms of public storytelling that extend research beyond the walls of academia.

Mimi George has a Ph.D. and is a cultural anthropologist and sailor who supports training youth to apply ancestral voyaging knowledge to current problems, including biodiversity loss and climate change. Mimi has documented voyaging traditions of Austronesian people in Papua New Guinea; a small, mixed-gender crew wintering a sailboat in Antarctic sea-ice, sea-hunter and reindeer herder networks and migrations across Bering Straits, and Polynesian islanders in SE Solomon Islands who build vessels and navigate by ancient designs, materials and methods, including weather modification and calling for ancestral lights that show the way to land. Mimi worked in response to requests for help by the people she worked with. She describes prominent roles of women and children in voyaging cultures, and how revival of ancestral voyaging networks creates sustainable and resilient communities and ecologies.

Lynn Harris is Professor of history at East Carolina University. Prof. Harris has a background in nautical and terrestrial archaeology and maritime history. She teaches courses in underwater archaeology methods, maritime material culture, maritime landscapes, watercraft history, coastal cultural resource management, African and Caribbean maritime history and archaeology. Teaching assignments have included directing summer abroad study programs and international field schools in Namibia, South Africa, Costa Rica and Dominican Republic. Harris also engages in research projects on the south eastern seaboard that integrate student researchers. Currently, two grant projects are underway. One to research and expand the African American maritime history on Portsmouth Island NC and another to conduct rapid site surveys of coastal heritage at risk using a variety of technologies and tools on diverse sites in NC, SC and GA. Each case study site has state or national historic significance, conservation management challenges and serves as an intellectual platform to segue between preservation of an historic icon and research questions. For the past four years, Harris has led a collaborative project with Costa Rican partners to study two shipwrecks and maritime legacies in Cahuita located on the Caribbean coast of Costa Rica. Most recently, Harris authored and edited two books, in addition to co-authoring articles published in *International Journal of Nautical Archaeology*, *Journal of Maritime Archaeology*, and *Coriolis: Interdisciplinary Journal of Maritime Studies*. Harris currently serves as faculty in the Program in Maritime Studies, Atlantic World Program, and Coastal Resources Management Doctoral Degree Program. She has served for several years as a member of the Advisory Council in Maritime Archaeology and is a Nautical Archaeology Society instructor who offered public workshops for diving stewardship groups locally and internationally with her colleagues. Harris welcomes graduate students who are interested in underwater archaeology, interdisciplinary research and applied maritime history projects (Source of biography East Carolina University, US).

Prof. Jeremy Hills is a Chartered Environmentalist with over 25 years of research and consultancy on the coastal and ocean. He has a M.Sc. in Ecology from Durham University (UK) and a Ph.D. focused on environmental modelling from Glasgow University (UK). Jeremy has worked in over 50 countries across the world. He has worked for many actors including international institutions such as the EU, UNESCO, UNDP, UNEP, UNESCAP, World Bank, Asian Development Bank, bilateral aid organisations and a range of other Non-Governmental (NGO) and commercial organisations. He has developed national and regional marine planning and policy and was a lead author on the Global Environment Outlook-6 (UNEP). He was Director of the Institute of Marine Resources at The University of the South Pacific and is presently a Professor at the same institution in the Office of the Deputy Vice-Chancellor (Research, Innovation & International). His research is on ocean policy and governance in developing countries, with a particular focus on development assistance, financing and blue economy.

Isabel Hofmeyr is Professor Emeritus at the University of the Witwatersrand and Global Distinguished Professor at NYU. She has published extensively on the Indian Ocean and oceanic studies. Her most recent book is *Dockside Reading: Hydrocolonialism and the Custom House* (2022). With Charne Lavery, she runs the Oceanic Humanities for the Global South project with partners from India, Mozambique, Jamaica and Barbados.

Elise Huffer is a culture, heritage and development consultant in Fiji and the Pacific Islands. For 10 years, Elise was the Culture Adviser at the Pacific Community (SPC, 2008–2017) providing technical advice on cultural policy, cultural industries and heritage to 22 Pacific Island countries. Prior to joining SPC, Elise was Associate Professor at the University of the South Pacific (USP) where she designed and taught the Pacific Studies graduate programme. She is Adjunct Associate Professor at the Oceania Centre for Arts, Culture and Pacific Studies at the USP and the Vice-Chair Oceania of the IUCN Commission on Environmental, Economic and Social Policy. She is a Fiji citizen and has been living in the Pacific Islands region for 35 years.

Confidence Joseph is an Oceanic Humanities for the Global South Ph.D. fellow at WISER and an African Literature doctoral candidate at the University of the Witwatersrand. Her current research is on the representation of water and water spirits in Southern African Literature. Her other research interests include Zimbabwean postcolonial literature and gender studies.

Taryn Pereira is an activist researcher and facilitator, with a background in urban water justice and social learning to support civil society networks. Taryn is currently a co-investigator on the One Ocean Hub, focusing on the building of solidarity between academic researchers and community-based activists working across disciplines, sectors and knowledge systems towards coastal justice. As a researcher and co-facilitator on the Lalela ulwandle project Taryn experienced the incredibly powerful role that Empatheatre can play in bridging many of the intractable dividing lines in our society—lines of race, class, gender, literacy, language—and in so doing to set the stage for the building of equitable relationships between different groups of people, based on critical, reflexive solidarity.

Jens Kruger is the Acting Deputy Director for the Ocean and Maritime Programme of the Pacific Community (SPC). Jens is a graduate of the University of the South Pacific and completed an MSc at the University of Waikato, New Zealand. He has over 20 years of experience leading multi-disciplinary applied research projects and has a particular interest in the use of marine science and technology in managing the development challenges of Pacific Small Island Developing States. Jens has worked in over 20 countries including several years as a seafarer on marine survey vessels. This work in the industry and with intergovernmental organisations has resulted in more than 80 publications including technical reports, maps, scientific papers and book chapters. Jens has shaped the Pacific regional approach to maritime boundaries for many years and is passionate about collaborative approaches to ocean governance.

Dr. Charne Lavery is Lecturer in the Department of English at the University of Pretoria and Research Fellow on the Oceanic Humanities for the Global South project (www.oceanichumanities.com) based at WISER, University of the Witwatersrand, South Africa. She explores literary and cultural representations of the deep ocean, the Indian Ocean, the Southern Ocean and the Antarctic seas, researching ocean writing of the global South in a time of environmental change. She completed her DPhil in English at the University of Oxford as a Rhodes Scholar, and her B.A. at the University of Cape Town. She is a South African Humanities and Social Sciences delegate to the International Scientific Committee on Antarctic Research (SCAR), co-editor of the Palgrave series *Maritime Literature and Culture*, and a board member of the journal *Global Nineteenth-Century Studies*. Her monograph *Writing Ocean Worlds: Indian Ocean Fiction in English* is forthcoming from Palgrave.

Dr. Laura Major is an anthropologist with international field research experience and expertise in environmental, political and legal anthropology, and the study of material cultures. As a research fellow with the GCRF One Ocean Hub based at the University of Strathclyde, Glasgow, UK, she developed collaborative transdisciplinary work with researchers based in the Caribbean, Ghana and South Africa.

Curtis Marean is the Foundation Professor of School of Human Evolution and Social Change, Arizona State University, Honorary Professor, Nelson Mandela University. Dr. Marean's research interests focus on the origins of modern humans, the prehistory of Africa, the study of animal bones from archaeological sites, and climates and environments of the past. In the area of the origins of modern humans, he is particularly interested in questions about foraging strategies and the evolution of modern human behaviour. Dr. Marean has a special interest in human occupation of grassland and coastal ecosystems. Dr. Marean conducts a variety of studies using zooarchaeology, the study of animal bones, and taphonomy, the study of how bones become fossils. He also is a dedicated field researcher and has conducted fieldwork in Kenya, Tanzania and Somalia, and since 1991 has focused his field efforts in coastal South Africa. He is the principal investigator for the South African Coast Paleoclimate, Paleoenvironment, Paleoecology, Paleoanthropology (SACP4) project based around Mossel Bay in South Africa at the field locality of Pinnacle Point. This large international project, funded by the National Science Foundation and the Hyde Family Foundation, employs a trans-disciplinary approach to modern human origins, climate and environment. Under his directorship, Pinnacle Point has become one of the world's most important localities for the study of modern human origins.

Dr. Dylan McGarry is an educational sociologist and artist from Durban, South Africa. He is a senior researcher at the Environmental Learning Research Centre (ELRC) at the University currently known as Rhodes. As well as the South African Director of the Global One Ocean Hub research network. Dylan is the co-founder of Empatheatre, and a passionate artist and storyteller. He explores practice-based research into connective aesthetics, transgressive social learning, decolonisation, queer-eco pedagogy, immersive empathy and socio-ecological development in South Africa. His artwork and social praxis (which is closely related to his research) is particularly focused on empathy, and he primarily works with imagination, listening and intuition as actual sculptural materials in social settings to offer new ways to encourage personal, relational and collective agency.

Evanthie Michalena has specialised in green growth and sustainable energy policies and approaches since 1993. Evanthie was awarded a national scholarship from the Hellenic State (Greece) to accomplish her Ph.D. research (in the Geography Department in the Sorbonne University, France) and a Marie-Curie International European Fellowship from the European Commission to pursue postdoctoral research. She is Adjunct Professor at Fiji University, a Senior Research Associate in Sorbonne University, France and in the Aegean University, Greece. Evanthie is also a reviewer for established academic journals such as Nature. Evanthie assists the European Commission in the evaluation of proposals under numerous energy and environmental Calls as an EC Expert, and is appointed at the Special Advisor (Secretariat) for the Hellenic Regulator for Energy (RAE), Greece, where she now works on the monitoring of the Greek electricity market with an emphasis on sustainable energy and energy efficiency. She is also appointed as an expert from international organisations such as United Nations (UN) and Global Green Growth Institute (GGGI), European ones (such as the European Energy Regulator-CEER) and has offered her consulting and academic services to more than 80 countries around the world.

Elisa Morgera is Professor of Global Environmental Law at Strathclyde University Law School, Glasgow, UK and Director of the UKRI GCRF One Ocean Hub. She specialises in international biodiversity law and its interactions with international human rights law.

Dr. David O'Kane is Associate of the Max Planck Institute for Social Anthropology, and a graduate of the National University of Ireland, Maynooth. He has conducted anthropological research in Eritrea and Sierra Leone, and has taught anthropology in both of those countries, as well as in the Republic of Ireland, Northern Ireland, mainland Britain, New Zealand and the Russian Federation. He is currently developing research projects that will look at the effects of the global coronavirus pandemic on Sierra Leone (and the role of the experience of the Ebola Virus Disease epidemic of 2014–2016 in that country's response to the pandemic), and the impact of the coronavirus pandemic on cross-border relations on the island of Ireland.

Dr. Georgina Yaa Oduro is Senior Lecturer with the Department of Sociology and Anthropology of the University of Cape Coast, Ghana. She also doubles as the Director of the Centre for Gender Research, Advocacy and Documentation (CEGRAD) of the same University. She is part of the Ghanaian team researching the UKRI/GCRF funded multi-disciplinary and multi-country One Ocean Hub (OOH) project entitled: *Re-focusing Multi-scalar, Integrated Ocean Management for SDG Synergies*. She is the lead researcher for the RP 5 component of the OOH research project in Ghana. As a Sociologist and Gender expert, she brings along nuanced socio-cultural perspectives in ocean research. She is positioned within the tangible and intangible heritage aspects of the project and has interest in women and children in the Blue Economy.

Dr. Varsha Patel is an independent social scientist. Her research: 'Memories, Royal Ports and Ruins of Sailing Boats: Sediments of Maritime Routes along the Bhavnagar Coast, Western India, 1900–2014' was supported by the Max Planck Institute for Social Anthropology between 2013 and 2016. She designed and taught seminars including 'Indian Ocean Histories', 'Nature, Science and Empire', 'Wildlife and Power in Colonial and Post-India', and 'Animals in Modern Indian Thought' at the University of Kassel until 2020. At present Varsha cultivates research interests in the Indian Ocean, Princely States in India and in Human Animal Studies.

Dr. Elena Perez-Alvaro is a postdoctoral fellow of the NRF Research Chair in Ocean Cultures and Heritage at Nelson Mandela University. She is also Blue Shield Representative for ICOM UK, acting as a liaison between organisations to protect cultural heritage during conflict, including humane and natural disasters. She is an accredited and authorised Associate Professor by the Minister of Universities of the Spanish Government and she works as a director of masters' dissertation for the Master of Cultural Management at the International University of La Rioja (Spain). She has large experience as a marine heritage consultant.

Ryan Pillay is the Deputy Director: Arts, Culture and Heritage at Nelson Mandela University. Pillay works as a researcher and project manager in the NRF Chair for Ocean Cultures and Heritage, Ocean Account Framework, and One Ocean Hub. His research areas include gender, heritage, memory and place, transformation, visual participatory methods and art-based methodologies. Through social enquiry he has designed, facilitated and documented processes in both the public, private and secular sectors. He currently holds the position of Trustee at the South End Museum and the Gcina Mhlope Trust in South Africa.

Anthea Christoffels-DuPlessis is an emerging researcher and an academic at the Nelson Mandela University, Faculty of Law Port Elizabeth, South Africa. She is a doctoral candidate under the supervision of Professor Patrick Vrancken and her research focus is on customary fishing rights in South African fisheries law.

Dr. Pedro Pombo is Assistant Professor at Goa University and Associated Researcher at the Centre for Research on Slavery and Indentured, University of Mauritius. He received his Ph.D. in Anthropology from ISCTE-IUL, Lisbon (2015) with an ethnographic exploration on space, belonging and local history in Southern Mozambique. Earlier, Pedro graduated in Decorative Arts and Design and developing research on Indo-Portuguese architecture. Dr. Pombo's research traces Afro-Asian circulations through aesthetic and anthropological approaches, focusing on archival and material traces, heritages, landscapes and memories in the Indian Ocean. He is co-author of the documentary *The Club* (2021), on Goan diaspora in Tanzania, with the Nalini Elvino de Sousa, funded by the RTP-Portuguese Television. He is also an associated researcher with the Southern Atlantic/Indian Oceanic Africa cluster of the Project Regions 2050, WiSER, Wits University, South Africa and a 2021 Fellow of The Africa Multiple Cluster of Excellence, Bayreuth University, Germany. He was invited to present in the 'AfricAsia: Overlooked Histories of Exchange' Symposium, organised by the National Museums of Asian and African Art, Smithsonian Institute (2020) and in the Webinar series 'Indian Ocean World Material Histories', George Mason University (2021). From 2022, he will also be a research associate of the Ocean Cultures and Heritage Chair Programme at Nelson Mandela University, South Africa.

Jale Samuwai has a Ph.D. in Climate Change from the University of the South Pacific in climate finance. He is an independent Pacific-based researcher with work interest in the Pacific and SIDS in general. Jale was the interim Economic Justice Lead for Oxfam in the Pacific. He is presently the Resilience Finance Analyst who supports the Framework for Resilient Development in the Pacific based at the Pacific Islands Forum Secretariat. His research areas include climate change, climate and ocean finance and equity and justice.

Dr. Dominique Santos is Senior Lecturer in Anthropology at Rhodes University and Course Co-Ordinator for the Postgraduate Diploma in Heritage Management. Her scholarly work explores the intersections of music, play, dreaming and heritage practices with intimate experiences of the self, space and social change. She is interested in the place of dreaming and Indigenous Knowledge systems as speculative methodology when working with life stories, archives and heritage sites, and is an initiated stick diviner in the Dagara tradition.

Rev Thebe Shale is a traditional healer who grew up in his grandmother's traditional Emaswati home, also including baSotho and amaXhosa lineages and is a member of the Bakwena clan. Prior to initiating as a traditional healer, he was a Catholic priest who served in Rome, and spent time in Lourdes and Germany researching sites of miraculous healing. His work as a traditional healer has taken him all over South Africa, and Kenya, where he studied hippo communication. He currently resides in Port St John in the Eastern Cape where he is establishing a school for preserving knowledge of indigenous herbal medicine.

Anita Smith is Associate Professor of Heritage and Archaeology at La Trobe University, Australia. She has more than 20 years of research experience in the Pacific Islands where she has worked with communities and governments, recording cultural landscapes and leading capacity building programs in heritage management. Anita is the cultural expert member of Australia's Delegation to the UNESCO World Heritage Committee and an Advisor to the UNESCO Pacific World Heritage Program. Her research in collaboration with First Nations communities has supported four successful World Heritage nominations. Anita is the lead author of 'Intangible cultural heritage, traditional knowledge and the protection of Pacific Ocean biodiversity through mechanisms of the United Nations' (*Historic Environment*, 2021) and co-author of the chapter 'Developments in Management Approaches' in the *United Nations Second World Ocean Assessment (WOA II)* launched by the UN Secretary-General in April 2021.

Katy Soapi is the coordinator for the Pacific Community Centre for Ocean Science at the Pacific Community in Fiji, a position she took up in January 2021. Before that, Soapi was the Manager of the Pacific Natural Products Research Centre at the University of the South Pacific (USP) in Fiji. Soapi grew up on the island of Rendova, part of the Solomon Islands, in the Pacific. Katy completed her Bachelor's degree at the USP, a Master's at the University of Sydney in Australia and a Ph.D. at the UK's University of East Anglia. Katy then returned to the Pacific, taking up a position at the USP as a lecturer and eventually becoming the Manager of the Pacific Natural Products Research Centre, housed at the USP.

Dr. Jackie Sunde is a researcher for the One Ocean Hub based in the Department of Environmental and Geographical Science, University of Cape Town. Jackie has worked as a research activist in the small-scale fisheries sector for the past two decades. Her Ph.D. focused on living customary law and the governance of small-scale fisheries in South Africa.

Dr. Jessica Leigh Thornton is a social anthropologist and postdoctoral fellow of the National Institute for Humanities and Social Sciences at the University of Pretoria, South Africa. Dr. Thornton is also a fieldwork manager of the NRF Research Chair in Ocean Cultures and Heritage. She has worked as Project Manager for multiple research projects such as the SAPS research project on gangsterism and for a South African LOTTO funded project entitled 'Moments in Time: Field Guides to Heritage in the Eastern Cape Province'.

Dr. Saskia Vermeylen is reader at Strathclyde University Law School, Glasgow, UK and a socio-legal property scholar.

Dr. David Wilson is Lecturer in Early Modern Maritime and Scottish History at the University of Strathclyde, Glasgow. His research interests include piracy, colonisation, maritime law and coastal communities. His first monograph, *Suppressing Piracy in the Early Eighteenth Century: Pirates, Merchants and British Imperial Authority in the Atlantic and Indian Oceans*, was published with The Boydell Press in 2021. He is a Co-Investigator on One Ocean Hub, focusing on colonisation, law and coastal communities in Ghana.

List of Figures

Fig. 6.1	View of the archaeological site from the opposite mountain (Photo by Dr. Evanthie Michalena, 2021)	58
Fig. 6.2	"Hesitant", and very small scale, archaeological excavations by a team of Professors of Archaeology and the Ministry of Culture (in the red circle) (Photo by Dr. Evanthie Michalena, 2021)	59
Fig. 6.3	View of the archaeological site from the opposite mountain: a half-submerged, decaying ship, beside an industrial shipyard inside an archaeological zone of absolute protection, type "A" (Photo by Dr. Evanthie Michalena, 2021)	60
Fig. 6.4	A view over designated archaeological site from the opposite mountain—and of the New Greek Shipyards, S.A. (Photo by Dr. Evanthie Michalena)	63
Fig. 6.5	View of the archaeological site from the opposite mountain—Shipyards on Kynosoura, an ostensibly absolutely protected "Grade A" archaeological site (Photo by Dr. Evanthie Michalena, 2021)	64

List of Figures

Fig. 9.1	Ocracoke inlet North Carolina 1775. An Accurate Map of North and South Carolina with Their Indian Frontiers, Shewing in a distinct manner all the Mountains, Rivers, Swamps, Marshes, Bays, Creeks, Harbors, Sandbanks, and Soundings on the Coasts; with The Roads and Indian Paths; as well as The Boundary or Provincial Lines, The Several Townships and other divisions of the Land in Both the Provinces; the whole from Actual Surveys. By Henry Mouzon and Others Public Domain: https://commons.wikimedia.org/wiki/File:Ocracoke_inlet_north_carolina_1775.jpg	112
Fig. 9.2	Fishermen on the Outer Banks land a mullet seine in this photograph, ca. 1884 (North Carolina Division of Archives and History)	119
Fig. 9.3	Shell Castle Pitcher made for John Wallace circa 1805–1810. Image bears the inscription "A North View of Governors Shell Castle and Harbour, North Carolina" (Image from NC Museum of History)	122
Fig. 9.4	Image of a typical black and white mullet gang at Brown Goode, ed. (1884). *The Fisheries and Fishery Industries of the Commission of Fish and Fisheries*, p. 562	122
Fig. 9.5	Survey Area 2019 at Portsmouth Island near Haulover Point and up Doctors Creek. Program in Maritime Studies, East Carolina University image	124
Fig. 9.6	Water survey team aligned 2 m apart on a tape, searching for submerged artifacts and structures on an eroding bank. Program in Maritime Studies, East Carolina University image	125
Fig. 9.7	Headstones of enslaved grandmother Rose Pickett (1836–1909) and her free African American grandson Henry Pigott (1896–1971). Program in Maritime Studies, East Carolina University image	126
Fig. 9.8	Menhaden Netting Operation (NC State Archives)	128
Fig. 9.9	NPS Signage on Portsmouth Island about Tom Piggott (Photo by Lynn Harris)	130
Fig. 9.10	Historical shoreline positions since 1848 for the Portsmouth Town to Ocracoke (Harris et al., 2018; Thieler & Hammer-Klose, 1999)	132
Fig. 13.1	Seychelles exclusive economic zone	204
Fig. 14.1	Monument to Commemorate the 170th Anniversary of Abolition in Mauritius (*Source* the author, 2016)	227
Fig. 14.2	Map of the villages and towns of Southeast of Mauritius (*Source* https://www.mapsland.com/africa/mauritius/detailed-road-map-of-mauritius-with-cities-and-villages accessed 23/08/2021)	229

List of Figures

Fig. 14.3	Ville Noire village. The village's slave history is indicated via its twinning with Ilha de Mozambique, and the allusions to efforts to overcome enduring inequality (*Source* https://www.lemauricien.com/actualites/ville-noire-ancre-dans-son-histoire-et-tourne-vers-lavenir/275203/ accessed 23/08/2021)	236
Fig. 14.4	Extent of the MV Wakashio Oil Spill, August 2020 (*Source* https://giscan.com/monitoring-the-impact-of-a-shipwreck-the-case-of-the-wakashio-bulk-carrier/ accessed 10/08/2021)	239
Fig. 14.5	Volunteers carry booms for the Wakashio oil spill (Photo by Daphney Dupré, 2020)	242
Fig. 14.6	A volunteer helps to clean up after the Wakashio oil spill. Photo by Daphney Dupré, 2020	242
Fig. 15.1	Image of disparities that can occur in the water level during neap tide (*Source* Author's own image)	253
Fig. 15.2	Pictures taken off the coast of Kilindoni during a flight from Mafia to Dar es Salaam on 15 February 2019	256
Fig. 15.3	Sign located next to the bus stop at Kilindoni. The Swahili phrase underneath the picture of the whale shark reads "let us protect the whale sharks. Their existence is our benefit" (*Source* Author's own photo)	261
Fig. 15.4	Whale shark statue located next to the airport. https://www.booking.com/hotel/tz/mafia-island-tours-amp-safaris.en-gb.html?activeTab=photosGallery (Public Domain) accessed 21/08/2021	261
Fig. 18.1	Community on the beach describing their customary rights and territory to the conservation officials at Hobeni, Dwesa-Cwebe (Photo source Jackie Sunde [co-author])	346
Fig. 18.2	Co-accused in the Gongqose customary rights court case—Co-accused fishers and their legal team and expert witnesses (Photo source Jackie Sunde [co-author])	348

1

Introduction: Blue Heritage, Human Development and the Climate Crisis

Rosabelle Boswell

In the Hollywood blockbuster, *The Day After Tomorrow*, scientists discover that a predicted environmental crisis is happening a lot faster than expected. In a matter of days, instead of years, Earth would change beyond recognition, and humans would be forced to adapt. The film highlights the globalised nature of the disaster and the minutiae of human relationships, thrown into relief against a backdrop of unimaginable catastrophe. This book does not sensationalise the unfolding disaster that is the climate change crisis. That crisis is the present reality of billions worldwide, even if there are disagreements presently about the speed with which it is happening, or the coming intensity of its effects in the coming years. What this book hopes to do, is to provide insight into historical and contemporary human cultural attachments to the sea (sometimes described as cultural heritage), and the impacts of either political decisions and/or environmental degradation on human relationships with the sea and coasts. An important claim of the book is that humans hold diverse perspectives of the sea and coast. Inclusion of diverse,

---

The research presented in this chapter is supported by an NRF UID Grant number 129662.

R. Boswell (✉)
Nelson Mandela University, Room 104, Port Elizabeth, South Africa
e-mail: rose.boswell@mandela.ac.za

often marginalised perspectives of the sea in development, as well as consideration of marine inspired values and cultural practices deepen democracy and advance environmentally sound practices of ocean management.

The importance of inclusion cannot be overstated. The globalised Covid-19 pandemic is revealing historical and enduring schisms which affect not only mitigation of the pandemic but also global effort to resolve the climate crisis. Millions worldwide are already being forced to endure climate changes affecting livelihood, health and habitat. The situation is even more dire for those living in the Least Developed Countries (LDCs), as these countries are further compromised by human development challenges.

In the midst of the Covid-19 pandemic, national governments in LDCs and more developed countries are prioritising the immediate challenge of saving their citizens' lives. In part, this has involved curtailment of basic rights and ultimately, access to independent forms of livelihood. At the time of writing this book, most national governments across the developed world, however, were seeking to 'normalise' life for their citizens. They had already ensured that their citizens are vaccinated against Covid-19 and that life, as it was, before the pandemic, could resume in limited forms. As this book starkly reveals, for LDCs, Covid-19 looms large, alongside the wicked problems of exclusion, poverty, poor access to livelihood and enduring pollution. It appears to some authors in the book, that these problems trump both the climate crisis and the Covid-19 crisis.

The book also shows that national governments in the LDCs need to do more, to prioritise and safeguard Earth's blue heritages, that is, the tangible and intangible artifacts and 'assets' of the oceans and coasts, as well as the intimate, cultural relations that people have with Earth's oceans and coasts, as the latter are very difficult to repair and/or restore. Moreover, important decisions are being made by both international bodies and these governments that put both the tangible and intangible 'blue' heritages at risk of terminal degradation. Of added concern to the authors, is the fact that, to date, there are 1154 World Heritage Sites the majority of which are territorial. Only 50 sites are World Marine Heritage Sites. Not one of these sites is a cultural or culturally associated site. And on the list appraised, there were only two natural World Marine Heritage Sites in Africa. Considering tangible and intangible cultural heritages and at the time of writing, there were 145 such sites in Africa but up to 39% of these contain heritages in danger. The poor regard for culture and/or its precarious situation in Africa is astounding. In Africa and among ordinary people, culture is considered a major social asset and source of human rights, cohesion, creativity, prosperity and livelihood.

A critical review of the Blue Economy (BE) imperative in Africa, states that, 'large scale BE initiatives prioritize economic gains at the expense of environmental degradation and the exclusion of local communities' (Ifesinachi et al., 2020). The exclusion of local communities from the ocean management sector is critical given the intractable, and rapidly intensifying problems of climate change and general, environmental degradation that are devasting Earth's oceans.

Humans rely on the oceans for work, subsistence, habitation, healing and leisure. Over the span of human evolution, they have engaged with the oceans and coasts, conceptualising the marine space, aesthetically depicting it, using it for naval and imperial ambitions, lamenting it as a grave for slave descendant brothers and sisters, sailed it for sport and dived into it for other worldly experience.

This book, as previously stated, seeks to offer further insight into the diverse human social and cultural engagements with Earth's oceans and coasts. In part, the book shows what is at stake, if the trend of environmental degradation and the exclusion of local communities from ocean management continues. The book, however, also hopes to nuance accounts of negative, causal anthropogenic impacts on the environment, to reveal complex, symbiotic and intersubjective human interactions with the oceans and coasts. A major contribution of the book is its high estimation and valuation of human and cultural contribution to the 'blue economy'.

In monetary terms, it is estimated that the global ocean economy is worth 24 trillion US Dollars and that it is the seventh largest economy on earth (World Wildlife Fund, 2018). While several international reports provide detail on the financial dividends expected from the ocean economy (OECD, 2016; World Wildlife Fund, 2017), there appears to be little focus on the human, social and cultural value of, contribution to, and benefits to be gained from the oceans.

At the inception of the Intergovernmental Panel on Climate Change (IPCC) in 1988, states approving the committee pledged to align their strategies for growth with the UN SDGs. But since then, it has become evident that facilitating economic growth that is inclusive, equal and environmentally sound is extraordinarily difficult to achieve. Today and despite the reality of climate change, many nation states are still not keen to decarbonise the economy or pursue environmentally sound 'growth'. Historical legacies of inequality and injustice, still unresolved for centuries, are adding to the problem. Making inclusive and equitable participation in ocean management and climate management difficult.

Mindful of these challenges and the experiences of those caught in the climate crisis maelstrom, the United Nations (UN) has set in motion various actions, declarations and global agreements to help nation states mitigate and/or adapt to the coming crisis. The UN interventions were apparent at the 2017 UN General Assembly where state parties pledged to support and implement SDG 14, the sustainable use and development of the oceans. The ensuing global ocean governance discussions, including COP21, culminated in the UN Decade of Ocean Science, which emphasised prioritisation of the health of marine ecosystems and the securing of ocean-based livelihoods.

The Post-2015 UN Development Agenda task team (UNESCO, 2012) however, identified that culture, 'a driver of development' was not explicitly stated in the SDGs and that effort must be made to include culture as a key metric. In this regard, the EU Parliament's report on actions for sustainability (2017), emphasised that culture, an essential resource and fourth pillar of development, cuts across all the SDGs. To this end, culture should be explicitly articulated as part of the EU's sustainable development goals. Similar sentiments were expressed in the African Union's Agenda 2063, where a major aspiration noted is the achievement of a culture-centric Africa, where sustainable development processes do not exclude culture. The AU indicated it aspires to achieve an 'An Africa with a strong cultural identity, common heritage, shared values and ethics' (African Union Commission, 2015, p. 2).

The inclusion of culture in development, however, goes beyond its relevance to the UN SDGs and ultimately to EU and AU goals. As stated further on, UNESCO's concept of culture and heritage (used to advance the conservation of biological and cultural diversity for decades), is evolving and has been at the forefront of globalised concepts of culture. While this understanding of culture is now included in discussions on human flourishing, belonging and diversity, governments in the Least Developed Countries noted in this book, have tended to portray culture and heritage as uncomplicated legacies and symbols of diversity. Heritage and culture appear to be for celebratory and recreational purposes.

But as this book shows, heritage and in particular 'blue' heritage, are important sources of cultural values, practices, sense of history, identity and solidarity. Understanding this is critical to the substantive implementation of the UN SDGs and global interventions for human development. Some of the authors writing here specifically mention the UN SDGs in the context of a discussion on the UN Decade of Ocean Science for Sustainable Development 2021–2030, the UN declaration on the Rights of Indigenous Peoples (UNDRIP) and the UN SDGs, specifically UN SDG 14 (Life below Water), UN SDG 1 (poverty alleviation) and UN SDG 5 (gender equality).

The authors in this book also discuss UNESCO's defined tangible and intangible cultural heritages and the relevance of this understanding of heritage to ocean health and the protection of biological and cultural diversity. They offer fascinating accounts of common or everyday heritage and cultural beliefs, those heritages not inscribed on either national or global heritage lists. So, for this book, the UN's decadal commitments are important to the story of better ocean health and climate, but so is the rich array of stories, interviews, encounters, readings and perspectives, of humans' holistic ocean management practices. These narratives enrich and diversify our understanding of human interaction with the essential resource that is the ocean.

A further contribution of the book is that it provides detail on the consequences of social and cultural exclusion from ocean management and development decisions. It also shows the extraordinary resilience and inventiveness of humans as they seek to cope with and overcome exclusion from important decisions regarding their environment.

This leads me to the title of the book. The title, 'blue heritage' is specifically chosen because it speaks to the rich legacy of Earth's oceans to humans, which the current generation must restore and safeguard for their own and the benefit of those to come. The chapters seek to subvert the prevailing focus on economic dividends to be obtained from the oceans. Some chapters in the book tell of how governments remain focused on the economic benefits of ocean exploration and exploitation, while others discuss the very long history of humans defining, engaging, living with and sharing the ocean and its resources across cultural and social boundaries. These are studies of a culturally rich engagement with the sea, from a literary and philosophical analysis of the medium, to the use of the sea for subsistence and health.

Finally, the book foregrounds the concept of heritage, in its narrower (institutional) sense. The aim is to show that the ocean is not merely the common heritage of humankind, as defined by UN Convention for the Law of the Seas (UNCLOS) but also that Earth's oceans and coasts contain and showcase a rich array of particularistic tangible and intangible cultural heritages as defined by UNESCO. To this end, conserving marine ecosystems and safeguarding unique coastal cultures and livelihoods, are key to protecting biological and cultural diversity, as well as the knowledge these can impart in a time of climate crisis.

Moving beyond a listing of World coastal heritages and their essential attributes (Claudino-Sales, 2019), the book also considers the intangible cultural heritage of local communities, as well as marine cultural heritage. Again, here the focus is not so much the 50 or so sites inscribed as World Marine Heritage, but the intangible cultural heritage that people hold. Ultimately, the authors in the book seek to make further audible, the voices of indigenous and autochthonous peoples, as well those holding transnational and transcultural identity, people who often share a common vision for the safeguarding of the oceans and coasts.

An Urgent Agenda

The urgency of integrated and inclusive development action, as well as timely and substantive effort to address climate change is now starkly revealed in the UN IPCC report on climate change discussed at COP26 in November 2021. The IPCC report outlines the catastrophic implications of climate change, wrought by rising sea levels and increasing temperature on Earth. Media reports preceding the event, confirm the dire situation, reporting on uncharacteristic flooding, excessive rainfall, devastating fires and rapidly diminishing biodiversity worldwide. Critiques of the UN IPCC reports point to earlier erroneous claims regarding climate change, conflict of interest in presenting climate change as a crisis and questions regarding the accuracy of the reports for the situation in the global South (Schiermeier, 2010, pp. 566–567).

However, and despite the critiques noted above, one cannot entirely deny climate change or the urgency with which it must be addressed. Urgent and collaborative action at national, regional and global levels is needed now to safeguard the health of marine ecosystems and other natural contexts (i.e. dunes and beaches) that support coastal 'health' and marine life. To this end, the European Union (EU), the African Union (AU) and the Pacific Regional Forum are separately seeking solutions to address the environmental threats posed.

In June 2021, the UN Secretary-General António Guerres described the human impact on ocean health as a 'war on nature'. Referring to plastic waste pollution, ocean acidification and sea-level rise as a result of temperature change and the melting of Earth's glaciers, he called for all states parties to renew their commitment to addressing environmental degradation on Earth. However, and as this book shows, addressing climate change and environmental degradation is not possible without the commitment from local communities, indigenous peoples and an array of coastal and ocean-involved stakeholders. National governments worldwide may not yet recognise this, believing that signing the various UN conventions and implementing 'benevolent' top-down approaches to ocean management will suffice in efforts to conserve marine biodiversity and ocean health.

Being researchers in the field of culture and/or ocean sciences, the authors of this book are aware of the complexities of inclusion, both of people and perspectives. They offer timely advice for an urgent agenda, one that requires significant governmental and human behavioural change if the climate change goals are to be met by 2050. Part I of the book offers a fourfold perspective on 'blue' heritages in the humanities and social sciences, opening with a piece on human evolution, another on islands and humanity, a third on oceanic humanities and a fourth on heritage itself.

Part II of the book offers several lenses through which human engagement with the coast has been perceived, while in Part III of the book the authors focused on contemporary challenges to blue heritage management. The authors adopt a critical view of heritage, analysing it alongside ocean management, a process that, as already noted, holds its own complexities. Finally the book also considers the science-to-policy interface, noting challenges inherent in the process of communicating and transposing scientific findings (especially complexly articulated intangible cultural heritages), for policy use and inclusion in the hope of advancing more democratic and inclusive ocean management.

While a rich literature and multiply situated discourse of the oceans is emerging in the oceanic humanities that is relevant to an incisive discussion on 'blue heritage' (see the orientation contribution by Hofmeyr and Lavery in this book), several perspectives and imperatives are 'colliding' in the field of climate change, as scholars press to work collaboratively in mitigation of a globalised climate crisis. Unfortunately, and as the UN itself has noted, not enough credence is given to the humanities and social sciences in global efforts to advance sustainable ocean management. The focus remains resolutely on natural science contributions, even though evidence of climate change and degrading ocean health point to anthropogenic factors, and specifically, behavioural trends, including prevailing (modernist and neoliberal) ideologies and values. Scientists concur, that an integrated approach to addressing climate change (and ultimately ocean health) is likely to yield greater success, and that decarbonising the economy should be the ultimate priority (Anderson et al., 2019, pp. 933–934).

Offering a historical perspective from the start, the book begins with a reflection on the impact of early forms of exclusion on coastal communities. In doing so, the book highlights the role of historical and enduring inequalities in shaping ocean management and ocean health. Curtis Marean's analysis of tangible and intangible cultural heritage among early modern humans along South Africa's southern coasts offers insight into prehistory, arguing that early sociality and intercultural exchange were key to the earliest human engagement with the sea and coasts in South Africa.

The orientation on islands, offered by Godfrey Baldacchino follows next, revealing the globalised existence of islands as rich sites of tangible and cultural heritage. From these emerged unique practices necessary for early sustainable use of the sea. The final section of the orienting chapters offers a discussion on heritage and the place and meaning of blue heritage in mitigation of the climate crisis and its effects on humans.

In Part II there are selected historical examples of human engagement with the sea, either under conditions of duress, as in Patel's paper on India under British Imperial rule, and in Pombo's chapter on architecture and heritage in Goa. This is followed by Harris' account of African American coastal heritage in the United States. Part II also includes Hills' account of historical blue heritage in Greece, as well as Joseph's literary analysis of identity and the sea and Calyx discussion on the poetry and politics of women divers in the Pacific.

Part III of the book eases into further case studies from the Pacific, Europe, Africa and Asia. While these case studies cannot claim to cover all the issues for all these complex locales, they do offer insight into the specificities of inclusion in ocean management processes. Erwin, Pereira, McGarry and Coppen's paper on the highly acclaimed *Lalela uLwandle* (Listen to Hear) theatrical performance, showcase diverse but interwoven narratives of ocean management, foregrounding the importance of listening to 'hear' in inclusive ocean management practices.

I take on the ongoing environmental disaster story of the MV Wakashio in Mauritius, While Choppy offers an account of the musical, coastal heritages of the Seychelles. Asse and Perez-Alvaro write on underwater cultural heritage, its manifestations and challenges, while Haan analyses blue heritage among fishermen in Mafia Island, Tanzania. The afterword is offered by George Abungu.

Conclusion

The chapters in this book do not offer an exhaustive analysis, or a definitive set of case studies regarding 'blue' heritage and its management. The chapters do not consider only LDCs as defined by the UN. And, finalising the manuscript, the editors became aware that a rich literature in humanities still needs to be foregrounded in the ocean sciences for ocean researchers and activists to obtain deep insight into the complex interactions of humans, oceans and coasts. This literature and their potential authors were difficult to come by in the period during which this book was being edited. Furthermore, the authors were all writing during the Covid-19 pandemic and the duress that the pandemic brought to fellow colleagues' lives and families. While authors across history have written under difficult circumstances, it is important for readers to remain aware of the conditions under which writers write. There is no point in being self-reflexive in the research process and ignoring the conditions under which we work and produce knowledge.

In this regard, the editors of this book have chosen not to overly edit the contributions of the authors. They have chosen instead to allow the authentic voice of each author to filter through and to allow their specific writing style to communicate knowledge of blue heritage. This allows for a multivocal and multi-layered process and written outcome, that acknowledges and makes space for the politics of knowledge and its unique forms of communication in the world. The authors concur that many voices, arising from diverse politicised and knowledge spaces are required to attend to the formidable challenges of climate change and human development.

An opportunity missed in the blockbuster Hollywood film, *The Day After Tomorrow*, is that of different voices. While we notice the protagonists, the supporting actors and those contributing to vignettes showcasing aspects of the disaster, we do not really get to hear their wisdom or thoughts on the situation at hand. Unlike the film, this book hopes to render audible, the voices of many different stakeholders in the ocean sustainability and climate change story. Our hope also, is that the chapters offered here will help to advance discussion on the role of humanities and social sciences in debates and interventions for sustainable ocean use.

References

African Union. (2015). *Agenda 2063: The Africa we want*. African Union Commission: Addis Ababa.
Anderson, C. M., DeFries, R. S., Litterman, R., Matson, P. A., Nepstad, D. C., Pacala, S., Schlesinger, W. H., Shaw, M. R., Smith, M., Weber, C., & Field, C. B. (2019). Natural climate solutions are not enough. *Science, 363*, 933–934.
Claudino-Sales, V. (2019). *Coastal world heritage sites*. Springer Nature.
European Union. (2017). *EU report on action for sustainability (Opinion of the committee on culture and education)* https://www.europarl.europa.eu/doceo/document/A-8-2017-0239_EN.html#title5. Accessed 27 April 2022.

Ifesinachi, O.-Y., Nelly, I. K., Nelson, A. F., Uku, M., Elegbede, J., Isa, O., & Ibukun. J. A. (2020). The blue economy–Cultural livelihood–Ecosystem conservation triangle: The African experience. *Frontiers in Marine Science*. https://www.frontiersin.org/articles/10.3389/fmars.2020.00586/full. Accessed 20 Sept 2020.

OECD. (2016). *The ocean economy in 2030: The ocean as a sustainable source of economic growth*. OECD.

Schiermeier, Q. (2010, February 4). IPCC flooded by criticism: Climate body slammed for errors and potential conflicts of interest. *Nature, 463*, 566–567. https://www.nature.com/articles/463596a.pdf. Accessed 18 Aug 2021.

UNESCO. (2012). *UN system task team on the post-2015 UN Development Agenda: Culture, an enabler and driver of sustainable development (A Thematic Thinkpiece)*. UNESCO.

World Wildlife Fund. (2017). *Reviving the Western Indian Ocean economy*. https://sustainabledevelopment.un.org/content/documents/13692WWF2.pdf. Accessed 2 Aug 2021.

World Wildlife Fund. (2018). *Principles for a sustainable blue economy*. http://awsassets.panda.org/downloads/wwf_marine_briefing_principles_blue_economy.pdf. Accessed 12 Aug 2021.

Part I

Humans and Humanities in Ocean Heritage

2

The Evolutionary Roots of Blue Heritage

Curtis Marean

Introduction

Unlike the rest of the animal kingdom, modern humans have culture as their primary form of adaptation. By culture I mean the system of customs and behaviors and technology that members of a society use to cope with their world and with each other. It is transmitted from person to person through observation and teaching, and much of that transmission is symbolically mediated through language and other symbols. The capacity for having culture is evolved, and there are three key evolved features of modern humans that make culture possible—an advanced cognition, a psychology for a special form of social learning, and a psychology for cooperating at large scales with unrelated individuals. Without these evolved capacities there can be no modern human culture (Marean, 2015).

Among paleoanthropologists, there is debate as to when these capacities arose, but there is increasing evidence that an advanced cognition and the modern human form of social learning was in place by around 200,000 years ago or earlier. The origin time and context of the third key feature, what I call

---

C. Marean (✉)
School of Human Evolution and Social Change, Arizona State University, Tempe, AZ, USA
e-mail: Curtis.marean@asu.edu

hyper-prosociality, is still a matter of debate but I have proposed its evolution is intimately bound to the movement of humans into the aquatic niche—coastal and riverine/lacustrine food resources (Marean, 2016). This proclivity to cooperate at large scales was a game-changer, allowing the formation of multi-scale society and ultimately very large-scale complex civilization. But at the same time, this ability to cooperate at large scale brought with it an innate tendency toward out-group bias that can be triggered in such a way that it results in societal division and conflict of the type so prevalent today.

Thus, it is possible that the waters of the world and our ancestor's adaptations to them may sit at the root of our ability to cooperate at vast scales and at the same time molded our tendency to divide into the bickering self-centered groups raging across society and social media today. The former is fundamental to our way out of the disaster of human induced climate change, and the latter is the impediment to that large-scale cooperation we need to solve the many real crises we face as the custodian species of our world. In this essay I summarize the evidence and theory that underlies this possible connection between life on the water's edge and humanity's great cooperative asset and tragic divisive failure.

What Is Unique About Aquatic Hunter-Gatherers?

The vast majority of human evolution occurred in a hunting and gathering economy. The study of hunter-gatherers began at the very origin point of anthropology with the field research of people like Franz Boas, Baldwin Spencer, and Alfred Radcliffe-Brown in the late 1800s. There was a slow accumulation of data on hunter-gatherers that, in my opinion, reached a phase around the 1960s where the diversity and quality of that data allowed useful and accurate generalizations to be made. This was the beginning of good theory building in anthropology on what structured the diversity of hunter-gatherer adaptations that were recorded through nearly 100 years of observations. It was at this stage that it became evident that hunter-gatherers who relied significantly on aquatic resources differed from terrestrial hunter-gathers in some key ways.

At first these observations were qualitative, but eventually some were quantified. In the discussion below, I will rely on summaries provided in the following (Binford, 2001; Erlandson, 2001; Erlandson & Moss, 2001; Kelly, 1995; Knauft et al., 1991; Marean, 2014, 2016). In comparison to terrestrial hunter-gatherers, these are some of the key characteristics of aquatic hunter-gatherers:

2 The Evolutionary Roots of Blue Heritage

1. Aquatic hunter-gatherers, as the name implies, get much or most of their sustenance from aquatic sources. These tend to be high-quality foods and are particularly well-known to be high in omega-3 fatty acids. This has many downstream impacts as noted in the following list.
2. Aquatic hunter-gatherers often rely more on storage. For example, fish are stored for long periods through drying or special forms of controlled rotting.
3. Aquatic hunter-gatherers have heavier technology in that it is made of more parts, more diverse raw materials, and thus is less subject to being carried around with ease.
4. Aquatic hunter-gatherers are more sedentary and less mobile. Terrestrial hunter-gatherers tend to make many residential moves, sometimes 20–30 a year or more. Aquatic hunters-gatherers have reduced residential mobility, often parking themselves near where the water-based foods are concentrated and/or where they have stored the food taken from water.
5. Aquatic hunter-gatherer bands are often larger, and since they are more sedentary, take on the form of villages that resemble what we see with food producers.
6. Aquatic hunter-gatherers have smaller territories, at both scales of sociality. So, the annual round of a band, and the size of the territory of the ethnolinguistic group, are smaller than terrestrial hunter-gatherers. A downstream impact of this is that aquatic hunter-gatherers have heavily packed landscapes with many small tribal territories, and a good example is shown in the Kroeber's ethnolinguistic map of California (Kroeber, 1925).
7. Aquatic hunter-gatherers often live at higher population densities.
8. Aquatic hunter-gatherers often have higher levels of economic and social differentiation, and thus the egalitarian ethic so common among terrestrial hunter-gatherers is relaxed.
9. Aquatic hunter-gatherers have greater levels of inter-group conflict and territoriality. This is well documented in both the ethnographic and archaeological record and is easily explained by a very powerful theory from behavioral ecology called the economic defendability theory (Davies et al., 2012) that predicts that territoriality, or defense of space, rises as resource density and predictability rises.

This list of key traits separating aquatic from terrestrial hunter-gatherers is the product of comparative ethnographic analysis from a wide range of scholars. But are there other ways that aquatic hunter-gatherers are different? I think there are, but at this stage, these are hypotheses that I will introduce here and explore more in future work.

First, I think it is likely that aquatic hunter-gatherers and their societies are hotspots for technological innovation. Lives on the water's edge pose interesting challenges for hunter-gatherers once they expand beyond the intertidal zone such as into fishing, and there one finds the need for twine and bone working to make hooks and grinding stone for sinkers and so on. This of course suggests that coastal adaptations have big downstream impacts on groups that are not aquatic—innovations can get diffused inland.

Second, coastlines are natural refuges during periods of climate change. We currently live in an interglacial, but for the majority of the last million years, the world was in a glacial climate. Glacials result in colder temperatures and, generally speaking, more aridity in Africa. Colder temperatures and drier climates generally result in reduced biomass and biodiversity in terrestrial ecosystems. Oceans also get colder during glacials, but unlike terrestrial ecosystems, colder ocean waters normally result in greater biomass, and thus the productivity of the inter-tidal is enhanced during colder climates. During the transition between glacials and interglacials, the coast may move, and this of course will cause disruptions, but hunter-gatherers can move with it. This means that once a tribe has shifted to a coastal adaptation, and that coastal adaptation has stimulated innovations because the food base is stable, there will be population and societal stability, and thus a greater likelihood for those innovations to be retained.

So, there is absolutely no doubt that aquatic hunter-gatherers differ significantly, on average, from terrestrial hunter-gatherers in some fundamental ways. This leads inexorably to an important question—when and where did humans move into the aquatic niche, and how did it effect human evolution? Could that new way of life create an entirely different selection regime for modern humans that changed the course of human evolution? To address that question, we need to link aquatic adaptations to sociality.

The Shared Sociality of Modern Humans

Modern human terrestrial hunter-gatherers tend to live in bands of roughly 20–30 people, often sometimes called the local group (Marlowe, 2005). This is the primary foraging and social unit throughout most of the year. This basic first level of sociality is shared with our closest living relatives the chimps where it is called the troop. But modern humans have a deeply shared sociality of multi-scale society, where we are networked together with kin and non-kin into tribes that share a language and other customs (Chapais, 2008). Here I use the words tribe and ethno-linguistic group interchangeably. Formal mate exchange is the primary glue that ties a tribe together, and members of a tribe tend to find mates outside the band but inside the tribe. Tribes not only mate together, but they also fight together and defend their territory.

Effective participation in multi-scale society requires a set of special cooperative behaviors on the part of humans (Fehr & Fischbacher, 2003). First, for large-scale cooperation to work there must be in place a psychology to cooperate with non-kin, which as I noted before is extremely rare in the animal kingdom. Second, there must be a complex cognition that facilitates language and rather strong memory storage. Both are required to remember and negotiate the complex relations required of multi-scale society such as agreements for mate exchange, and kinship terms to keep track of relationships that exist as a function of descent, affinity through things like marriage, and fictive kin. I used the term "hyper-prosociality" to refer to this combined cognitive-psychological package unique to modern humans among living animals (Marean, 2015). In my opinion, a complex cognition had to evolve first before multi-scale society could evolve and before the conditions for hyper-prosociality could evolve. It is a matter of debate as to whether or not other closely related humans like Neanderthals had a similar hyper-prosociality—I don't think they did, and this is why they are extinct.

But the evolution of hyper-prosociality and multi-scale society also brought psychological proclivities toward out-group bias rooted in the new social regime of tribal structure. Out-group bias refers to negative emotions toward "others" such as distrust and prejudice, and is also referred to as "parochial altruism" (Bernhard et al., 2006). These can be triggered by certain stressors, such as resource scarcity. Political leaders and social media influencers can prey on this innate tendency toward outgroup bias, fanning it with rhetoric and false information. Social media algorithms that push negative posts to the top have an amplifying impact and are a conduit for brain viruses like sectarianism. The extraordinary tribalism that we see today is, in my opinion, an expression of these evolved psychological proclivities triggered by stressors effecting so many and fanned by social media AIs and influencers.

This raises an obvious question—when and where did the "troop to tribe" (Ambrose, 2010) transition occur in human evolution. In my opinion, this is one of the great unsolved paleoanthropological questions. I think as a starting point we can safely say that the primary psychological complex for hyper-prosociality was in place prior to the exit of modern humans out of Africa around 70,000 years ago. The reason for this is at all modern humans have this shared sociality, so it is just simpler to argue that it evolved once in one place and was carried onward and outward, rather than having evolved multiple times among different populations of modern humans. So that leaves a long possible time span from 70,000 to anything back to the shared common ancestor with chimps, which clearly did not have hyper-prosociality.

There are a variety of hypotheses as to when and where the "troop to tribe" transition occurred. But I think there are three that have articulated a strong theoretical basis for their argument. A well-known discussion is developed in the wonderful book "Mother and Others" by Sarah Hrdy (2011). We can think of this as the "Cooperative Breeding Hypothesis". A simplification of it is that among early hominin groups, group members started to alloparent the children of others because as encephalization occurred, the demands for provisioning them increased, and this created a pathway for the evolution of a socially intelligent ape that started the process toward hyper-prosociality. A second hypothesis was developed by E.O Wilson in his book "The Social Conquest of Earth" (Wilson, 2012) and we can call it the "Nest Hypothesis". Wilson uses his expertise in eusocial insects to make the case that all highly eusocial ants build and defend nests, and these nests are expensive and filled with resources. This leads to group cohesion. He argues that when humans first began to build campsites these would have had similar value as the eusocial nests, and thus selected for cooperation to defend those nests. Both these hypotheses, in my opinion, posit a deep time depth to the origins of high levels of cooperation, likely back to early *Homo*, because that is when encephalization and the use of campsites began.

I have drawn on the theory developed by Bowles and Gintis in their book "A Cooperative Species" (Bowles & Gintis, 2011). Using a combination of math and computer simulation, they show that when competition develops between cultural groups, those groups with individuals more prone to within-group cooperation in conflict with other groups will expand at the expense of others. Thus, competition and intergroup conflict in a cultural species create a selection regime for the spread of prosocial individuals and genes. I joined this theory to the theory of economic defendability which posits that inter-group conflict is triggered when organisms rely on dense and predictable resources (Marean, 2014, 2016). My review of African terrestrial resources showed that terrestrial resources on the African continent tend to not be dense and predictable, and this explains why intergroup conflict is uncommon among known African hunter-gatherers. The only dense and predictable resources are in aquatic contexts, and therefore it follows that when humans moved into aquatic niches intergroup conflict was triggered or elevated, creating a strong selection regime for the spread of highly prosocial behaviors.

How and When Did the Aquatic Revolution Occur?

The Terrestrial Hunter-Gatherer of Our Origins

The vast majority of human evolution that resulted in modern humans took place on African land as hunter-gatherers. There have been multiple exits out of Africa, and these resulted in radiations of hominin evolution that resulted in human species similar to us in some ways, such as Neanderthals and Denisovans. But it is in Africa where modern humans evolved, and they left Africa around 70,000 years ago and replaced these other hominin species. The explanations for this replacement include competitive exclusion, outright warfare, usurpation of their land, or a combination of all of these. Along the way, there were multiple mass extinction events, particularly of megafauna. I think the great modern human diaspora, and the extinction of hominin species and megafauna was a single related event made possible by the evolution of hyper-prosociality and the "troop to tribe" transition. This gave modern humans the ability to out-compete hominins that had just single-level societies and also allowed the formation of the large cooperative tactical hunting strategies that allowed humans to overhunt megafauna.

But let's step backward in time and consider what these early modern humans in Africa were like around 300,000 years ago. They were hunter-gatherers that focused primarily on terrestrial resources. The archaeology shows us that the animals they exploited were primarily large mammals—antelope of all sizes, pigs, zebra, and so on. Smaller animals such as hares and dune mole rats and lizards and snakes were sometimes taken but not as a regular food source. They had relatively light but effective technology. Wooden and stone-tipped spears were present, and at this stage it seems like these would have been stabbing and hand-cast spears; bows and arrows were not present (Lombard, 2021), but the jury is still out on atlatls. Clothing made of skins was almost certainly present though the earliest definitive evidence we have is about 120—90 ka (Hallett et al., 2021)—I suspect it dates far earlier. These early modern humans were hunter-gatherers who lived in bands of roughly 20–30 people. They had high residential mobility meaning they moved their camps many times each year; 30 or more times a year. One pattern that is evident in the anatomical record is that Pleistocene early modern humans were rather robust—big-boned, sometimes quite tall people—heavily muscled formidable humans, but note that such a body is expensive in calories. I think this robust body form was a consequence of being highly mobile with very large territories caused by the unstable nature of Pleistocene climates, thus stimulating a need for very large territories. This was largely a shared adaptation between modern humans and other late Pleistocene hominins of the world and represented a parity of sorts such that when they encountered each other, as sometimes happened in the Levant, neither had any distinct advantage over the other. This then changed in Africa, and I think it changed first in coastal South Africa.

Enter the Aquatic Hunter-Gatherer

There is no debate that the ethnographic record on hunter-gatherers shows us that aquatic hunter-gatherers are fundamentally different compared to terrestrial hunter-gatherers in a number of important ways, as I have discussed above. Thus, it stands to reason that when hunter-gatherers in our evolutionary past made this shift to aquatic resources there were downstream impacts on their lives, and these would have been significant. This illustrates the power of the comparative ethnographic approach in anthropology—it provides us with a frame of reference upon which to build expectations and theory in human evolutionary studies. While human behavior is not subject to lawlike predictability from theory, the range of variation we see in modern ethnographic accounts provides us with expectations, and when theory intersects in a way that articulates and explains that variation, we can use it to help illuminate the past. That being so, we must ask the question "when did humans enter the aquatic niche?".

I have reviewed the evidence for the origins of aquatic resource use, and for the most part that review is still up to date (Marean, 2016). The earliest evidence we have for the use of aquatic resources is at Pinnacle Point near Mossel Bay dating to just before 160,000 years ago. That evidence shows us that people were exploiting the upper and middle intertidal zone as well as exploiting beached whales. Interestingly, this is the earliest best example of people using and modifying ochre (Marean et al., 2007). We were careful in our reports of that evidence to point out that while aquatic resources were being exploited, it is unclear how intensive that exploitation actually was. It is unclear if this was a full-blown coastal adaptation, where the life revolves around the sea and its rhythms, or rather a more modest but systematic use of coastal resources. This matters a lot, but like is so often the case, more research is needed. The challenge is that the site is very old, and the deposits dating to this time are not well preserved and rare in the cave. We absolutely need more evidence from this time period, but it has proven very difficult to find deposits dating to earlier than 125,000 years ago on the coast, largely because there was a high sea level of +5–6 m just before that and in many cases, it washed away older sediments.

Around 110,000 years ago in South Africa we see recurrent evidence for a coastal adaptation at multiple sites, and this is signaled by (1) dense shell middens, (2) species ranging from wider intertidal zones, and (3) evidence for symbolic systems imbedded in the sea such as seashells, tools made from seashells, and seashells being used as ornaments (Marean, 2014). There is also emerging evidence for coastal resource use in North Africa roughly 110,000 years ago, such as the presence of ocean mollusks being used as beads, but there is yet to be reported a shell midden at this age. In eastern Central Africa, there is evidence for freshwater fish exploitation at around 90 ka (Brooks et al., 1995; Yellen et al., 1995), but this remains a single observation done when luminescence dating was still in its infancy. We need more research into these riverine and lacustrine contexts.

So at least by 110,000 years ago, probably earlier but not yet discovered, there is on the shores of South Africa a population of modern humans embedded in a coastal adaptation. I think this population was spread largely from the rich and species diverse shores of the east coast to the cold species-poor but high biomass shores of the west coast. There was likely some diversity in adaptation across this biomass and diversity gradient, but they shared the hallmarks of the coastal hunter-gatherer. Their coastal adaptation was focused mostly on the intertidal zone, and they did not begin intensive fishing until much later, closer to the Holocene. They still practiced significant amounts of large mammal hunting and exploited the rich plant foods of this region. So, this was a coastal adaptation with substantial input of terrestrial foods, not unlike other coastal hunter-gatherers documented in California and Australia. On the south coast, they had the rich Palaeo-Agulhas Plain and its large herds of grassland species to hunt, providing a trifecta of plant, terrestrial mammal, and seafoods perhaps unique to this place (Marean et al., 2020). They had reduced residential mobility, and this set up the conditions for the production of the rich material record we see in these archaeological sites. This then acted as a hot spot for innovation and refugia for its preservation, which then allowed the retention and diffusion of innovations outward from this region to other areas in Africa. And perhaps the most consequential of innovations was conflict over these resources, which then set up the selection regime for the evolution of hyper-prosociality and the "troop to tribe" transition. The cooperative and quarrelsome human had evolved.

Conclusions

Here I have outlined a particular sequence of steps in the evolution of modern humans. It is a hypothetical model that proposes that large-scale cooperation was the final addition to a suite of traits, along with advanced cognition and a special form of social learning, that provided the capacity for the modern human cultural adaptation. Clearly, there is no consensus that this model is correct, but what it does have going for it is a strong theoretical basis and good articulation with our currently available empirical record. It is also testable, but the required empirical records will take many years to develop. But even if it is ultimately proven incorrect, I think it will remain the case that the first foundations of humanity's Blue Heritage occurred on the shores of South Africa (see Boswell 2022 in chapter 1), and those first coastal people had the unique coastal lifestyle we see in the hunter-gatherer ethnographic record.

In 2011, Andrew Hall, who at that time was Chief Executive Officer of Heritage Western Cape, Jenna Lavin, at that time Assistant Director: Research, Policy and Planning at Heritage Western Cape, and I welcomed to Mossel Bay a large group of South African scientists and stakeholders. There we made the pitch for a serial World Heritage Nomination for Pinnacle Point and several other important sites on the South African coast. Our argument was that these sites documented the beginnings of Blue Heritage and at the same time the origins of culture, and thus were a special heritage belonging to the world. We felt that this story needed to be told locally and worldwide, while at the same time leveraging this resource as a way to draw local and foreign visitors to the coast for the purpose of job creation. Over the years, studies have been done, and reports written, and leadership changed—Dr. Mariagrazia Galimberti has been for the last several years leading the charge (she did her Ph.D. on Pinnacle Point materials). That application is now complete and soon to be submitted, so it is my hope that in the near future we will have a World Heritage Site on the South African coast, telling the story of the deep antiquity of Blue Heritage in this very special place, where I have been so fortunate to be a visiting scientist for nearly 40 years, and have been shown so much friendship and hospitality from my South African hosts. I thank all of them so much for that.

References

Ambrose, S. H. (2010). Coevolution of composite-tool technology, constructive memory, and language: Implications for the evolution of modern human behavior. *Current Anthropology, 51*(S1), S135–S147.

Bernhard, H., Fischbacher, U., & Fehr, E. (2006). Parochial altruism in humans. *Nature, 442*(7105), 912–915.

Binford, L. R. (2001). *Constructing frames of reference: An analytical method for archaeological theory building using ethnographic and environmental data sets.* University of California Press.

Boswell, R. (2022). Introduction: Blue heritage, human development and the climate crisis In R. Boswell (Ed.), *Blue heritage: Global perspectives on ocean histories and cultures.* Palgrave.

Bowles, S., & Gintis, H. (2011). *A cooperative species: Human reciprocity and its evolution.* Princeton University Press.

Brooks, A. S., Helgren, D. M., Cramer, J. S., Franklin, A., Hornyak, W., Keating, J. M., Klein, R. G., Rink, W. J., Schwarcz, H., Smith, K. N. L., Stewart, K., Todd, N. E., Verniers, J., & Yellen, J. E. (1995). Dating and context of three Middle Stone Age sites with bone points in the upper Semliki Valley, Zaire. *Science, 268*, 548–553.

Chapais, B. (2008). *Primeval kinship.* Harvard University Press.

Davies, N. B., Krebs, J. R., & West, S. A. (2012). *An introduction to behavioural ecology.* Wiley.

Erlandson, J. M. (2001). The archaeology of aquatic adaptations: Paradigms for a new millennium. *Journal of Archaeological Research, 9*(4), 287–350.

Erlandson, J. M., & Moss, M. L. (2001). Shellfish feeders, carrion eaters, and the archaeology of aquatic adaptations. *American Antiquity, 66*(3), 413–432.

Fehr, E., & Fischbacher, U. (2003). The nature of human altruism. *Nature, 425*(6960), 785–791.

Hallett, E. Y., Marean, C. W., Steele, T. E., Álvarez-Fernández, E., Jacobs, Z., Cerasoni, J. N., & Dibble, H. L. (2021). A worked bone assemblage from 120,000-90,000 year old deposits at Contrebandiers Cave, Atlantic Coast, Morocco. *iScience*, 1–21. https://doi.org/10.1016/j.isci.2021.102988

Hrdy, S. B. (2011). *Mothers and others: The evolutionary origins of mutual understanding.* Belknap Press.

Kelly, R. L. (1995). *The foraging spectrum.* Smithsonian Institution Press.

Knauft, B. M., Abler, T. S., Betzig, L., Boehm, C., Dentan, R. K., Kiefer, T. M., Otterbein, K. F., Paddock, J., & Rodseth, L. (1991). Violence and sociality in human evolution [and comments and replies]. *Current Anthropology, 32*(4), 391–428. https://doi.org/10.2307/2743815

Kroeber, A. L. (1925). *Handbook of the Indians of California* (Vol. 78). Smithsonian Institution, Bureau of American Ethnology.

Lombard, M. (2021). Variation in hunting weaponry for more than 300,000 years: A tip cross-sectional area study of Middle Stone Age points from southern Africa. *Quaternary Science Reviews, 264*, 107021.

Marean, C. W. (2014). The origins and significance of coastal resource use in Africa and Western Eurasia. *Journal of Human Evolution, 77*, 17–40.

Marean, C. W. (2015). An evolutionary anthropological perspective on modern human origins. *Annual Review of Anthropology, 44*, 533–556.

Marean, C. W. (2016). The transition to foraging for dense and predictable resources and its impact on the evolution of modern humans. *Philosophical Transactions of the Royal Society of London B: Biological Sciences, 371*(1698), 1–12.

Marean, C. W., Bar-Matthews, M., Bernatchez, J., Fisher, E., Goldberg, P., Herries, A. I. R., Jacobs, Z., Jerardino, A., Karkanas, P., Minichillo, T., Nilssen, P. J., Thompson, E., Watts, I., & Williams, H. M. (2007). Early human use of marine resources and pigment in South Africa during the Middle Pleistocene. *Nature, 449*, 905–908.

Marean, C. W., Cowling, R. M., & Franklin, J. (2020). The Palaeo-Agulhas Plain: Temporal and spatial variation in an extraordinary extinct ecosystem of the Pleistocene of the Cape Floristic Region. *Quaternary Science Reviews, 235*, 106161. https://doi.org/10.1016/j.quascirev.2019.106161

Marlowe, F. W. (2005). Hunter-gatherers and human evolution. *Evolutionary Anthropology, 14*(2), 54–67.

Wilson, E. O. (2012). *The social conquest of earth*. W. W. Norton & Company.

Yellen, J. E., Brooks, A. S., Cornelissen, E., Mehlman, M. J., & Stewart, K. (1995). A middle stone age worked bone industry from Katanda, Upper Semliki Valley, Zaire. *Science, 268*, 553–556.

3

Islands and the Sea

Godfrey Baldacchino

When I teach study units relating to islands, I always dedicate one session expressly to discussing the sea. It is not much, considering the enormous role and contribution that the sea plays in all our lives—including my own, something that I found an opportunity to write about (Baldacchino, 2015)—and then especially in the lives of island residents, whose countries, economies, societies and environments are always girded, and sometimes dictated, by that aquatic medium. Over 600 million people live on islands today, which is about one out of every ten. 250 million of these in Indonesia alone. The African coastline is generously girded with islands; from Djerba (Tunisia) in the north; Soqotra (Yemen) in the east; Dyer (South Africa) in the south; and Eticoga (Guinea-Bissau) in the west. Aldabra (Seychelles), Kuntah Kinteh (Gambia), Robben (South Africa) and Stone Town (Zanzibar, Tanzania) are some African island sites ensconced as UNESCO World Heritage. Six African nations are exclusively enisled: Cape Verde, Comoros, Madagascar, Mauritius, São Tomé y Príncipe and Seychelles.

Many islanders would have specific episodes etched in their memory that speak to particular significant events in their own personal, or their island, or their country's history that would have an intimate connection with the sea.

G. Baldacchino (✉)
University of Malta, Msida, Malta
e-mail: godfrey.baldacchino@um.edu.mt

Mention August 15th to the people of my own country: Malta, a small European archipelago state, the world's tenth smallest country by land area. The feast of Saint Mary in the Christian calendar, this is the day that, in 1942, the remnants of a British naval convoy limped into Malta, then a British colony, providing vital food and fuel to a nation under siege from the Axis forces during the Second World War and avoiding its likely surrender. Mention March 11th to the Japanese, and they will recall the 2011 Tōhoku earthquake and tsunami, the strongest ever recorded in Japan, and resulting in some 20,000 casualties. And there is no need to mention any date to the citizens of the small atoll states of Kiribati, Maldives, Nauru and Tuvalu—along with the many residents of islands in such disparate locales as the Brahmaputra delta (Bangladesh) and eastern Papua New Guinea. Their men and women are quite aware of the livelihoods they procure from the sea and the shore; the ravages of occasional floods or king tides; and the more foreboding prospect of slow but inexorable sea level rise that is yet to force their permanent relocation. West Africa's coastline is especially prone to a similar catastrophe, with settlements like Keta City (Ghana), Saint Louis (Senegal), and Lagos (Nigeria) especially in peril (Ahedor, 2021).

There are various ways in which one can approach the study of water from an 'island studies' perspective. There are myriad methodologies that allow us to connect with water: and this book is itself testimony to this epistemological richness and diversity. In this brief essay, I will highlight two, extreme interpretations of the role of the sea and the ocean. In their own contrasting ways, their insights illuminate the myriad interconnectivities that *Homo sapiens* has with the sea. These bonds become deeper and more intimate when dealing with islands and islanders. Moreover, as we progress towards smaller and smaller islands, the role of the sea becomes more salient. In small islands, life is dictated and determined by the ferry schedule. What goes on aboard ferries, and during that ever necessary downtime when one must wait to board, is an integral part of small island life: ferry tales, indeed (Vannini, 2012).

First, the Facts

So: first, I invite you to ask your friends, or students in a classroom, of any age, to volunteer at least ten reasons why the sea, and water in general, is so important to us. The 65 students following my 'Island Life' course in summer 2021 at the University of Prince Edward Island, Canada, offered the following answers. They said the sea can be associated with:

1. Water Cycle (Evaporation/ Transpiration, Condensation, Precipitation)
2. Site for Fish and other Marine Food (shellfish, seaweed …)
3. Site for Seabed Minerals
4. Site for Health & Beauty Products
5. Transport Medium
6. Jobs in Maritime and Marine Sectors
7. Venue for Relaxation and Tourism
8. Attenuator of Extreme Temperatures
9. Carbon Sink
10. Source of Oxygen
11. Weather Regulator
12. Universal Dump

There are many other connections; those that are so obvious that we take them for granted (water is essential for life, for example) and other associations that are known and can be added to this list (water as the medium for rituals of purification in many religions, for example). But there are also possibly many other linkages that we may yet discover or that we do not as yet acknowledge. Water has been recognised as one of four basic elements (along with air, wind and fire) as well as home to three Zodiac signs within Western astrology: Cancer, Scorpio and Pisces. Water governs the Pig and the Rat in the Chinese Zodiac; and it represents wealth and money in Chinese geomancy (or Feng Shui).

The above listing corresponds to a factual approach to the question as posed. It follows from a rational engagement with the world around us; and where we pursue a positivist rigour that corresponds to the scientific method. This methodology is most at home in the hard or natural sciences, and follows from observation and fieldwork that seek to test out initial hypotheses about 'how the world works'.

There are other possible approaches, however. I have been criticised for falling into this trap, and presenting 'island studies' as if it were indeed a purely scientific endeavour, in the sense described above. Instead, island studies, and its interrogation of the role of the sea from this narrow scientific angle, would have 'some distance to go' (Fletcher, 2011). I have accepted this criticism as fair and seek to make amends.

Then, the Fiction

Hence, the second approach, where we resort to the rich trove of fictional literature to inform our understanding of islands and their relationship with the sea. I could dwell on Robinson Crusoe and the ensuing genre of Robinsonnades where heroes (and rarely heroines) arrive accidentally on deserted islands and survive the various ordeals and challenges thrown at them, eventually leaving the island as transformed persons. But this is a path that has been trodden too often. Let me instead focus on three select works of fiction that allow me to make a few observations that are relevant to this collection.

Take *Gould's Book of Fish*. Written by Tasmanian writer Richard Flanagan (hence an islander). Factually, it offers a fascinating history grounded in multiple insular geographies: that of Sarah Island, home of a notorious nineteenth-century penal colony in Van Diemen's Land (now Tasmania), itself an island off Australia, the world's island continent. Additionally, it is a book that tells us much about life, love, death, and the value of truth and how slippery (like a fish) this proves to be. The beautifully rich and captivating language throws up an entire sea of woe and wonder (plus various aquatic metaphors). The plot itself is woven around the real character of William Gould, sentenced on a small island off an island continent for a crime he did not commit: he narrates his tortuous journey of survival and tenuous hold on sanity, and reaching for meaning in his life by writing his story, and painting his fish. Twelve of them; one for each chapter; or, better, one chapter for each of the painted fish actually painted by the actual Gould (Flanagan, 2001).

Or take *The Old Man and the Sea*. Another novel, set on another island (Cuba); but this time written by an American novelist (not an islander), and equally celebrated in its movie form with the same name, with Spencer Tracy in a captivating rendition of Santiago, an old fisher and the main character. Tracy was nominated for the best actor award for his part in the movie. Regretfully, we do not have a similar accolade for the giant blue marlin that was as equal a protagonist in this drama. For this is not just a narrative of an old man's epic battle with a fish over three days; but of a fish's epic battle with a seasoned fisher who, in spite of his many years and long run of bad luck, is intent on catching it (Hemingway, 1952).

Meanwhile, this piece is appearing in a book about Africa, so let me add an 'African island' example to my select shortlist: *Brothers of the Sea*, by D. R. Sherman. There is an uncanny similarity to Hemingway's masterpiece: another old fisher, also without catching a single fish for too long; also set on an island (but this time, it is the Seychelles and not Cuba). Then a bottlenose dolphin saves the fisher's adopted son Paul from a shark attack. A friendship develops, but a dangerous triangle sets in when a beautiful young girl joins in, with predictable outcomes (Sherman, 1966).

All three stories happen to be written in English; and all three lead characters are male. But I cite them here as representatives of the world's rich and multilingual literatures that engage with the complex relationship between all people, the sea, and its creatures. A certain admiration goes out to our characters, heroes in their own way, as they tease out the story of their lives *in, with* and *by* the sea. They are all, somehow, 'coming of age' in these tales. Are not all islanders, in some obvious or perverse way, like Gould or Santiago or Paul, with their lives drenched and stitched in water? And, taking the cue from Flanagan, would not many islanders feel comfortable in being called amphibious, when their lives depend so deeply on that liquid medium and what it portends. Hay (2013, p. 209) goes so far as to argue that 'to be girt by sea creates distinctive island psychologies'. And, while hard to prove, 'the sea remains the thing – the key factor in the construction of island identity – and its almost infinite biophysical (and cultural) richness is replicated in the complex particularity within island cultures' (ibid., p. 229).

Meanwhile, the dolphin, the blue marlin and the convict's twelve 'fish' are also supreme characters of the protagonists' stories. They too inspire agency. Gould is metamorphosed into a seahorse; all that is left of the marlin is a bony carcass, but it rekindles Santiago's sense of vitality and purpose; and the dolphin comes with human qualities: endearing; shy yet teasing.

And so, via fact or via fiction, we must conclude that we owe the sea, and its living contents, a grave and great respect. In the pages of this book, is a rich sample of testimonies, narratives and ethnographies of multiple engagements with the sea.

References

Ahedor, J. (2021, September 24). West Africa is sinking. *Earth.Org*. https://earth.org/sea-level-rise-west-africa-is-sinking/

Baldacchino, G. (2015). There is so much more to sea: The myriad aquatic engagements of humankind. *Etnofoor, 27*(2), 179–184.

Flanagan, R. (2001). *Gould's book of fish: A story in twelve fish*. Grove Press.

Fletcher, L. (2011). '... some distance to go': A critical survey of Island Studies [Paper in special issue: The literature of postcolonial islands. DeLoughrey, Elizabeth (Ed.).]. *New Literatures Review* (47–48), 17–34.

Hay, P. (2013). What the sea portends: A reconsideration of contested island tropes. *Island Studies Journal, 8*(2), 209–232.

Hemingway, E. (1952). *The old man and the sea*. Charles Scribner's Sons.

Sherman, D. R. (1966). *Brothers of the sea*. Little, Brown & Co.
Vannini, P. (2012). *Ferry tales: Mobility, place, and time on Canada's west coast*. Routledge.

Filmography

The Old Man and the Sea. (1958). Directed by John Sturges.

4

Oceanic Humanities for Blue Heritage

Isabel Hofmeyr and Charne Lavery

The Humanities Perspective on the Oceans

Rising sea levels require new styles of humanities research that speak to environmental and decolonial themes. While earlier traditions of oceanic research focus on the sea as a conduit for human activity, more surface than volumetric depth, a new critical oceanic study grapples with how to go below the waterline. This work encompasses an expanding field of tracing how, by what media and genres, with what effects, the unseen ocean is mediated to human audiences. Whether speculative fiction, underwater photography, aquariums, coral or shipwrecks, how do these forms mediate the undersea and how do they deal with representational problems of scale, depth and visibility?[1] Using

---

The research presented in this chapter is supported by an NRF Grant UID: NRF 129219 'Antarctica, Africa and the Arts' and NRF grant Southern African Literature: Hydrocolonial Perspectives, UID 120813.

I. Hofmeyr
University of the Witwatersrand, Johannesburg, South Africa
e-mail: isabel.hofmeyr@wits.ac.za

C. Lavery (✉)
University of Cape Town, Cape Town, South Africa
e-mail: charnelavery@gmail.com

© The Author(s), under exclusive license to Springer Nature Switzerland AG 2022
R. Boswell et al. (eds.), *The Palgrave Handbook of Blue Heritage*, https://doi.org/10.1007/978-3-030-99347-4_4

four rubrics, this first orientation provides an overview of trends in the field known variously as oceanic humanities, blue humanities or critical ocean studies.

Wet Ontologies and Immersive Methodologies

There is now an emerging and rich body of scholarship that seeks to make visible the deep-seated land- and human-orientations of much research. Terming these "dry ontologies" or "dry technologies", this work aims to "immerse" concepts and theories to produce new modes of analysis. For human geographers Philip Steinberg and Kimberley Peters (2015), this has meant going beyond the reanimation of space through highlighting its "verticality, its materiality, and its temporality" to centralizing the ocean as the spatial foundation for theorization. The ocean, as "indisputably voluminous, stubbornly material, and unmistakably undergoing continual reformation," produces "wet ontologies" which allow us to move beyond debates shaped by terrestrial limits (p. 248). Taking the "oceans' nonhuman scale and depth as a first critical position and principle" in literary studies, Hester Blum and others have demonstrated the ways in which texts can be read through the "material and ontological flux of the ocean" (2013, p. 153). Moreover, as Steinberg and Peters (2019) have more recently pointed out, the ocean is not only liquid but also solid, as ice, and vaporous, as air. This means that other modes of oceanic perception should be added to immersive methods, embracing "more-than-wet ontologies".

Immersion takes different analytical forms. Some scholars literally go underwater, using diving to relativize land-based epistemic perspectives (Jue, 2020, pp. 1–3). Life experiences of diving, swimming, surfing and sailing produce new insights into well-known archives, such as slave narratives whose expression of a "black aquatics" has long been overlooked (Dawson, 2018, p. 6). Others examine histories of submersive technologies and the shifts these produce in visuality (Alaimo, 2011; Cohen, 2017). Ann Elias (2019) connects coral to imperial histories in her *Coral Empire,* demonstrating the ways in which underwater photography in the early twentieth century present coral as a "modern spectacle". Some scholars travel underwater analytically (rather than actually) in order to "conceptually displace" technologies like writing, photography or databases to estrange "dry" ideas of inscription and archive (Jue, 2018, 2020), or explore methods of reading under water to connect maritime and literary archaeologies (Lavery, 2020).

Jason DeCaires Taylor's submerged sculptures off the coast of Cancun have become a well-known focus for investigating submarine aesthetics, a field explored by a rapidly growing number of underwater artists, using performance, visual and sonic media (Cohen, 2017; Critchley, 2018; DeLoughrey, 2017). His sculptures are located just offshore in shallow water where visitors can snorkel to view the exhibits. Designed to become artificial coral reefs, they are placed close to existing coral formations so that seeding can take place. As Elizabeth DeLoughrey's account of DeCaires Taylor's corpus indicates (2017), the experience of viewing these installations is shaped by wind and current and hence differs from "normal" viewing in an art gallery. She continues, "impressions are informed by light, the viscosity of the water, the age of the sculptures, and the presence of marine species. While the exhibits are 'permanent,' the sculptures are not; they change every day based on their occupation by bacteria, algae, and, eventually, coral" (37). These various experiential and analytic methods of reading, thinking and seeing through seawater are producing novel insights about memorialization, submerged histories and multispecies being-with (Haraway, 2016).

Bodies of Water and Amphibious Aesthetics

Several methods embrace an elemental reading of water itself, along with the typologies, topographies and geologies of the ocean and related water bodies. The coast, for instance, is imagined as the site of human evolution itself, and hence one of the most enduring and productive artistic terrains (Gillis, 2012). Postcolonial literary critics have variously analysed the littoral as an ecotone, a place of "fractal multiplicity" and amphibiousness, which writers use to complicate orthodoxies of all sorts (Allan et al., 2017). As Meg Samuelson indicates, "littoral literature and coastal form … muddle the inside-outside binary that delineates nations and continents, and which has been particularly stark in framing Africa in both imperial and nativist thought" (2017, p. 17). Coastal morphology and its associated water formations or "waterside chronotopes" in Margaret Cohen's formulation, like lagoon, estuary, delta, shoal, white water and brown water constitute literary micro-regions (2007, p. 649). As climate change increasingly buffets coastlines, these regions like the Ganges–Brahmaputra delta, the Lagos lagoon or the Bombay archipelago become more prominent, producing narratives of coastal life and its perilous terraqueous futures (Ghosh, 2004; Martel, 2003; Okarafor, 2016).

Water as elemental media is elaborated on by scholars like Melody Jue (2020) and Steve Mentz (2020). Sarah Nuttall (2020) employs the hydrological cycle to link oceans to land via rainfall, highlighting the temporal implications of what she calls "pluvial time". The link between bodies of water and the body in water is turned by Astrida Neimanis into a theory of feminist watery embodiment (2017). This and other feminist theorization draw on Stacy Alaimo's "transcorporeality" (2010), the co-constitution of microbial and elemental life across corporeal boundaries. The skewing by the water of ordinary modes of perception and individuality, along with amphibious territorialities, has also been taken up by queer studies. The "queer blue sea" takes as a model and critical context the shifting qualities of waves or the "queer life" under the sea, as part of a "query ecology" as a "mode of relating to the nonhuman that is erotic, pleasurable, multiple, and shot through with world-making potentialities as well as with memories of trauma, violence, and loss" (Winkiel, 2019, p. 143; see also Huggan & Marland, 2020).

Multispecies Approaches

Multispecies frameworks have increasingly been adopted across a range of disciplines. Likewise in oceanic humanities, multispecies methods have been deployed by a range of scholars. These methods have sought to move away from anthropomorphic understandings of the seas, so prominent in popular culture as well as the focus on what Stefan Helmreich calls charismatic "marine mascots" (Helmreich, 2009, p. 13). Whether whales (a nineteenth-century "symbol of work, trade, and natural history" [5]) or dolphins ("an emblem of the twentieth-century sea" [5]), these have promoted human-centred views of the oceans with the mammal simply a platform for extending bipedal imaginings beyond land. In his work on the microbial ocean, Helmreich demonstrates how microscopic organisms not only promote twenty-first century "technoscientifically fueled imaginings" (6) of the ocean but encourage scholars to think about new frontiers of what constitutes "life" in a way that shifts us away from human-centred understandings and across an expanded range of scale to the viral and microbial.

Literary scholars have also ventured into marine multispecies methods as a way of generating new conceptual possibilities, for instance through an emphasis on marine plant ecologies. One recent trend has been to critique the model of the rhizome which has long been influential in the radical humanities as a way of thinking about "underground" and "out of site" forms of connection and interaction. Critiquing these as terra-based and presupposing a more or less fixed environment, literary scholars have turned to forms like coral and sargassum to generate new methods and approaches. Ottmar Ette (2017) has demonstrated how Mauritian poet, Khal Torabully uses coral as a radical ontological schema. As Torabully notes, coral provides an "agglutinating form of being together, consisting of stratification, densification, sedimentation, not unlike the palimpsest, instead of an erratic form of being together. It maintains the egalitarian aspect of being together but is open to all currents" (p. 115). Alternatively, in discussing a range of Caribbean-related texts, Aaron Pinnix (2019) centres sargassum (free-floating seaweed). As an ecosystem independent of land, sargassum entangles, rather than connects, "a circulating reservoir of meaning that figuratively surfaces violences inflicted on enslaved Africans during the Middle Passage" (p. 424). Sargassum "makes the ocean as material critically present, while also refusing to ever be attached to one location" (p. 426).

Multispecies thinking has further promoted intersectional, feminist and queer approaches, especially through cultural studies' attention to the way in which sea creatures can radically relativize human sensoriums and sense-making systems. Discussing the jellyfish, Eva Hayward asks: "Does face-to-face 'seeing' matter for organisms of a radically different scale and for whom 'eyes' are light receptors rather than picture makers?" (2012, p. 164). Stacey Alaimo also addresses the radical alterity of the jellyfish as constituting a challenge to the overlapping fields of science and aesthetics, while potentially motivating a more aquatic environmentalism (2013). These works form part of a broader trend of thinking-with other species, be these starfish, coral or octopus (Ette, 2017; Hayward, 2008). In the southern African context, Lesley Green (2016) examines the West Coast rock lobster, *Jasus lalandii*, or "kreef", as an iconic, complicating species, while Meg Samuelson (2018) thinks with sharks towards both species extinction and racial terror. Historian Marcus Rediker (2008) similarly uses sharks as a way into transatlantic slave histories, while literary scholar Joshua Bennett expands multispecies shark histories to propose the notion of a "black hydropoetics" (2018).

Black Hydropoetics

The black radical tradition has always been engaged with the undersea, as generations of scholars have sought to reconfigure the ongoing legacies of the Atlantic slave trade and the Middle passage. Ancestral and "aquafuturist" (to use Suzanna Chan's term [2017]), these hydropoetic explorations have been pursued across a rich range of creative and scholarly media that revisit the Atlantic undersea as a realm of speculative diasporic histories. Notable examples include artist Ellen Gallagher's mixed media explorations of the black Atlantic submarine (Gallagher, n.d.), the meditations of Christina Sharpe (2016) on the molecular remains of enslaved bodies and their "residence time", both of which draw on the electronic music of Drexciya and the underwater realm it imagines, where the children of drowned captives have adapted to submarine living (Wikipedia, n.d.). Together these constitute black hydropoetics as a major focus of diasporic scholarship and constitute the undersea as a potent source of ancestral memory and imagination.

Putting black hydropoetics in relation to southern African creolized waters opens up suggestive submarine cartographies. These might map how southern African oceanic ancestral traditions relate to the drowned communities of both the Atlantic and the Indian Ocean, the arena from which Cape slaves were drawn.[2] Once one considers this enlarged realm, the dramatis personae expand, taking in the djinns and genies of the Indian Ocean; the ancestors of the African oceans; the submerged deities of Indian indentured communities; and the drowned of both the Middle Passage and the Indian Ocean.

Indigenous Epistemologies and Creolized Water

Postcolonial theory has always sought to decolonize knowledge by making visible the colonial contours that shape existing curricula and canons. One method for relativizing such constructions is to examine older indigenous epistemologies of water and the oceans while paying attention to the ways in which these pre-colonial and colonial hydro-epistemologies interact and creolize.

The southern African case provides an apposite example of such creolization where maritime imperialism, "indigenous" hydrocosmologies, slave understandings of water, and settler hydrologies converge to produce inspirited views of water which intersect and creolize each other. For imperial ideologies, the ocean is infused with the spirit of empire apparent in the thrill of the maritime adventure novel, tales of naval heroes and the romance of the sea. For colonial settlers, water resources direct land dispossession, apparent in the "water-related" endings of names that frequently characterize colonial farms (-fontein, [fountain], -spruit [stream]) (Guelke & Shell, 1992, p. 820). Settlers were avid dam builders and as David Hughes (2006) has demonstrated for Zimbabwe, white farmers constructed dams as much for agricultural as aesthetic reasons. In the semi-arid interior, the dam with its reflective surface of water and indented shoreline approximated ideals of the British landscape, derived from glacier-sculpted waterscapes and mild year-round rain.

As regards vernacular hydrocosmologies, southern African waterworlds (both fresh and saline) bustle with congregations of deities, generally snake- and mermaid-like beings, ancestors and rain creatures who exercise control over rainfall (Bernard, 2010). Under Dutch and British imperialism, enslaved communities were transported from South-East Asia bringing with them Muslim ideas of water djinns (Die Djin-Vrou, 1939). These various hydrocosmologies interacted with Khoisan (First Nation) beliefs about water creatures producing the idea of the *watermeisie/watermeid* (water girl/maid) as Mapule Mohulatsi's research (2019) has recently demonstrated. This figure is now widely believed to occupy large bodies of water and is a staple of vernacular mythologies found in most South African townships (segregated black residential established under apartheid).

Yet, as with any situation of creolization, some forms have greater purchase on power, and in this case, more traction on the ocean than others. As submarine infrastructures in colonial port cities began to colonize water and create artificial land in the harbour precinct, they gave imperial theologies a firmer foothold, rendering the purchase of other spiritscapes on the ocean weaker. Harbour development erased coastal sites that would have been points of ancestral access, characterized by "living water", that was active and moving. The Durban beachfront itself was increasingly defined as a whites-only space, limiting access to the ocean for ritual or festival purposes. By supporting the proponents of maritime manliness while also defining the ocean as an infrastructural space, harbour engineering helped sideline submarine cosmopolitanism. On land, much the same process occurred as the cosmopolitanism of the port city was brought under the control of white officials, increasingly invested in a fortified and racialized maritime boundary.

Notes

1. On speculative fictions, see Chan (2017); on underwater photography, Cohen (2019a); on aquariums and coral, Elias (2019, pp. 125–126); on shipwrecks, Cohen (2019b).
2. Sibeko (2020) explores intersections between African ancestral beliefs and Atlantic slavery, a theme also touched on in Putuma (2017); see also Baderoon (2009).

References

Alaimo, S. (2010). *Bodily natures: Science, environment, and the material self*. Indiana University Press.

Alaimo, S. (2011). New materialisms, old humanisms, or, following the submersible. *NORA-Nordic Journal of Feminist and Gender Research, 19*(4), 280–284.

Alaimo, S. (2013). Jellyfish science, jellyfish aesthetics: Posthuman reconfigurations of the sensible. In J. MacLeod, C. Chen, & A. Neimanis (Eds.), *Thinking with water* (pp. 139–164). McGill-Queens University Press.

Allen, N., Groom, N., & Smith, J. (Eds.). (2017). *Coastal works: Cultures of the Atlantic edge*. Oxford University Press.

Baderoon, G. (2009). The African oceans: Tracing the sea as memory of slavery in South African literature and culture. *Research in African Literatures, 40*(4), 89–107.

Bennett, J. (2018). 'Beyond the vomiting dark': Toward a black hydropoetics. In A. Hume & G. Osborne (Eds.), *Ecopoetics: Essays in the field* (pp. 102–117). University of Iowa Press.

Bernard, P. S. (2010). *Messages from the deep: Water divinities, dreams and diviners in southern Africa* [Doctoral thesis, Rhodes University].

Blum, H. (2013). Introduction: Oceanic studies. *Atlantic Studies, 10*(2), 151–155.

Chan, S. (2017). "Alive…again": Unmoored in the aquafuture of Ellen Gallagher's "Watery Ecstatic." *Women's Studies Quarterly, 45*(1/2), 246–263.

Cohen, M. (2007). The chronotopes of the sea. In F. Moretti (Ed.), *The novel* (Vol. 2, pp. 647–666). Princeton University Press.

Cohen, M. (2017). Seeing through water: The paintings of Zarh Pritchard. In N. Allen, N. Groom, & J. Smith (Eds.), *Coastal works: Cultures of the Atlantic edge* (pp. 205–224). Oxford University Press.

Cohen, M. (2019a). The underwater imagination: From environment to film set, 1954–1956. *English Language Notes, 57*(1), 51–71.

Cohen, M. (2019b). The shipwreck as undersea gothic. In M. Cohen & K. Quigley (Eds.), *The aesthetics of the undersea* (pp. 155–165). Routledge.

Critchley, E. (2018). *Pursuit of beauty: Art beneath the waves*. https://www.bbc.co.uk/programmes/m00013nr. Accessed 31 May 2021.

Dawson, K. (2018). *Undercurrents of power: Aquatic culture in the African diaspora*. University of Pennsylvania Press.

DeLoughrey, E. (2017). Submarine futures of the anthropocene. *Comparative Literature, 69*(1), 32–44.

De Prada-Semper, J. (2016). "A partial clue": The genesis and context of Qing and Orpen's conversation. In J. De Prada-Semper et al. (Eds.), *On the trail of Qing and Orpen* (pp. 29–96). Standard Bank of South Africa.

"Die djin-vrou" (The djinn woman). (1939). In *Uit die slamse buurt: deel 1: Kaapse sprokies, fabels en legendes*, retold by I.D. du Plessis (pp. 75–78). Nasionale Pers.

Elias, A. (2019). *The coral empire: Underwater oceans, colonial tropics, visual modernity*. Duke University Press.

Ette, O. (2017). Khal Torabully: "coolies" and corals, or living in transarchipelagic worlds. *Journal of the African Literature Association, 11*(1), 112–119.

Gallagher, E. (n.d.). https://www.hauserwirth.com/artists/2783-ellen-gallagher. Accessed 31 May 2021.

Ghosh, A. (2004). *The hungry tide*. Penguin.

Gillis, J. R. (2012). *The human shore: Seacoasts in history*. University of Chicago Press.

Green, L. (2016). Calcalemus Jasus Lalandii: Accounting for South African lobster. In B. Latour & C. Leclerc (Eds.), *Reset modernity!* MIT.

Guelke, L., & Shell, R. (1992). Landscape of conquest: Frontier water alienation and Khoikhoi strategies of survival, 1652–1780. *Journal of Southern African Studies, 18*(4), 803–824.

Haraway, D. (2016). *Staying with the trouble: Making kin in the chthulucene*. Duke University Press.

Hayward, E. (2008). More lessons from a starfish: Prefixial flesh and transspeciated selves. *Women's Studies Quarterly, 36*(3/4), 64–85.

Hayward, E. (2012). Sensational jellyfish: Aquarium affects and the matter of immersion. *Differences, 25*(3), 161–196.

Helmreich, S. (2009). *Alien ocean: Anthropological voyages in microbial seas*. University of California Press.

Huggan, G., & Marland, P. (2020). *Queer blue sea: Sexuality and the aquatic uncanny in Philip Hoare's transatlantic eco-narratives*. ISLE: Interdisciplinary studies in literature and environment. https://doi.org/10.1093/isle/isaa161

Hughes, D. M. (2006). Whites and water: How Euro-Africans made nature at Kariba Dam. *Journal of Southern African Studies, 32*(4), 823–838.

Jue, M. (2018). Submerging Kittler. *Social Science Information, 57*(3), 476–482.

Jue, M. C. (2020). *Wild blue media: Thinking through seawater*. Duke University Press.

Lavery, C. (2020). Diving into the slave wreck: The São José Paquete d'Africa and Yvette Christiansë's Imprendehora. *Eastern African Literary and Cultural Studies, 6*(4), 269–283.

Martel, Y. (2003). *The life of Pi*. Canongate.

Mentz, S. (2020). *Ocean*. Bloomsbury.

Mohulatsi, M. (2019). *Black aesthetics and the deep ocean* [Unpublished Masters Thesis, University of the Witwatersrand].

Neimanis, A. (2017). *Bodies of water: Posthuman feminist phenomenology*. Bloomsbury.

Nuttall, S. (2020). Pluvial time/wet form. *New Literary History, 51*(2), 455–472.

Okarafor, N. (2016). *Lagoon*. Saga Press.

Pinnix, A. (2019). Sargassum in the Black Atlantic: Entanglement and the abyss in Bearden, Walcott, and Philip. *Atlantic Studies, 16*(4), 423–451.

Putuma, K. (2017). *Collective amnesia.* uHlanga.

Rediker, M. (2008). History from below the water line: Sharks and the Atlantic slave trade. *Atlantic Studies, 5*(2), 285–297.

Samuelson, M. (2017). Coastal form: Amphibian positions, wider worlds and planetary horizons on the African Indian ocean littoral. *Comparative Literature, 69*(1), 16–24.

Samuelson, M. (2018). Thinking with sharks: Racial terror, species extinction and other Anthropocene fault lines. *Australian Humanities Review, 63,* 31–47.

Sharpe, C. (2016). *In the wake: On blackness and being.* Duke University Press.

Sibeko, O. (2020). *Bottled seawater: A sea inland* [Unpublished Masters Thesis, University of the Witwatersrand].

Steinberg, P., & Peters, K. (2015). Wet ontologies, fluid spaces: Giving depth to volume through oceanic thinking. *Environment and Planning D: Society and Space, 33,* 247–264.

Steinberg, P., & Peters, K. (2019). The ocean in excess: Towards a more-than-wet ontology. *Dialogues in Human Geography, 9*(3), 293–307.

Wikipedia. (n.d.). *Drexciya.* https://en.wikipedia.org/wiki/Drexciya. Accessed 31 May 2021.

Winkiel, L. (2019). A queer ecology of the sea: Reading Virginia Woolf's The Waves. *Feminist Modernist Studies, 2*(2), 141–163.

5

World and Marine Heritages

Rosabelle Boswell

UNESCO has largely shaped the modern concept of heritage since the early 1970s. The task is not easy because there is extraordinary cultural diversity and social diversity in the world, making the process of shaping heritage, defining culture and recognising heritage challenging. Previously, heritage was previously conceived as an unproblematic universally manageable and largely celebratory legacy to be handed from one generation to the next (Garcia-Canclini, 1995). The conservation of heritage was to safeguard cultural and biological diversity across the globe.

Today, heritage is recognised as both cultural artefact of deep value to local communities, as well as a potential source and symbol of the power for elites in global society (Kirshenblatt-Gimblett, 2004; Winter, 2013). Heritage is also perceived to be a part of humanity's tangible and social evolution. Therefore, it cannot be merely discussed as a political tool. Tangible and intangible heritage are key to human evolution and to intimate human connection with the sea. Blue heritages are to be found in the artefacts and habitats remaining in coastal areas worldwide, (as discussed by Curtis Marean in this book's

---

The research presented in this chapter is supported by an NRF UID Grant number 129662.

R. Boswell (✉)
Nelson Mandela University, Port Elizabeth, South Africa
e-mail: Rose.boswell@mandela.ac.za

set of orienting discussions) and they are to be found in behaviours, values, practices and rituals associated with the sea.

In the African context and certainly in other parts of the world with experience of European colonisation, tangible and intangible heritage remains largely complex and contested, not just because of colonisation itself but because neo-colonial values emerging after independence from colonial rule have encouraged hierarchical and exclusionary heritage practices. In the context of South Africa, for example, post-apartheid leaders continued to prioritise tangible heritage, especially monuments to anti-apartheid heroes (Manetsi, 2019; Nettleton & Fubah, 2019). These commemorations and monuments served to celebrate and elevate anti-apartheid heroes in the newly formed nation state. The dismantling of colonial heritage and elevation of liberation heritage forms part of a complicated process of identity reconstruction in decolonisation (Nyamnjoh, 2016). As Long (2000) explained a while ago, the heritages of the colonised became important to new narratives of identity and belonging after colonial rule and could no longer be portrayed as the (marginal) heritage of the Other.

A major part of the decolonisation and democratisation process also included the excavation (Foucault & Sheridan, 1972), recognition, and restoration of local and indigenous heritages (Masoga, 2017). These have the function of not merely redressing the past or dealing with the trauma caused by colonisation; they became important to sharing new epistemologies among a citizenry hungry for new and empowering knowledge forms.

Heritage scholars writing from Africa and about Africa (Abungu, 2019; Petersen et al., 2015) have noted that Africa features poorly on UNESCO's voluminous list of World Heritages Sites. Moreover, the rich intangible cultural heritage of Africa, and in this case, its marine ICH, remains largely uncatalogued and unrepresented (Boswell, 2011). When considering the coastal and marine tangible and intangible heritage of Africa and other Least Developed Countries (LDCs) of the world, one finds an equally dire situation. As noted previously, there are 50 World Marine Sites, of these, only two are to be found in Africa. There are no cultural sites associated with World Marine Sites, even though and as this book proves, there is a rich tangible and coastal intangible cultural heritage in a diversity of 'less developed' nation states.

When considering marine and maritime heritage, one finds that most of the sites inscribed on the World Marine Heritage list are natural marine heritages. In the discussions offered in this book, however, a rich maritime heritage is revealed, suggesting the need to recognise maritime cultural heritage (and all its complexities), in a more distinct manner. Maritime heritage would include all the tangible cultural heritages in the marine sphere, overlapping with underwater cultural heritage, which is recognised by UNESCO.

In 2021, UNESCO stripped Liverpool city in England, of its World Heritage Status. Apparently, this is the only third incidence ever, of UNESCO removing World Heritage Status from a site. The reasons given included the fact that unhindered development continued within the buffer zone of the mercantile city and that this compromised its status, as a unique and authentic site. Some of the chapters in this book note how intangible cultural heritage conservation continues (and remains valuable) despite non-recognition of such heritages in global society.

Culture and Heritage in Least Developed Countries

Literature on cultural heritage notes that World Heritage Sites are important to LDCs, as these sites generate income for both national government and the communities where the sites are located. Considering the location of LDCs, one finds that of the 48 LDCs identified, 34 are in Africa, nine are small islands and the remainder are in Asia and the Caribbean, the latter being countries in which there was slavery and colonial rule for centuries. If World Heritage Sites are important to LDCs and World Heritages are deemed important to a wide array of socioeconomic activities, why are there not more of them in Africa, which contains most of the LDCs?

UNESCO states that it has an urgent development agenda for LDCs and that the mobilisation of cultural heritage and the output of cultural goods play a critical role in the process. In the chapters to follow, there are many examples of intangible cultural heritage in LDCs. The organisation (UNESCO, 2016, p. 10) states that,

> Cultural and natural heritage, including UNESCO inscribed World Heritage sites, is an important asset for sustainable economic development, which attract investments and ensure green and locally-based jobs. Activities associated with the stewardship of cultural and natural heritage often make use of local resources and skills and preserve local knowledge systems and infrastructures. Furthermore, sustainable cultural tourism is an economic driver which enhances the international profile of destinations, enables resource creation to increase competitiveness and strengthens local communities.

Furthermore,

> Cultural and creative industries (CCI) are one of the most dynamic and fastest-growing sectors of the world economy, contributing to sustainable economic growth, income generation and the creation of decent jobs. New data produced by the UNESCO Institute for Statistics (UIS) show that the export of cultural goods worldwide reached approximately US$ 212.8 billion in 2013, which more than doubled from 2004. CCI also generate US$2.2 billion a year, i.e. 3% of world GDP, and employ 29.5 million people (1% of the world's active population).9 CCI revenues exceed those of telecom services and employ more people than the car industry of Europe, Japan and the USA combined (29.5 million jobs vs. 25 million).10 In Mali, the culture sector accounted for 5.8% of employment in 2004 and 2.4% of GDP in 2006, including in the informal component (accounting for 57 % of the national economy).

However, LDCs only accounted for only one per cent of the total, global output of cultural goods, a situation that may be explained by the fact that few of them even have World Heritage Sites. Another point that must be made is that when one considers that Africa has only two Marine World Heritage Sites and that globally (and to date), only 50 of these have been identified, compared to more than a thousand terrestrial sites, one begins to understand that (a) coastal, cultural heritages are either poorly understood by the UN or (b) that these heritages are not perceived as important to the big questions of ocean sustainability and climate change. Given the research conducted thus far in South Africa at least (See this book's chapters by Erwin et al., and Thornton & Pillay), it is becoming evident that LDCs and certainly developed countries do not have sufficient information on the cultural and heritage dimensions of coastal existence.

Setting aside the potentially lucrative business of cultural industry outputs, which serve to provide (in the first instance) consumable goods to a mostly elite market, culture has value in and of itself, and so does heritage. Neither has to be as Edson (2004) stated, a product or service. Moreover, the casting of LDCs as potential producers of cultural products arising from recognised cultural heritages, can sinisterly feed into an existing, highly unequal (and yet still complex) dynamic of LDCs as producers and developed countries as consumers. It would be interesting to see another conversation about LDCs involved in cultural industry outputs for the development of LDCs.

Identifying and Maintaining World Heritages

Presently, however, both within and beyond LDCs, heritage is publicly discussed in relation to World Heritage. When considering sites recognised for World Heritage status by UNESCO, it is noted that these need to be maintained in accordance with the requirements set by the World Heritage Committee. Maintenance is both expensive and often counterintuitive, as they are perhaps located in sites used intensively by human populations, such as in Stone Town, Zanzibar. People living in such places may desire modern conveniences and building alterations, even though UNESCO has identified the unique historical architecture of a place, as that which is worthy of universal valuation and conservation. Moreover and as documented by those focusing on vulnerable heritages, it is noted that in countries where there is civil conflict or the increasing effects of climate change, it has become difficult to conserve and efficiently manage heritage. Many LDCs are facing drought, civil conflict and the legacies of long-term inequality. These facts mean that it is difficult to manage terrestrial heritage (natural and cultural, tangible and intangible), and, it is likely to be equally difficult to identify, manage, foreground and integrate marine and coastal cultural heritage in efforts towards ocean sustainability.

However, even before getting to the stage of management or upkeep, those LDCs seeking to elevate national heritages (terrestrial or marine) to World Heritage status, must have sufficiently competent scribes to assist in the preparation of nomination documents for the World Heritage Committee. And, before that, they need to convince their national government (and in particular, the national commission for UNESCO) that the chosen site is worthy of special recognition, above other places of cultural value. In a context where colonialism has dramatically altered both physical and social landscapes, it is incredibly difficult to achieve consensus. Not least because, there are often, after colonialism, competing claims for recognition and assertion of identity.

Finally, when considering the issue of heritage and of 'blue' heritage, there is also the problem of historical schisms within identifiable groups. Slave owners and colonials may not have annihilated cultural values and practices, but they did foster ideas of ineluctable difference and, in the colonial period in Africa at least, there were also efforts to divide and rule. The point being made here is that, not only did colonials seek to silence and erase/marginalise local heritages, they advanced ideas of difference and distinction, reducing knowledge of cultural commonalities and shared values.

Thus, women across differently identified ethnic groups may would have shared in similar life-cycle experiences and perhaps, experiences of oppression in patriarchal settings. However, the larger focus remained on gross differences between groups and the distinctive cultural contributions of each.

Diaspora Connections

Before there were thoughts on climate change in the social sciences and humanities, diaspora studies reveal remarkable continuities and commonalities between Africa and those who were either forcibly relocated or migrated beyond Africa. These continuities appear in language use, music, food, aesthetics and, in direct relevance to the title of this book, 'blue' heritage. Thus, in between important discussions on the horrors of slavery and the resilience of those seeking to adapt to new worlds, there are accounts of continuities in fishing, sailing and navigational techniques. Thus, and as we learn from this book, both the islanders of Mafia Island (c.f. Hasan's chapter) and Maori sailors navigate the sea using the moon and the stars.

My earlier fieldwork in Tanzania (2004–2008), secondary data research on Mozambique and observation of fishing techniques in Mauritius, also reveals that many of the people in these settings use the same kind of fishtrap. Accessing seafood sources is also gendered and these social divisions are apparent across both Tanzania and Mauritius. In brief, studies of the intangible cultural heritage of coastal communities, show that such communities have heritages that are not recognised, nominated or inscribed on the World Heritage List. The heritages are also not considered as valuable sources of knowledge for survival and livelihood generation. The heritages are also under threat from pollution, coastal commercial development, sea level rise and urbanisation.

Conclusion

Kirshenblatt-Gimblett (2004) has usefully described heritage as a mechanism for metacultural production. However, it is also useful to consider Smith's more recent view of heritage as important to social meaning and belonging or Herzfeld's idea of heritage and cultural intimacy (Byrne, 2011). In a time of climate crisis, of disjunctures and deterritorialisations, it will become ever more important for people to have a sense of certainty, or, they will have to learn to live with uncertainty (Boswell & Pillay, 2021). Even so, social meaning and belonging have long provided a source of certainty in times of distress. Reflecting on marine and cultural heritages, it may be important to shift attention from global processes of heritage management to consider local level, community based, endogenous cultural values in specific environmental settings. In shifting from the simple perspective of heritage and heritagisation, to understand how local communities have and are developing close and culturally intimate relations with the sea and coasts, we may be in a better position to understand the value of these resources to humanity, as well as how to ensure their long-term sustainability.

References

Abungu, G. O. (2019). Museums: Geopolitics, decolonisation, globalisation and migration. *Museum International, 71*(1–2), 62–71. https://doi.org/10.1080/13500775.2019.1638030

Boswell, R. (2011) *Re-presenting heritage in Zanzibar and Madagascar*. Eclipse.

Boswell, R., & Pillay, R. (2021). Changing heritage, changing the world? The case of a South African university. *Journal of Contemporary African Studies*. https://www.tandfonline.com/doi/abs/10.1080/02589001.2021.1945550. Accessed 7 Aug 2021.

Byrne, D. (2011). Archaeological heritage and cultural intimacy: An interview with Michael Herzfeld. *Journal of Social Archaeology, 11*(2), 144–157.

Edson, G. (2004). Heritage: Pride or passion, product or service? *International Journal of Heritage Studies, 10*(4), 333–348.

Foucault, M., & Sheridan, A. (1972). *The archaeology of knowledge*. Pantheon Books.

Garcìa-Canclini, N. (1995). *Hybrid cultures: Strategies for entering and leaving modernity*. University of Minnesota Press.

Kirshenblatt-Gimblett, B. (2004). Intangible heritage and the metacultural production of heritage. *Museum International, 561*(2), 52–66.

Long, D. L. (2000). Cultural heritage management in post-colonial polities: Not the heritage of the other. *International Journal of Heritage Studies, 6*(4), 317–322.

Manetsi, T. (2019). Heritage denunciation and heritage enunciation. In A. Nettleton & M. A. Fubah (Eds.), *Exchanging symbols: Monuments and memorials in post-apartheid South Africa*. SUN Press.

Masoga, M. (2017). Critical reflections on selected local narratives of contextual South African indigenous knowledge. In P. Ngulube (Ed.), *The handbook of research on theoretical perspectives on indigenous knowledge systems in developing countries* (pp. 310–331). IGI Global.

Nettleton, A., & Fubah, M. A. (2019). *Exchanging symbols: Monuments and memorials in post-apartheid South Africa*. SUN Press.

Nyamnjoh, F. (2016). *#RhodesMustFall: Nibbling at resilient colonialism in South Africa*. RPCIG Langaa.

Peterson, D. R., Gavua, K., & Rassool, C. (2015). *The politics of heritage in Africa: Economies, histories, and infrastructures.* Cambridge University Press.

Rankin, E. (2013). Creating/curating cultural capital: Monuments and museums for post-apartheid South Africa. *Humanities, 2,* 72–98.

UNESCO. (2016). *Sustainable development in the least developed countries, towards 2030.* UNESCO.

UNESCO. (2020). *The future we want: The role of culture in sustainable development.* http://www.unesco.org/new/en/culture/themes/culture-and-development/the-future-wewant-the-role-of-culture/. Accessed 19 Feb 2021.

Winter, T. (2013). Clarifying the critical in critical heritage studies. *International Journal of Heritage Studies, 19*(6), 532–545.

Part II

Historical, Cultural and Literary Perspectives

6

Evolving Hegemonies of Blue Heritage: From Ancient Greece to Today

Evanthie Michalena and Jeremy Hills

In this chapter we unpack the shifting hegemonies of blue heritage over time, and through the contests which have shaped their past and their present. We do so by drawing on a study of contrasting influences on a specific coastal area of an island, separated by over twenty-five centuries, a period that extends from today back to the time of the ancient Greeks. This area is part of an island called Salamis, which lies in the Saronic Gulf of Greece. The area contains the protected archaeological site of a famous and decisive historical event, the sea battle of Salamis, in which the Greeks defeated the Persians 2500 years ago. More recently, this site has been exploited by the ship repair and recycling industry, with grave consequences for the preservation of its heritage. Our chapter, therefore, considers two battles: the real, historical sea battle that took place just off the island's shore, and the contemporary battle

---

The research presented in this chapter was not conducted with direct funding.

E. Michalena (✉)
Regulatory Authority for Energy, Athens, Greece

J. Hills
University of South Pacific, Suva, Fiji
e-mail: jeremy.hills@usp.ac.fj

between heritage preservation and industrial exploitation. The narrative of this chapter concludes at a point of interface: at this point are joined the differing and conflicting uses of this site, and the governance system that stands over them.

Islands and Salamis

Islands are at the forefront of the global fight against climate change, as they are among the first sites to experience the devastating impacts that global warming has on local ecosystems and livelihoods. Indeed, islands host locally most of the infrastructures needed for the management of their resources, while the (often intense) demands of seasonal tourism exact a heavy toll on both island infrastructures and resources.

Many islands in Greece face challenges derived from insularity and seasonality, and experience permanent handicaps with regard to energy production, transportation, natural resources management, access to markets and economic diversification. However, Salamis is an island located very close to both the mainland (it lies less than two kilometers from the mainland) and the Greek administrative capital of Athens. Salamis, thus, is a case where these inherent insular and seasonal characteristics, located as they are near to the national center of governance, can create opportunities for the island to function as a laboratory of technological, social and financial innovation. Salamis could, in the near future, rise to be seen as the ideal test-bed for the deployment of smart, integrated solutions, which maximize the synergies between energy, transport, water and waste management through the use of cutting-edge technologies (including Information and Computing Technology [ICT]), and which can be replicated in other small island municipalities, mountainous, rural and generally geographically isolated areas.

Technology and infrastructure create opportunities for Salamis, but its natural and cultural resources are another strong asset. It is blessed with many archaeological sites, including one of the most important archaeological sites of Greece, that of the battle of Salamis. Its importance is indicated by the fact that it is listed as the highest grade "A" site by the Ministry of Culture of Greece, a status equivalent to that held by the Parthenon of the Acropolis in Athens. The combination of locality and heritage provides huge potentials to this island, and also strengthens the capacity of island authorities and communities to ensure the optimal use of infrastructures and resources, and thus create an enabling environment for sustainable economic activities to flourish on the island.

The relationship between islands, sustainability and the struggle to counter climate change is recognized at the European Union (EU) level, and even above it. The EU energy and climate policy framework for 2030, the Covenant of Mayors (an EU initiative under which municipalities agree to curb their greenhouse gas emissions: see Christoforidis et al., 2013, p. 644) and recent agreements under the United Nations Framework Convention on Climate Change (UNFCCC),[1] all emphasize the role of local authorities in tackling climate change by reducing emissions, building resilience and establishing cooperation platforms at local, national and international levels. Actors at these levels—local and regional authorities, the business sector, academia, and civil society—are key for devising and implementing a place-based, transformative development agenda that exploits islands' competitive advantages and generates sustainable local growth and prosperity for citizens and visitors. Salamis is thus in a strong position in relation to its blue economy, as it has the potential to engage in a participatory, inclusive and diverse fashion with well-established governance regimes at all levels: at island level, at regional and national levels and at the level of the EU as a whole.

Culture and nature are, according to the United Nations Educational Scientific and Cultural Organization, two sides of the same coin (UNESCO, 2021). Culture is rooted in time and place, and helps people frame how they relate to nature and their physical environment, to the earth and to the cosmos. Or, as some scholars put it, "the way individuals relate to the natural environment is culturally patterned" (Milfont & Wesley Schultz, 2016). Rather than existing in separate and parallel realms, biological and cultural diversities are closely interdependent: they have developed over time through mutual adaptation between humans and the environment. Environment and culture interact with and affect one another in complex ways in a proxy co-evolutionary process. This suggests (again according to UNESCO) that any local policy aiming to protect the natural environment and achieve sustainable development will necessarily have to take the culture of the affected communities into consideration, and act upon the implications of that consideration (UNESCO, 2021).

The close linkage between nature and culture emphasized by UNESCO is, in turn, strongly aligned to sustainable development. This concept refers to that form of "development that meets the needs of the present without compromising the ability of future generations to meet their own needs" (World Commission on Environment and Development, 1987), and rests on three pillars, economic, environmental and social. Conceptually, sustainable development includes both "development", viewed simply as economic change, and "development" as conceived in the broadest possible manner, including, among other things, social and environmental development. Such development is oriented to future generations and progressed in a manner that does not reduce present options and capacities but ensures generational equity (Mitoula & Kavouras, 2020). Equity is not the only thing that should span generations: so, too, should the lessons of history, and of historical events like the Battle of Salamis.

The Battle of Salamis in 480 BC

The second Persian invasion of Greece had seen significant victories for the Persians at the battles of Thermopylae and Artemisium, which they followed with advances into the Greek heartlands of Boetia and Attica. Then, off the island of Salamis in 480 BC, the Greeks defeated the Persian naval fleet, in what was the largest naval battle ever fought in the ancient world (Krentz, 2020). In this battle, the Greeks were outnumbered, but still decisively overcame Xerxes and his empire.

The ancient Greek historian Herodotus relates that, around 490 BC Darius, the King of Persia, sent ambassadors to many of the Greek city-states. Their mission was to demand "Earth and Water" from these cities: according to Persian custom, the rendering of these symbols was a sign of allegiance to the Persian empire (Herodotus, *Hist.* [Volume III], pp. 193–195). Some Hellenic city-states were surrounded and one (Thassos) demonstrated its submission to Persia by demolishing its own city walls. In Sparta, however, Persian ambassadors were thrown into a well, as a sign of sarcasm, as this was where they would find "earth and water". In Athens, meanwhile, the ambassadors were executed. Herodotus informs us that the Athenians threw the Persian ambassadors off a cliff, because this was the manner in which they were used to executing "criminals". Seeking revenge for the killing of its ambassadors, Persia sought to take revenge on Athens in 490 BC, at the Battle of Marathon. There, the soldiers of Athens crushed the Persian troops and saved their city.

Ten years later however, the Persians returned, this time with seemingly insurmountable military power. The Hellenic leader Themistocles reasoned that the Persians would not proceed further through to the Peloponnese if they could be lured to Salamis Island by the Greek fleet. Themistocles also threatened the Spartan Hellenes who were hesitant to support him, noting that if his plan was not followed then he would lead the Athenians to leave Greece, and thus transferring the naval power of Athens to lower Italy. This threat, the last ace of Themistocles, was what made the leader of the Spartans align his city with the Athenians. The Spartan leader Evryviadis knew well that without Athenian ships the Greeks would not be able to fight Persians, and that their absence would mean surrender. This is how the decisive battle between Greeks and Persians came to take place at Salamis.

As Herodotus reports, Themistocles also resorted to cunning stratagems: he sent a messenger to the Persians, to inform them that "the Greeks have lost heart and are planning flight, and that now is the hour for you to achieve an incomparable feat of arms, if you suffer them not to escape. For there is no union in their counsels, nor will they withstand you anymore, and you will see them battling against each other, your friends against your foes" (Herodotus, *Hist.* [Volume IV], p. 73).

The Persians, he said, would win an unprecedented victory over the Greeks, because the latter were beginning to quarrel and disagree among themselves. The Roman author Plutarch (writing several centuries later) reports that Xerxes, the leader of the Persians, believed that these words were said by a friend and, after listening with joy, immediately instructed all Persia's ships to sail to the Straits of Salamina, in order to force the Greek fleet into a naval battle (Plutarch, *Lives*, p. 37).

Meanwhile, the exiled statesman Aristides sailed from Aegina (a nearby island in the Saronic Gulf) to meet the Athenians and offer his services against the enemy. He saw that he could not enter the Straits of Salamis because they were barred by the loitering Persian fleet. He landed at a nearby village on Salamis called Selinia and then went on foot to the Bay of Ambelakia where the Greek fleet was massed. The enclosed nature of the Bay of Ambelakia, with its surrounding hills, makes it one of the best protected anchorages in the region. On meeting with the Greeks, Aristides informed them that they were surrounded by Persian vessels waiting in the Straits outside the Bay. Convinced by their predicament, the Greeks took up their positions for battle.

According to Herodotus, 380 Greek ships took part in the battle, along with 76,000 Greek fighters from Athens, Sparta, Corinth, Megara and many of the islands. Strauss notes that ten years after Marathon, the Greeks were now a unified force, which "ranged from the richest person to the poorest, from the cavalier to knave, from Panhellenic champions to losers at the childhood game of knucklebones...." (Strauss, 2004, p. 144). It now seemed that "suddenly, there were no more Athenians, no more Spartans, no more Corinthians. There were only Greeks" (Strauss, 2004, p. 144). The

Persians took part with a more substantial force of 1207 ships (according to the dramatist Aeschylus, who fought at Salamis) and about 517,000 men (Aeschylus, *Pers.*, p.11).

A golden throne for the Persian King Xerxes was placed at the foot of Mount Aigaleo. From this place, it was easy for him to observe his fleet in the Strait of Salamina and supervise his men, certain that the victory would be his. Many centuries later, a contemporary Greek author would recount this scene, drawing on the English poet Lord Byron's evocation of the event[2]:

On the back of the rock sits a king and gazes at Salamis, daughter of the sea. It's all his. At his feet were his uniform and his warriors. Did he count them at dawn and when they were at sunset? (Mpithizis, 2010).

On this day 2500 years ago, surprise was the most effective and economical of weapons. When the Persians saw the sizeable and unified Greek fleet emerge from the Bay, they must not have known or believed that the Greeks were ready to fight. The first sign of concern for the Persians was an unexpected sound from all Greek ports—the singing of a war song, a pagan Greek practice, which is described by Aeschylus as one that instils courage and hope in hearts of those who use it, by spreading fear among enemies. Then, the Persians heard the loud sound of the special trumpets, an undeniable call to arms. After that, came the sound of paddles: they were starting to hit the water hard, and the Battle of Salamis began.

On the right (east) flank were the Spartans with Evryviadis, who led the advance; the Athenians held the left (western) edge; the Aeginian ships were probably next to the Athenians and the other Greeks were in line with each other. The Greek fleet spread widely over a front two and a half kilometers wide (from Kynosoura to Cape Tropea or the Islet of Saint Georgios). The triremes of the back row were ready to fight back against any Persian ship that would attempt to penetrate the front row. The two rivals stood up to each other, but then the Greeks pretended to retreat, leaving the enemy at their back and pointing their bows toward the coast of Salamis. Through this "retreat" they led the Persians close enough to the shoreline of Salamis to allow the Athenians massed there to strike at them from the coast.

As is often the case in the history of the battles, the first blood was shed not at the behest of the generals, but on the initiative of a subordinate who was tired of waiting. Someone from the Municipality of Pallini charged ahead with his trireme, and, with its sharp bow, he pierced a Persian vessel. This gave the Greeks the first victory of the day, although his bow had been stuck too deep inside the enemy vessel and the men could not detach it (Mpithizis, 2010).

The Persian ships were designed for greater maneuverability, but lacked that ability in the confined space into which they had been enticed, especially in the fickle winds blowing close to shore (Krentz, 2020). The Greek trireme ships were heavier and sturdier than the Persian vessels, and won by ramming the confused and constrained Persians. The confusion was exacerbated by the heroism of the Greeks in body-to-body combat, termed "hirodikia" ("I raise my hand"). The Greek crews whose ships were sunk could still come out of the battle alive from their sunken ships and survive the battle, since they could swim to the shore and be saved on a friendly coastline. This was not the case for the Persian soldiers: few of them knew how to swim, and many of them drowned. Within hours the Greek victory was complete. The straits were filled with debris and wrecks and, from his throne, King Xerxes had clearly seen the humiliation of his fleet. The Persian casualties were 500 ships and 46,000 men. The Greeks lost 40 ships, but Themistocles' plan had triumphed. On that day, any further expansion of the Persian kingdom was prevented.

Salamis would become a global symbol of heroism. For some historians, scientists and scholars there is little doubt of the superiority of the ancient Greek spirit and that of the Athenian Municipality in particular, which reached its apogee after the victory over the Persians in the Straits of Salamis. Greece became a beacon after the battle of Salamis, one that shed its rich spiritual light on all the countries that had lived, until then, in spiritual darkness. Professor Llewellyn Jones states: "In many respects, for Xerxes the Greek campaign was mission accomplished. He succeeded in his objective of sacking Athens and bringing the Athenians low. Yet it cannot be denied that the subsequent defeat of the Persian navy at Salamis, even if it was not a matter of overarching concern for the Persian empire which continued to flourish for 150 years, was a turning point in the Greek conception of 'self' and the beginning of the Greek ideology of freedom" (Bragg, 2017). Such, at least, is one popular reading of Salamis and its consequences.

Scholars still debate just about every aspect of the battle, from the reliability of the ancient sources to the nature of the wooden warships involved, from the numbers of these ships to the topography of the Salamis strait at the time of the battle, from the credibility of Themistocles' trick to lure Xerxes into fighting, to the reconstruction of the fighting itself and its last act, in which land troops played a role (Krentz, 2020).

According to ancient geographers and historians (first and second century CE), the Bay of Ambelakia is a place one can find preserved submerged antiquities, which are gradually submerged before again re-emerging, according to the cycles of tidal and sedimentary processes around the bay and in the

Straits. This has been validated by contemporary underwater archaeological research around the bay: during 2016, the first detailed underwater reconnaissance research was carried out under a three-year program on the eastern coast of Salamina, in particular in the area of Ambelakia—Kynosoura and, more particularly, in the inner (western) part of the Bay.

The Bay was considered, in ancient times, to be the commercial and military port of the classical state, the Municipality of Salamis. At that time, it was one of the four most important ports around the Athenian city-state. This Bay is also a neighbor to the most important monuments of the victory: the polyandrion (tomb) of Salaminomachus and the Trophy, on Kynosoura (Fig. 6.1).

Archaeological excavations in the bay have been taking place for a long time, and still continue today, if only on a very small scale (Witchmann, 2021). The excavations are the result of an interdisciplinary underwater research program that, at the time of writing (2020–2021), is entering its sixth year (Archaeology Newsroom, 2021). Many magnificent findings from ancient times have been made, even if the most recent ones cannot yet be found in any museums or any published scientific papers, but are, rather,

Fig. 6.1 View of the archaeological site from the opposite mountain (Photo by Dr. Evanthie Michalena, 2021)

Fig. 6.2 "Hesitant", and very small scale, archaeological excavations by a team of Professors of Archaeology and the Ministry of Culture (in the red circle) (Photo by Dr. Evanthie Michalena, 2021)

just photographed and published on the internet (Archaeology Newsroom, 2019) (Fig. 6.2).

The Bay of Ambelakia at Present

The epic battle of Salamis is now two and a half millennia old. Today, debris and destruction are also seen in the Bay of Ambelakia, but these are neither the human remains of soldiers, or the physical remnants of triremes, those ancient galley-type vessels. They come from an entirely different source. The shores of the Bay of Ambelakia are not decked out with statues and pathways, as would befit an "Elysian fields of memory" where humanity should come to remember, and children should go to learn to dream (Mpithizis, 2010). The vista the Bay presents today is one of half-submerged and decaying ships, strewn around industrial shipyards. Together, these complete the unkempt image of a modern industrial shipwreck (Fig. 6.3).

The whole site of Ambelakia Bay, where the Battle of Salamis took place, is designated as a protected "historical site" by the Hellenic Ministry of Culture.

Fig. 6.3 View of the archaeological site from the opposite mountain: a half-submerged, decaying ship, beside an industrial shipyard inside an archaeological zone of absolute protection, type "A" (Photo by Dr. Evanthie Michalena, 2021)

By way of further support for the archaeological treasures of the site, in 2001 the Ministry of Culture defined zones of absolute protection type "A" (where the construction, alteration of the soil, construction and operation of ship repair units are strictly forbidden) and type "B" (construction is permitted, but only under strict conditions).

However, the visitor who has grown up learning about the battle of Salamis and the victory of Greeks against Persians, and who would like to visit the site, will come face to face with a very different reality. The peninsula of Kynosoura (the south bank of the Bay of Ambelakia) is three and a half kilometers long and 300–700 meters wide, and its total area is about half a square kilometer (Mpithizis, 2010). It now includes residential buildings and commercial and industrial complexes where shipbuilding and ship repair activities are carried on.

After the definition of the archaeological zones, the shipyard companies submitted cases to the Supreme Court of Justice (SCJ), seeking to cancel the definition. The Supreme Court overruled their cases and confirmed the supremacy of the archaeological designation. However, in spite of this, the environmental operating conditions of the ship industries were authorized in 2005 for 5 years by the Hellenic Ministry of Environment, the Ministry of Development and the Ministry of Commercial Maritime. This operating permit was not approved by the Ministry of Culture and successive requests for approval/renewal of the environmental conditions were rejected.

In 2013, a decision by the Hellenic Ministry of Maritime Affairs (MMA) determined the conditions of construction and use (shipbuilding and repair zone) in Kynosoura and Ambelakia. At the same time, a study from the National Technical University of Athens predicted that the coming years would see increased pressures in the Salamina Straits from the operation of the Port of Piraeus in the coming years (Belavilas, 2018). Among the suggestions of the study are the consolidation and the promotion of the historical points of this site of public use, the gradual shifting of the shipyards of Kynosoura to other locations and the prohibition of the expansion of the shipyards of Punta.

In 2020, forty-three parliamentary deputies belonging to the opposition party SYRIZA (the Coalition of the Radical Left) challenged the government's decision to grant a 15-year extension of the operation permit of the Kynosoura Shipyards, which are located in the archaeological zone "A" (Web press Koinoniki, 2020). The decision[3] of the Ministry of Environment and Energy ignored this zone "A" status, and the shipyard's permission to work in the area was extended for a further 15 years.

The protest by the SYRIZA deputies appeared to overlook the fact that during the period of the SYRIZA/ANEL[4] coalition government (2015–2019) the shipyards had already been operating without having the necessary permits; a fact apparently unknown to the protesting SYRIZA deputies. Therefore, both the government and the now opposition (then SYRIZA government) were entangled in the intrigue, consisting of half-submerged ships, ruins of German fortifications, urban waste and an abandoned quarry, alongside the shipbuilding activities (Ioannidis, 2018).

A few years ago, civil society actors, who were seeking to challenge the government's licensing of illegal facilities on the historical peninsula, sought legal advice through the Salamis Municipality. The lawyer from whom they sought that advice is now a minister in the current Greek government of Kyriakos Mitsotakis. The problem on which he advised remains unresolved. Worse still, the present Minister of Culture, Mrs. Mendoni, replies only with silence when the Ministry of Environment grants a license for ship-breaking activity to take place inside the archaeological site "A".

The Four Shipbuilding Units at the North of the Bay

Opposite the "Tomb of the Salamis fighters", to the north, at the location Kato Pounta Ambelakia, within the Second Protection Zone (Zone "B") of the archaeological site of Ambelakia, four shipbuilding firms are carrying on operations: "Panagiotakis Shipyards Ltd.", the "Theodoropoulos Group", "Naval Repair Salaminos S.A. (Koros Nikolaos)" and "New Greek Shipyards S.A." (formerly "Spanopoulos S.A."). Of these, the last shipbuilding company began its operations in 2004 as a machine shop for the repair of marine engines and machinery, having obtained a first installation license from the Greek State in 2003, after which it was granted further approval, in 2006, for the execution of embankment works (Belavilas, 2018). However, with a total area of 12,081 m^2, some of them were built without the consent of the Ministry of Culture (Belavilas, 2018). The Spanopoulos company then constructed a port facility made of reinforced concrete, which is extended until it reached Zone "A" of Supreme Protection (Belavilas, 2018).

Despite State licensing that provides for limited and specific works to be done in the area and under strictly specific environmental terms decreed by the state, large-scale ports and earthworks are found in the area. These include the construction of platforms, the emplacement of reinforced concrete blocks in the sea, excavation work at the bottom of the sea, large-scale excavations, the placement of boulders in various parts of the Bay and the use of huge quantities of reinforced concrete for the construction of working floors on embankment surfaces (Belavilas, 2018) (Fig. 6.4).

A sizeable area inside the Bay of Ambelakia was purchased by the shipbuilding company "Arcadia", of the Diamantis brothers, in 1971. The Diamantis brothers were (and are) related to Mr. Konstantinos Karamanlis (1907–1998), who was four times Prime Minister of Greece and twice the President of the Hellenic Republic, or, as some call him, "the First of the Nation".

The operation of the company Arcadia is not without state approval, as the company was granted a license by the Ministry of Culture in 1971 for the establishment of a makeshift machine shop on the site. Later, in 1977, the same ministry granted further permission for the installation of a shipyard, on the condition that a small hill (the "Tomb of Salamis fighters") would be excluded from the shipyard's operations. However, despite this legal obligation, the shipbuilding company has used the wider area as a waste dump, and has made efforts to establish businesses (including a coal transit station and a refrigerator factory) in that area. The business continues to operate to this day (Belavilas, 2018). Locals maintain that, in order for the shipyard to be

Fig. 6.4 A view over designated archaeological site from the opposite mountain—and of the New Greek Shipyards, S.A. (Photo by Dr. Evanthie Michalena)

built, it was necessary to destroy many of the ancient graves, where ancient fighters had been interred. Their contents were, it is said, then thrown in the sea.

In 2001, the State decided to declare Kynosoura a Zone of Ultimate Protection (Zone "A") under Government Gazette 1459/B/2001. In response, the two shipbuilding companies appealed to the Council of State, against the relevant ministerial decision[5] (Ministry of Culture/ARCH/A1/Φ02/54404/3270), but the Higher Courts of Justice (STE, dismissed the appeals [Belavilas, 2018]).

A big battle between the Hellenic Ministry of Culture and the local citizens, on the one side and the shipowners (and the politicians that support them) on the other side, began. The course of this strife was always in favor of the shipbuilders. The current minister of infrastructures, Mr. Konstantinos Karamanlis (a nephew of the former "First of the Nation" referred to above) has done little to save the reputation of his uncle. Meanwhile, the Chinese Shipping Company COSCO, which has been buying more and more shares from the then Hellenic Piraeus Port (OLP), continues to ignore legal decisions that rule out any claim to the whole region as belonging to OLP, and

Fig. 6.5 View of the archaeological site from the opposite mountain—Shipyards on Kynosoura, an ostensibly absolutely protected "Grade A" archaeological site (Photo by Dr. Evanthie Michalena, 2021)

which find, therefore, that this area should be annexed to the plans for OLP's extension (Fig. 6.5).

The New Decision Over the Ships Dismantlement Unit

By now, we have started to realize that the Greek workers employed in these commercial ventures, the very descendants of the ancient Greeks to whom those workers are indebted for the opportunity to inhabit and experience this sanctuary, have turned the Bay into a place of extensive ship repair activities and a degraded dump of concretized surfaces. Today, next to the tombs of the heroes of the Battle of Salamis, we find rusty ships dying of old age, buildings whose very legal existence is disputed, enormous floating ship repair platforms, and over 15 huge cranes which dominate the skyline of this epic place.

Just when (in May 2020) the island of Salamis was getting ready to celebrate the 2500th anniversary of the epic the Battle of Salamis, the Greek state chose to characterize part of the Bay of Ambelakia as a "Ship Repair

Site". Then, in July 2020, the then Greek Minister of Energy and Environment Kostis Hatzidakis, gave authorization for ships to be dismantled on this site, even though the European Commission does not permit ship dismantlement sites or activities in Greek territory. The Municipality of Salamis has appealed to the High Court of Justice against this decision, as have (independently of the Municipality) local citizens and SYRIZA members of the Greek parliament.

Two Clashing Hegemonies

The Salamis puzzle is a complex problem of polyonomy, bureaucracy and contradictions (Ioannidis, 2018). This is reflected in a relevant evaluation carried out by the Urban Environment Laboratory of the National Technical University of Athens (NTUA) in June 2018 (Belavilas, 2018), in response to a request from the civil society of Salamina. This "theatre of the absurd" starts from a real contradiction: the continuation of an economic activity, ship repair, which was first developed in the early 1970s, under the military dictatorship that ruled Greece at this time, in a zone that was subsequently declared "one of absolute protection" by Greece's democratic government.

The Greek state has never dealt properly with the archaeological value of the battle site. Excavations have been carried out there since 1965, but the progress of archaeologists has been slower than the creeping expansion of the shipbuilding industry. Ironically, archaeological protection has been invoked by each competing ship company, one against the other. Thus, the Kynosoura Shipyards, next to the Tymvos of Salamina, invoked archaeological protection when they saw the dizzying pace of development that has been taking place, since 2016, at the site of the rival Spanopoulos Shipyard (now operating under the name "New Hellenic Shipyards"), on the opposite peninsula of Salamina (which itself is officially under "Zone B" protection).

A crucial issue is that, as far as the Kynosoura Shipyards (owned by the Diamantis brothers) are concerned, it occupies a much larger area (approximately 58 acres) than the one in which it has been able to develop its facilities. In 2002, Tymvos (the site where the fighters of Salamis are buried) was excluded from all other forms of business development, so as to ensure the smooth implementation of the program for the promotion of the archaeological site (Belavilas, 2020).

In this way, the development of these facilities has completely distorted the coastal landscape of the historic naval battle site. "Since the landscape of the battle is marine, the main role in its perceptual dimension is played by the sea horizon and the coastline", write the NTUA scholars (Belavilas, 2018, 49). "The geographical measurement shows that the surrounding coasts of Perama (the town on the mainland) and Salamina, including the islets, have a development of about 24 km. Of these, 7.5 km are kept relatively unchanged, of which in their perfectly natural form only 5 km of the southern coast of Kynosoura and Atalanta" (Mpithizis, 2010). The Hellenic Environment Company in its letters in recent times has expressed its concern about the situation in the region to all co-competent bodies, and the Vice-President of the Hellenic Environmental Society, Costas Stamatopoulos, together with the President of the Architectural Heritage Council, Eleni Maistros, calling for the activation of those bodies on the occasion of the 2500th anniversary of the Battle in 2020 (Ioannidis, 2018).

The area was recently visited by the Minister of Culture, Mrs. Lina Mendoni, who was informed of the situation. However, sadly, the site of Tymvos (acknowledged by herself as a site of high archaeological significance a few years before she became a minister), has now been redesignated by the minister as "the so-called Tymvos", an expression that elicited adverse reactions from all interested political and regional administration parties (Typos Peiraios team, 2020)!

Blue Economy of Greece

According to the European Commission's report of 2018, blue economy sectors in the European Union are growing steadily (European Commission, 2018). More specifically, the blue economy in the EU produces a total turnover of €566 billion, generates €174 billion of value added, and creates jobs for nearly three and a half million people (European Commission, 2018). In several EU member states the blue economy has grown faster than the national economy in the last decade. During the global financial crisis that began in 2008, the blue economy proved more resilient in those member states, softening the effects of the downturn on coastal economies (European Commission, 2018).

Greece ranks among Europe's top five blue economies, with related sectors greatly contributing to its GDP and employment rates, in spite of the economic recession of recent years. According to the EU Blue Economy report UK (as of 2018, when that country was still an EU member state), Spain, Italy, France and Greece have Europe's biggest blue economies. When it comes to jobs and employment, Spain accounts for one-fifth of total employment in the sector, followed by Italy, the United Kingdom and Greece.

Combined, these four Member States account for more than half of the total blue economy-related jobs. At national level, the contribution of the blue economy to total national GDP significantly exceeds the EU average in several Member States, including Greece, where it accounted for 4.7% of the country's gross value added (GVA) in 2016.

The European Commission's Blue Economy Report states that, in Greece, the blue economy employs over 333,500 people and generates around €7.2 billion in GVA. It is dominated by the coastal tourism sector, which has contributed 76% of blue economy-related jobs, 67% of GVA and no less than 82% of overall profits from all related sectors in 2016. Maritime transport is also a large contributor, generating 16% of GVA and 11% of overall profits. The living resources sector (that sector based on activities such as fishing and aquaculture), on the other hand, generates 11.5% of the jobs and contributes 4% to GVA (European Commission, 2018). The blue economy has had a significant positive impact on Greek GDP and employment: while the national GDP fell strongly (28.5%) between 2009 and 2016, the amount of GVA contributed by the blue economy rose, over the same period, by 21%. Additionally, the percentage that the blue economy contributes to overall national GVA reached 4.7% in 2016, which is a 70% increase on the 2009 figure, when it stood at only 2.8%. Likewise, when the national levels of employment fell overall, the number of blue economy-based jobs grew by approximately 56%. The share of jobs covered by the blue economy now amounts to around 9.2% (a 93% increase compared to 2009).

This brings us back to the central theme of this chapter. The site of Ambelakia Bay on Salamis Island is important to the blue economy of Greece. The industrial ship complexes in the Bay provide jobs and income, especially to their owners who are not resident on the island. However, the opportunity cost of forcing archaeological tourism to co-exist with industrial decay has not yet been determined, even by the Municipal government which is influential in decisions on development on the island.

Over three-quarters of the blue economy of Greece is related to coastal tourism, and the economic potential that could be realized by reverting the area to a sustainable archaeological-focused visitors' area is, therefore, significant. In particular, there is strong potential for multiple locally owned small businesses to support such tourism, and, thereby, directly support the local island economy. The present situation, in other words, reflects a very narrow understanding of the blue economy concept. Voyer et al. (2018) have identified a range of themes and sub-themes embedded in that concept which bring together economic, environmental and social dimensions, and which have implications for innovation, technical capacity and governance.

At present, on Salamis Island, the blue economy seems to reflect only the economic dimensions associated with economic growth and employment. Inclusion of aspects such as restoration and protection of the environmental dimension, and equity and wellbeing of the social dimension are missing. Innovation related to the archaeological heritage (which is indelibly attached to the Bay) is absent, and governance does not display necessary aspects such as co-ordination and integration, and has not produced an effective regulatory framework. The dominance of the ship construction and repair industry is, in many senses, blocking the development of the blue economy in this area.

From a wider, regional perspective, however, one that includes the ship repair industry on the mainland around Piraeus port, we can see signs of a more detailed and comprehensive appreciation of the blue economy. The Municipality of Piraeus has produced a Blue Growth Strategy for the port, which offers a vision of a Piraeus empowered "to become an innovative, competitive, and resilient center for the organization and development of blue economy related activities with international orientation and scope" (Municipality of Piraeus, 2018). The municipality's Blue Plan, in this case, foresees economic expansion of traditional industries, but also the promotion of "blue entrepreneurship" and the enhancement of research, development and innovation in blue economy activities. This will be pursued by means such as micro-funds for venture labs, raising awareness of the blue economy in relation to culture, and the branding of Piraeus as a "destination". The depth and richness of identity, place and innovative futures invoked in the Piraeus Blue Plan contrasts markedly with the disregard for unique and place-based heritage exhibited by the present, and persistent, monotonic blue economy of the Bay of Ambelakia.

Concluding Remarks

The Battle of Salamis two and a half millennia ago was one where two competing sides fought, and where one victor emerged. The present-day battle between archaeology and industry in the same area is not so simple and has had, so far, no final, definitive outcome, except for the ongoing degradation of that area's irreplaceable archaeological treasures. The Greek government has advanced governance systems, has put provisions in place, and has adequate capacity, manpower and technological support to form and implement a solution. This is in line with the statements of those scientists who believe that "Re-learn" and "Re-value" are the keywords for a sustainable exploitation strategy of cultural heritage, combined with industrial development (Di Ruocco et al., 2017).

However, there seems to be an inability for a solution to emerge due to the inability for the archaeological and environmental proponents to reach an integrated decision (WWF 2019). It appears that the shipyards' licensees ignore the archaeological designations of their sites, and that the archaeological designation ignores the licenses for industrial operations on the same sites. In this ongoing battle neither side seems to see its enemy. Even comprehensive strategic planning for the blue economy on the neighboring mainland (also dominated by ports and the ship industry) has failed to influence planning for the Bay.

A narrative of a high priority for archeology is pronounced by the Greek state in its international statements. Greek Prime Minister Kyriakos Mitsotakis told the *Observer* (a British newspaper) that he would ask his British counterpart, Boris Johnson, to support the loan of the Parthenon Marbles[6] to Greece as a first step toward their permanent return (GTP Editing Team, 2019). At the same time, in another zone "A" archaeological site, the site of the Acropolis (where the Parthenon Marbles belong), construction of a new concrete pathway to facilitate wheelchairs, has been proposed. While this represents a new and positive enhancement of accessibility for disabled visitors, the fact that it was built on (and thus covering) the ancient stones of the Acropolis fueled a public row against national authorities in the summer of 2021 (Smith, 2021).

Although the pathway was placed over a synthetic membrane to stop damage to the underlying stones, the SYZRIZA parliamentarian and leader of the opposition Alexis Tsipras demanded that the government "stop abusing our cultural heritage" (Kyvrikosaios, 2021). Such apparent consternation on Tsipras' part is inconsistent with recent facts, as it fails to acknowledge his own party's complicity in the incremental concreting of another zone "A" site, the one which is the subject of this manuscript.

Moreover, the celebration of the concept of the blue economy, which has become "victorious" in Greece (Greek News Agenda Team, 2021), has been accompanied by little or no comment on the fact that it has failed, so far, to preserve the cultural, environmental and social side of an important archaeological site, and has been unable to check the creeping industrialization of another zone "A" site at Salamis Island. The opportunity costs of alternatively created economies are not considered, questioning the value of the blue economy narrative. The battle site and associated archaeological values are strictly place-based values, whereas ship-based industries are geographically transferable: therefore, the economic argument is not one-or-the-other but one that could evolve to recommend appropriate relocation of economic activities on the Greek coast, allowing both economic themes to flourish. The fragmentation of the Bay site into a zone of narrow permits and legal cases, however, seems to block any attempt at strategic management.

The failure of the government to restrict the ongoing industrial degradation of one of its highest priority archaeological sites has international consequences. Failure in this case makes the ongoing request by the Greek government, for the return of the Parthenon Marbles from the UK, look like a persistent joke—and a bad one. This is because the Parthenon Marbles look, now, as if they would clearly be in a safer place in the British Museum, rather than following the fate of other antiquities, consigned to the depths of a dirty and polluted sea. This narrates the story of a permitted and relentless destruction of an archaeological site and seascape where one of the most pivotal events in Greek and European history took place, a site that should have been one of the most highly protected examples of its type in the world.

Notes

1. The most recent of these is Paris agreement on climate change: as this chapter was being prepared for publication, in November 2021, the Conference of Parties was meeting in Glasgow (United Kingdom), under the auspices of the UNFCCC (Tobin & Barritt, 2021, p. 5).
2. Lord Byron's original lines, in his poem "the Isles of Greece", are as follows:
    A king sate on the rocky brow.
    Which looks o'er sea-born Salamis.
    And ships, by thousands, lay below,
    And men, in nations; – all were his!
    He counted them at break of day.
    And when the sun set, where were they? (Farrar, 1883, p. 215)
3. See decree YPEN/DIPA/68813/4143, 'Approval of environmental conditions for the operation of the facilities of the shipyards company "KYNOSOURAS S.A." in the area of Kynosoura and Ambelakia'.
4. ANEL, the junior partner in the Greek coalition government of 2015–2019, is a party of the radical populist right, whose acronym stands for 'Independent Greeks' (Aslanidis & Rovira Kaltwasser, 2016, p. 1078).
5. Ministry of Culture/ARCH/A1/Φ02/54404/3270.

6. The Parthenon Marbles, sometimes referred to as the Elgin Marbles, are a collection of marble sculptures that originally adorned the top of the exterior of the Parthenon in Athens, Greece. In 1801 a British nobleman (Lord Elgin) removed these sculptures to England; they are now in the British Museum in London. Greece considers the Parthenon Marbles to be stolen goods, and has frequently demanded that they be returned.

References

Aeschylus. (1909). *The Persians* (C. E. S. Headlam, Trans.). George Bell and Sons.

Archaeology Newsroom. (2019, June 14). New finds from underwater research on Salamis. Retrieved from https://www.archaeology.wiki/blog/2019/06/14/new-finds-from-underwater-research-on-salamis/ (Accessed 06 October 2021).

Archaeology Newsroom. (2021, April 16). Results of the underwater excavations off Salamina in 2020. Retrieved from https://www.archaeology.wiki/blog/2021/04/16/results-of-the-underwater-excavations-off-salamina-in-2020/ (Accessed 04 November 2021).

Aslanidis, P., & Rovira Kaltwasser, C. (2016, September 18). Dealing with Populists in Government: The SYRIZA-ANEL Coalition in Greece. *Democratization, 23*, (6), 1077–1091.

Belavilas, N. (2018). The archaeological landscape in Kynosoura of Salamis. Mapping of existing situation and exploration of the potential of its repair [in Greek]. National Technical University of Athens. Retrieved from https://www.docdroid.net/xl22sMp/kynosoura-salaminas-envlab-ntua-june-2018-pdf#page=29 (Accessed 21 October 2021).

Belavilas, N. (2020). The celebration of 2,500 years of the Battle of Salamis starts with a new decision for the operation of ships dismantlement unit on the archaeological land and sea space for the next 15 years [in Greek]. Retrieved from https://www.lifo.gr/blogs/almanac/biasmeno-arhaiologiko-topio-kai-oi-giortes-tis-naymahias-tis-salaminas (Accessed 04 November 2021).

Bragg, Melvyn (Host). (2017, March 23). The Battle of Salamis [audio podcast episode]. In Our Time, BBC Radio 4, https://www.bbc.co.uk/programmes/b08j99jl (Accessed 04 November 2021).

Christoforidis, G. C., Chatzisavvas, K. Ch., Lazarou, S., & Parisses, C. (2013, September 1). Covenant of Mayors Initiative—Public Perception Issues and Barriers in Greece. *Energy Policy, 60*, 643–655. https://doi.org/10.1016/j.enpol.2013.05.079

Di Ruocco, G., Sicignano, E., & Galizia, I. (2017). Strategy of sustainable development of an industrial archaeology. *Procedia Engineering, 180*, 1664–1674. Retrieved from https://www.sciencedirect.com/science/article/pii/S187770581718362

European Commission. (2018). *The 2018 annual economic report on EU blue economy*. EU Publications. Retrieved from https://op.europa.eu/en/publication-detail/-/publication/79299d10-8a35-11e8-ac6a-01aa75ed71a1 (Accessed 13 September 2021).

Farrar, F. W. (1883). *With the Poets: A Selection of English Poetry*. Funk & Wagnalls.

Greek News Agenda Team. (2021). *Greece among EU's "big five" in blue economy*. Retrieved from Greek News Agenda: https://www.greeknewsagenda.gr/topics/business-r-d/6761-blue-economy-2018 (Accessed 12 April 2021).

GTP Editing Team. (2019, September 6). *Greece to Officially Request Parthenon Marbles' Temporary Return*. Retrieved from https://news.gtp.gr/2019/09/06/greece-officially-request-parthenon-marbles-temporary-return/ (Accessed 10 June 2021).

Herodotus. (1922). *The Histories*. In Four Volumes. Volume III, Books VIII–IX (A. D. Godley, Trans.). Loeb Classical Library. William Heinemann.

Herodotus. (1969). *The Histories*. In Four Volumes. Volume IV, Books VIII–IX. (A. D. Godley, Trans.). Loeb Classical Library. William Heinemann.

Ioannidis, S. (2018). "Gordian Knot" the Battle of Salamis [in Greek]. Retrieved from https://www.kathimerini.gr/society/977213/gordios-desmos-i-naymachia-tis-salaminas/ (Accessed 03 May 2022).

Krentz, P. (2020, September 24). *Battle of Salamis: 480 BC*. Oxfordbibliographies.com. https://doi.org/10.1093/OBO/9780199791279-0196

Kyvrikosaios, D. (2021, June 9). *Greece faces row over wheelchair pathway at Acropolis*. Retrieved from https://www.reuters.com/world/europe/greece-faces-row-over-wheelchair-pathway-acropolis-2021-06-09/ (Accessed 24 October 2021).

Milfont, T., & Wesley Schultz, P. (2016). Culture and the natural environment. *Current Opinion in Psychology, 8*, 194–199.

Mitoula, R., & Kavouras, S. (2020). *Urban Development: Re-thinking City Branding. The role of Health and Safety*. Retrieved from Urbanistica Informazioni: https://www.academia.edu/44775619/Urban_development_Re_thinking_city_branding_The_role_of_health_and_safety (Accessed 07 May 2021).

Mpithizis, EA. (2010). *Ē naumachia tēs Salaminas: Tote kai tōra (The Battle of Salamis: Then and now)*. Association for the projection of the Salaminian culture.

Municipality of Piraeus. (2018). *Piraeus Blue Growth Strategy 2018–2024*. https://smartports.gr/wp-content/uploads/2018/05/KOKKALIS-BG-English-Presentation.pdf (Accessed 09 November 2021).

Plutarch. (1914). *Plutarch's Lives, Volume II: Themistocles and Camillus, Aristides and Cato Major, Cimon and Lucullus* (Bernadotte Perrin, Trans), ten volumes. Loeb Classical Library. William Heinemann.

Smith, H. (2021). 'Acropolis Now: Greeks Outraged at Concreting of Ancient Site'. *The Guardian.* Thursday, 10th June. https://www.theguardian.com/world/2021/jun/10/acropolis-now-greeks-outraged-at-concreting-of-ancient-site (Accessed 18 November 2021).

Strauss, B. (2004). *The Battle of Salamis. The Naval Encounter That Saved Greece - and Western Civilization.* Simon & Schuster.

Tobin, Paul, and Joshua Barritt. (2021, September 1). Glasgow's COP26: The Need for Urgency at "The Next Paris". *Political Insight, 12*(3), 4–7. https://doi.org/10.1177/20419058211044997.

Typos Peiraios team. (2020, 11 4). Λίνα Μενδώνη: Ο λεγόμενος Τύμβος των Σαλαμινομάχων. Retrieved from https://www.typospeiraiws.gr/%CE%B1%CE%BD%CE%B1%CF%83%CF%84%CE%AC%CF%84%CF%89%CF%83%CE%B7-%CE%B1%CF%80%CF%8C-%CF%84%CE%B7-%CE%B4%CE%AE%CE%BB%CF%89%CF%83%CE%B7-%CF%84%CE%AE%CF%82-%CF%85%CF%80%CE%BF%CF%85%CF%81%CE%B3%CE%BF%CF%8D/ (Accessed 12 July 2021).

UNESCO. (2021). Culture for Sustainable Development. Retrieved from http://www.unesco.org/new/en/culture/themes/culture-and-development/the-future-we-want-the-role-of-culture/the-two-sides-of-the-coin/ (Accessed 07 November 2021).

Voyer, M., Quirk, G., McIlgorm, A., & Azmi, K. (2018). Shades of blue: What do competing interpretations of the Blue Economy mean for oceans governance? *Journal of Environmental Policy & Planning, 20*(5), 595–616. https://doi.org/10.1080/1523908X.2018.1473153

Web press "Koinoniki". (2020). A story of craziness in Salamis Island [in Greek]. https://koinoniki.gr/2020/10/mia-istoria-trelas-sta-stena-tis-salaminas/ (Accessed 03 May 2022).

Witchmann, A. (2021, September). Archaeological Finds Shed Light on Battle of Salamis. Retrieved from https://greekreporter.com/2021/09/22/archaeological-finds-battle-of-salamis/ (Accessed 23 October 2021).

World Commission on Environment and Development. (1987). *Our Common Future.* Oxford University Press.

WWF. (2019, September 9). WWF: Two years from the sinking of Aghia Zoni II and still no punishment for the environmental crime [in Greek]. Retrieved from https://www.naftemporiki.gr/story/1513002/wwf-duo-xronia-apo-to-nauagio-tou-agia-zoni-ii-kamia-timoria-gia-to-periballontiko-egklima (Accessed 23 October 2021).

7

Tales of Ocean, Migration and Memory in Ancestral Homespaces in Goa

Pedro Pombo

Social differentiation and caste stratification existed in Goa, as in other regions of what is today India and across South Asia and integrated the Catholic conversion practices from the sixteenth century onwards. Under the Portuguese, the conversion of the local population to the Catholic faith happened with the maintenance of the vernacular village administration, the *gaunkari* from the Konkani *ganv* or *gaun* (village) and *kari* (administration) also known in Portuguese as *Comunidade* (community), which meant keeping generally intact the social status of the converts and the social codifications of the administration of the territory and of the sources of economic and social power. This historical process of religious conversion incorporating indigenous caste social and economic structures provided the background for the development of a Goan Catholic society, translated in specific aesthetic

---

It is an honour and a privilege to be part of such an exciting publication on this seductive field of research. I am thankful to Tejeswar Karkora, with whom I rediscovered Goa, visiting many villages across the state and exploring the beauty of lesser known Goan heritages. Many stories of empty houses and migration to East Africa were explored with Nalini Elvino de Sousa, who directs our documentary on Goans in Tanzania, and with whom I visited Dar es Salaam and Zanzibar in search of places, stories and recollections of the Goan diaspora.

P. Pombo (✉)
Goa University, Taleigao, India
e-mail: pedromanuelpombo@gmail.com

and cultural productions as well as particular social, economic and religious administrative systems. The conversion experience equally enabled the development of syncretic religious practices over time, explored in relevant anthropology researches on local religious and social structures (Henn, 2014; Newman, 2001; Perez, 2012).

The presence of caste and class-based hierarchies was, and is, part of the diasporic experiences from Goa (as well as from other regions of India) and directs us to pay attention to the diversity of backgrounds of Goan diaspora communities across the globe.

During the nineteenth century, Goans started to migrate firstly to cities as Bombay or Karachi, dynamic coastal cities of the British Raj, and, from the last decades of the nineteenth century, started to settle in Portuguese, German and British East Africa, being particularly active in the development of the largest cities in the region as well as in the countryside reached by new railroad networks. At the turn of the twentieth century, from the Portuguese colonial cities of Lourenço Marques (now Maputo) and Beira (in Mozambique) to the German city of Dar es Salaam or the territories administrated by the British, as the cities of Mombasa, Nairobi, Kampala and Stone Town in Zanzibar, Goan communities would grow in number and relevance. This diversity of circulation patterns and contexts has been explored in a growing body of work, from seminal stories of Goan migration at large (Frenz, 2014) to oral histories and generational accounts of diaspora experiences (Carvalho, 2014) which enquire about histories of specific Goan communities and modes of connecting with the homeland through festivals and traditions (Gupta, 2009), or the sensorial memories and belongings of diasporic lives (Rosales, 2009).

These social structures were translocated from Goa to the new settlement cities via several Goan clubs which reflected the social and economic status of local Goan families. During research on Goans in Tanzania, the social divisions of Goan communities in the diaspora are often mentioned as something that existed in the past or as something that slowly disappeared around the mid-twentieth century. Goan clubs in the diaspora therefore became fundamental places of community building and spaces for the maintenance of cultural heritages, social interactions and family life.

Goan communities in East Africa developed and maintained a variety of skills as: cooks, butchers, traders, clerks and tailors, photographers, musicians or architects, doctors, lawyers (Carvalho, 2016) and even consuls of Portugal (Frenz, 2013). These skills and forms of social visibility would be critical not only to survival but also important to family stories and memory-making between Goa, colonial South Asian and East African territories.

Reference must be made to the pioneering activity of Goan photographers in opening photographic studios in Zanzibar or Dar es Salaam and who are the authors of some of the oldest photographic records of this part of the world at the turn of the twentieth century (Gupta, 2016; Meier, 2013). At the same time Goan musicians would fill the most important Jazz clubs in Bombay and the social clubs spread along East Africa (Fernandes, 2012) as well as steamships, where Goan cooks and butlers were preferred for their upbringing closer to the European rules.

Besides the scholarly works already cited, the Goan publisher *Goa 1556*, led by the journalist and culture enthusiast Frederick Noronha, has been publishing a curious set of autobiographies and memories of diaspora Goans which highlights not only very personal experiences as well as the role that many Goan communities had in shaping colonial city life and the consequences of the African independences for the diverse Goan communities across East Africa.[1]

The relevance of this contextualization is that these social and economic dimensions of the diaspora would have consequences in the homeland and, more importantly, in the migrants' home villages and so-called ancestral homes. In a further section we will see how homespaces are perceived as places of display with almost a museographic sensibility, and observe how histories of migration are literally hanged on the wall and kept inside travel luggage in ancestral homespaces spread across Goan villages. For now, we will move towards the Goan houses, a central heritage element in the Goan cultural landscape.

House and Landscape: Verna Sensibility

The domestic architecture of the Catholic Goans, usually simply mentioned as Goan architecture, is the fruit of centuries of Portuguese influence on vernacular spatialities and building techniques. The particular cultural history of Goa provided the development of architectural models of different scales and richness that were adopted both by wealthy and socially important families belonging to the *Brahmin* and *Chardo* castes as well as by other social classes mostly connected with fishing and agricultural activities (Carita & Sapieha, 2002; Silveira, 2008). The development of an Indo-Portuguese, or Goan, architecture, would also be perceptible in the Hindu temples built or renovated during the eighteenth and nineteenth centuries in the so-called New Conquests, Goan territories that became under Portuguese rule during the second half of the eighteenth century (Pereira, 2001; Pombo, 2006).

Besides their particular aesthetics, a fundamental characteristic of this architectural heritage is the connection between architecture and ecology. Deeply rooted on land ownership, larger manor houses are mostly found scattered along the coastal lowlands, in close vicinity of the rice fields that sustained the families inhabiting them. The villages are usually spatially organized along the agricultural lands, paddy fields, coconut trees and cashew tree forests, and each neighbourhood has its own social composition, in a spatial organization of the social landscape of each village.

The cultural landscape of Goa is made, thus, of a fluid and intimate integration of vernacular architecture and environment (Pombo, 2019). The tiled roofs frame fields of growing rice or line up under the high and extremely slim coconut trees, in visual compositions that are profoundly rooted in the local culture and environment. The construction materials and colours translate the sensorial and chromatic qualities of the environment, while verandas, courtyards and sloppy tiled roofs allow the houses to breathe and be attuned with the hot dry season and the strong monsoon months.

This strong association between home and landscape has been central to the relations between diaspora and homeland. Home, landscape and family are seen and felt as intimately connected, and longing for Goa is often recalled by migrants as longing for the local landscapes, the experience of the monsoon or a Goan meal shared with the relatives over long meals.

The Ocean at the Window

One extraordinary way of bringing the ocean to mainland Goa is the unique tradition of building oyster-shell windows. Known as *carepa* windows in Portuguese, this uniquely Goan decorative feature is, with the *balcão*, one of the most common architectural and decorative features of Goan houses.

Estuary beaches as Siridão, Goa Velha or Chicalim are made of fine and dark sand and extremely vast low tides, due to the high sedimentation levels brought by the mighty Zuari river building over centuries of flat and slightly muddy coastal lands. Here, oysters grow in rich numbers, and the beaches are usually full of flat and translucid oyster shells in shades of white, yellow or mother of pearl chromatic reflections.

There is not a clear knowledge where the use of oyster shell on windows comes from, but the Portuguese windows with glass panes met, in Goa, with the traditional wooden shutters and the craftmanship of working with the nacre and mother of pearl, creating something unique and profoundly beautiful and poetic. At the same time, the lime used for construction commonly included crushed oyster shells, beneficing its properties in contact with the extreme humidity of the monsoon.

The extraordinary aesthetics of the *carepas* windows have been mentioned in many travelogues written by foreign visitors to Goa, and is without doubt one of the most recognizable and characteristic elements of the local architecture and aesthetic sensibility. Cut in small pieces, the oyster flat shells are inserted in wooden frames, composing double panes of square compositions. With time they gained small rectangles of glass in the middle of each pane, allowing a clear view from the interior of the house to the outside (Silveira, 2008).

The translucid and diaphanous light that the oyster shells produce will gentle illuminate interiors spaces, in an extremely soft, and filled with poetic sense, transition between the outdoors and the inner homespaces. Gaining curved and artistically decorated frames, *carepas* windows are omnipresent in the Goan landscape and are an astonishing testimony to the local skilled craftmanship and to the evolution of architectural styles that merge Portuguese and European features with local sensibilities and modes of living (Dastkari Haat Samiti, 2019).

The ocean, in its material remains of the oyster shell, becomes part of homes and heritages in the mainland, as the vernacular architecture of Goa has literally incorporated the ocean in its physical and aesthetic qualities. Through the translucid and textured light produced by these oyster shells we areindeed able to sense the ocean filtering the light, and to perceive its watery qualities and unclear horizons while looking at the produced colourful reflections that cover furniture and inner walls. The ocean, here, is literally composing homespaces and integral part of the material and intangible qualities of houses and family spaces: the ocean enters Goan homes in a beautiful and unexpected way.

Home, Memory and Belonging: The Centrality of the Ancestral House in Goa and the Goan Diaspora

This profound connection between house and landscape translates a local economy deeply dependent of the soil and its productivity, which historically coexists with oceanic economies in a territory that was central for maritime trade long before the Portuguese conquered the city of Goa to the Bijapur sultan Adil Shah in 1510. Despite several famous families tied with oceanic trade over the centuries became extremely wealthy, the local elites under Portuguese administration had their economic and social positions based on the administration and possession of rich plantation fields. The *gaunkari* system privileged the *bhatkars*, landowners usually from privileged castes, while the majority of the village inhabitants were tenants, *mundkars*, toiling a soil that would never belong to them. Because occupations tend to be hereditary, villages have seen, for centuries, the reproduction of social and economic systems, translated in the spatial organization of the territory.

This is an important aspect, since large manor houses would become central in developing a Goan and localized architecture, in the creation of spatial solutions and aesthetic solutions that would be reproduced in smaller buildings, embodying a truly Goan domestic architecture populating local villages in a diversity of forms and scales.

The growing migratory movements to South Asian and East African cities at the turn of the twentieth century would provoke a significant, and still understudied, architectural activity in coastal Goa's villages, translating flows of remittances and social changes sustained by the diverse influences that diaspora communities would bring to their ancestral villages. The monetary benefits brought back home by the diaspora would allow many Goan families to slowly increase their economic status. Remittances from relatives settled across the Indian Ocean colonial territories would allow a wide spread construction activity in ancestral homes, visually marking social and economic changes brought by migration to the state.

Many, or even most, of the house we see while traversing this region of Goa were built or renovated in this period. This fact translates to an architectural culture of the growing migratory movements, while the ancestral house, as it is commonly mentioned in Goa, would be not only a central place of personal belonging but also a privileged location of showing to the village society the success of the migrants in their new life.

This is a perceptible cultural feature of the Goan diaspora and the Goan built landscape: the house, its aesthetics and material qualities, would stay as fundamental spatial, social and emotional reference not only in life experiences but equally in cultural production and heritage attachments.

Selma Carvalho, a dedicated researcher on Goan literatures and diasporas, mentions this particular aspect, affirming that there has been a "consistent compulsion of the Goan Catholic writer to employ 'home' as the centripetal location" (Carvalho, 2018), reinforcing the place that home occupies in the sensorial (Boswell, 2017), emotional and imaginary attachments to the homeland and family histories. The ancestral home, thus, is central to those who migrate as it is to the ones who stay, and the house, understood in its material and aesthetical elements, is, of course, the repository of concepts of home and family history. Decorative and architectural influences brought from cosmopolitan port cities across the Indian region would be integrated in the Goan built heritage, turning the houses in repositories of geographic and cultural circulations.

Large landlord houses would grow in scale and richness, in a continuation of what succeeded in the late eighteenth century, becoming locally mentioned as palaces or mansions. New objects and technologies, decorative motifs and materials would fill large reception halls: Venetian or Belgian chandeliers suspended from high ceilings, Chinese porcelain from Macau invading the halls in blue and white dining sets or coloured monumental vases surmounting chest of drawers or side tables, gramophones filling the space with new music, hot water pipe systems serving modern toilet spaces, European made mosaic floor under the feet of dancing parties. The exuberance and ostentation of the nineteenth-century European art history would reach Goan villages and transform their architectural compositions. In much modest scales, smaller homes would equally reflect these influences. Typical architectural elements of Goa houses, as the *balcão* (balcony, a space at the entrance of the veranda with seating arrangements that serves both as a pleasure space and as a reception antechamber) and the veranda, would systematically incorporate small details that denote this period: the changing profile of the oyster-shell windows, the enlargement of proportions or the increasing relevance of the public façade of the house. In their inner spaces, we can usually observe the development of more formal reception room, often with large windows opening to the main veranda, creating a sense of scale and decoration that many times was not present in the more private areas of the house. The house would be enlarged in relation to the roads, in visual and material representations of family histories spread across vast distances. This would be accompanied by the visit of the migrants to their ancestral homes and villages, sustaining the incorporation of a whole universe of music, furniture, fashion senses, habits and cultural references that were mostly from outside a Portuguese sphere of influence, originated in globalizing cultural languages enabled by technologies as the steamship and the increasing colonial appropriation of the African continent.

The house, then, while integrating local social hierarchies and material elements of a "traditional" Goan life and heritage, would also embody traces of social mobility, economic and cultural changes, allowed by maritime migratory routes. The centrality of oceanic circulations in the coeval migratory routes and the development of diasporic cultures is perceived in the way they leave marks, physical and cultural, in these rural landscapes. We may not see the ocean from these balconies, but at the same time that we feel the diaphanous light coming from the traditional oyster-shell windows, we can still perceive a certain cosmopolitan modernity while seating among furniture and objects brought from other continents, observing old photographs and reading letters sent by those who left to settle in new cities, faraway shores and different landscapes.

Another layer where oceanic crossings are perceptible in Goan houses and their importance as literary locations is in the still blurred history of the slave trade to Goa during the long colonial period and the contemporary presence of Afro-descendant populations in the state.

While the role of the Portuguese in the Atlantic and Indian oceans slave trade for centuries is widely acknowledged, the presence of African slaves in Goa, as well in the other Portuguese colonial territories in the Indian subcontinent is still to be recognized in its full dimension (Pinto, 1991; Souza, 1989). The fact that the economic power of trading families and businesses in Portuguese India was, among other activities, sustained by the slave trade is an historical element that has been slowly but steadily obliterated from the local social memory and cultural references. It is known that before the abolition many maroon slaves settled in the forested hills just outside the Portuguese border, in the actual state of Karnataka, where among the Siddi communities (communities of Afro-descendants) we find many Christians with Portuguese origin names (Alpers & Catlin-Jairazbhoy, 2008). The settling, however, of African descendants in Goa after the abolition is, on the contrary, an unclear history which we can perceive through diverse archival dimensions connected with the home as simultaneously a family space and a location of historical processes. The contemporary existence of Afro-Goans, perceptible when crossing south Goa, can be curiously observed in photographic and literary dimensions, directly connected with the space of the house and its familiar dimension.

Research projects using family photographic archives, as the one developed by Savia Viegas, a writer, educationist and researcher (2015) or the book on Goan fashion by the late Wendrell Rodricks (2012), are important to unveil untold stories. In both projects we can see Goan extended families posing in front of their homes, translating the senses of self-fashioning and family atmosphere of the time. The peculiarity of these images is that they are accompanied by their Mozambican servants. At the turn of the twentieth century, Goan families with ties in Mozambique, which had been definitely conquered by Portugal in 1885, would often bring children as house servants to live in their Goan ancestral homes. These two examples of photographic archival research constitute important testimonies of this less known layer of the long maritime dimensions in Goa and its domestic spaces.

Complementing this visual archive, novels have been bringing this theme to literature and acknowledging African presences that have been for long invisible. The novel *Skin*, by Margaret Mascarenhas (2011), is a influential example, as Benedito Ferrão analyses (2014), linking the slave trade of the past with the contemporary presence of Afro-descendants, the village life revolving around social hierarchies and the domestic space where the history is developed.

From informal interviews and my own personal observation, in several areas of Salcete taluka, in South Goa, we can still see afro-descendent inhabitants, but socially they became part of the so-called lower castes dedicated to fishing activities, agricultural or domestic services.

The house as a repository of diaspora stories and maritime histories, in its material and human dimensions, is part of the Goan cultural landscapes, and at the same time that it can be understood as built heritage it can equally be seen as a space where past and present dialogue through personal stories, visual archives and literary works. In the next three sections we will visit three houses which keep in themselves three different modes of sensing oceanic heritages and histories in this coastal state.

The City of Zanzibar in a Goan Village

An unusual example of how stories of migration are materialized in domestic architecture in Goa is the Albuquerque House, in the coastal village of Anjuna, North Goa. Here, is the architectural language itself that tells of distant places, in an aesthetic universe that does not follow the common Goan features.

Dr. Manuel Francisco Albuquerque, Goan from the village of Anjuna, studied medicine in Bombay, Edinburg and Brussels before settling in Zanzibar in 1898. Besides becoming a royal palace's physician, his role in preventing and epidemic in the city awarded him the sultanate's highest honour: *Third Class of the Order of the Brilliant Star of Zanzibar*, in 1903 (Carvalho, 2014). He acted twice as Consul of Portugal in Zanzibar in the first years of the twentieth century, among other few Goans who were chosen by Portuguese government to represent Portugal in the archipelago's capital of Stone Town (Frenz, 2013). For this service to Portugal government he received one of the highest Portuguese honours, the *Ordem de Cristo*. This reflects how well integrated the Goan community was in Zanzibar at the dawn of the twentieth century, with a well-established population known for their important trading and professional role, as hoteliers, bakers or tailors and artistic activity, as photographers and musicians. Indeed, the relevance of Goan photography studios in the whole East Africa at the end of the nineteenth century is a story finally being studied and acknowledged (Gupta, 2016).

The uniqueness of the Albuquerque House is that, before returning to his home village of Anjuna, Dr. Albuquerque requested the sultan for an authorization to build a replica of the royal palace in Goa. This is how the story was preserved across several generations, and how it is told by Dr. Abuquerque's daughter-in-law, Ms. Ruth Abuquerque, with whom I met in 2019 and who painstakingly takes care of a section of the house. This unusual story is what explains the chosen peculiar aesthetics, which are not common in Goan domestic architecture.

Built in the vicinities of the cliffs overlooking the ocean, on what was, in 1919 (when the construction was completed) a largely open field terrain, this house became a visual marker in the village. With its two floors and towered octagonal volumes on each side of the main façade, the ground floor halls and reception rooms with stucco work and marble floors overlook the gardens in a succession of openings that link the interior with the environment. We do not observe the usual verandas and oyster-shell windows, neither the *balcão* nor the spatial distribution of the large Goan family homes, but we find a sense of the architectural culture of the nineteenth-century Indian Ocean port cities, in a fusion of influences that translated cosmopolitan and diverse cultural belongings. For this, as the story is still told, craftsmen and artisans from Zanzibar were brought to carve wood ceilings, design stucco works that decorating walls and ceilings and to lay down the patterns of ceramic and marble floors.

The result is a house outside of the traditional Catholic wealthy family house archetypes, and even if one is not aware of the story behind its construction one can feel there is something different and unexpected in its volumetry and general profile.

The Albuquerque house demonstrates that besides campaigns of renovation and the introduction of innovative decorative details, maritime histories of Goa and its people can equally be seen in architectural references aiming at bringing the life of an East African port city to a rural village on the other margin of the ocean.

In this home, Dr. Albuquerque's life in Zanzibar is, literally, reproduced in brick and mortar. This is a notable example of how the place of departure carries traces of lives lived across the ocean and of how we can find traces of migratory experiences in the ancestral villages and towns of those who departed.

The life story of Dr. Manuel and the story of his house are an enlightening example of the possibilities opened by a research looking at multiple and the relevance of the diverse materialities that Goan diaspora has left in Goa.

The Star of Zanzibar awarded to Dr. Albuquerque can be seen today in the Xavier Center for Historical Research, in Goa, included in the objects donated by his daughter-in-law to this institution. While observing this small medal with an inscription in Arabic script, we are indeed testifying one of the many traces that oceanic crossings have left in Goa.

Memories of a Shipwreck

Another unexpected testimony of the history of the Goan diaspora is not a house, as the previous example, but a compound wall with a miniature ship, built as memorial to the shipwreck of the steamship S.S. Britannia in 1941.

Hidden in one of the neighbourhoods of the village of Aldona, we find a house that presents a most uncommon compound wall: a replica of the British passenger steamship S.S. Britannia, one of the many steamships crossing the oceans in the first half of the twentieth century. The story behind this unexpected historical archive of sorts is an extraordinary example of how oceanic histories flow hinterland through living experiences and family networks and become part, in this case literally, of homes and family spaces.

One of the enduring connections of the Goan diaspora with the sea was the habit of employing Goans in the steamship services used to cross the Indian Ocean linking colonial territories in Asia and Africa in the first decades of the twentieth century. Goans were acknowledged by their skills as cooks, stewards and musicians in the British passenger ships connecting Indian Ocean port cities with Western Africa on its way to London.

While researching for private photographic archives and family stories in the village of Aldona, for the 2019s Goaphoto project "Aldona, Through Family Eyes",[2] I met the Goan journalist Melvyn Mesquita, who unveiled the most extraordinary history connecting this village with the global geographies of the Second World War.

The S.S. Britannia was one of the steam liners that connected South Asia with Europe through the Suez Canal, built by the company Anchor Line in 1925 and where several men from the village of Aldona were employed. In 1941, off the coast of Dakar, the steamship was attacked by a German cruiser, and after being warned by the German vessel, the crew and passengers were put in lifeboats. While many of them died, a small lifeboat reached the coasts of Maranhão, in Brazil, after a month in the sea, with some Goan seamen on board. Melvyn's grandfather was on this lifeboat, boat number 7, and unfortunately died before reaching shore, as many others. Out of 82 passengers drifting in this boat, only 38 reached Brazil alive. They were rescued and spend few days in the hospital in the city of São Luís do Maranhão. Years later, one of the survivors, frank West, published a book recalling his experience, titled *Lifeboat number 7*.

Aldona's inhabitants were still mourning their deceased when news of the few survivors reached the village and as told by some, one widowed woman ended up discovering that her husband was alive after all. Melvyn's work also brought to light a *mando* (Goan musical style) that was composed in the village as a memorial to the deceased ones and was performed in the religious celebrations for the souls of the departed seamen.[3]

We understand now why a village house in a small path amid thick forest has such an impressive miniature of the S.S. Britannia at its entrance: the grandson of one of the deceased seamen lives here, and decided to honour his family history with this sculptural memorial as well as with a collage of his grandfather's photographs and scans of newspaper news from the time.

Visiting this house and interviewing Melvyn Mesquita functioned as an opening window to larger landscapes and the astonishing life and worlds hidden in plain sight across many Goan villages.

Integrating the Goaphoto's exhibition in one of the village houses, the festival curators, Lola Mac Dougall and Akshay Mahajan, organized a display of old photographs, travel documents and books as a small museum dedicated to this event and to the memory of those who perished in this shipwreck. Melvyn Mesquita presented his research with the talk "Uncovering a Goan story on the S.S. Britannia" before local musicians played and sang the old *mando*.

This was an emotive endeavour, and significant for the purpose of this text, to find hidden stories of migration, biographies of houses and villages, events connecting a Goan village and the wider world. A notable aspect was the constant presence of the house, as a material heritage and emotional space, as a private museum where portraits of relatives and other geographies are kept, as elements of material culture and heritage, as mnemonic repositories of Goan society over the ages and as the centres of family tales and intimate storytelling.

Dhows and Travel Passports

Precious travel documents from the past constitute definitive memorabilia of family histories lived across the sea. Houses can become archives repository where old leather folders and cardboard boxes keep passports, travel documents and letters that tell private longings and new lives across the ocean.

In the South Goa village of Chinchinim, while looking for stories of Goans in Zanzibar for a documentary that the filmmaker Nalini Elvino de Sousa and me have been working on, we had the opportunity to examine fragmented travel documents from many decades ago, written in Portuguese and English. We enjoyed an evening seated on a veranda with columns painted in a tone of blue evoking the large ocean nearby, conversing about generations of migrants, houses and travels, while our interlocutor, Mr. Menezes, enthusiastically explained how his grandfather sailed in a dhow from Bombay to Zanzibar in the last years of the nineteenth century. After decades in Tanzania and Kenya, he too decided to return to his Goan house with his wife, preserving the family's history and securing documents and objects gathered in East Africa during the course of three generations living abroad.

Holding old passports issued by Portuguese and British colonial administrations, Mr. Menezes recounted the travels of his grandfather, the store his family owned with the Souza family in Dar es Salaam, the role Zanzibar played as the main port of East Africa at the turn of the twentieth century and the difficulties of sailing back home during the Second World War. Through torn papers with ancestors' names, passport photographs and few objects brought from East Africa, this beautiful house with its Goan *balcão* and veranda gains life with the stories told. Significant, thus, is its veiled presence in such dispersed family histories: after being vacant during a long time, this ancestral home is the anchoring port of a return to the homeland. The ancestral house was always there, and when all the other houses are emptied and transported in a return to the family village, it will be the old Indo-Portuguese furniture to receive its new inhabitants, who will open the newly painted *carepas* windows and lighten rooms and walls where old family portraits hang waiting for who looks at them.

In the stories told, documents carefully folded and the intense blue of the long balcony, this house is testimony of one among numerous tales of circulation that entangles the soil of this village with the vast oceanic horizons.

Doors Closed, Windows Shut, Living Memories

The examples here presented are inquisitive testimonies of the ways that houses, in their architectural language, affective values and composing elements, can indeed be seen as extraordinary museums and repositories of maritime histories behind and beyond the material and sensorial dimensions of the region.

Goan houses from the predominantly Catholic coastal regions have been recognized as a seductive heritage due to their aesthetic characteristics and a general feeling of nostalgia enacted by the state of abandonment of many of them (Gupta, 2009). Closed doors, falling ceilings and fading colours are, indeed, a common sight while traversing village neighbourhoods and wandering along rural secondary roads. It is inevitable that this landscape dotted with empty and ruining houses does not convey feelings of fading pasts and a curiosity for the vast number of stories and anecdotes that we imagine are kept behind doors.

The conjunction of this peculiar architecture, the history of Goa and its diaspora and the sense of nostalgia that pervades the rural environment are behind a growing understanding of the centrality of the house as archives of their own right, to be explored by scholars and artists. A striking visual example that opened Goan domestic heritage to wider audiences was the body of work made by the Indian photographer Dayanita Singh,[4] exploring the neighbourhood of the Goan village where she moved to years ago, Saligão. Her exhibition "Demello Waddo", titled after the name of her *waddo* (Konkani word for ward, which in this case gained its name from an important local family, De Mello) presented a selection of intimate black and white portraits of people and homespaces, translating the seductive characteristics of the local sensorial world. With the texture of her photographs, Dayanita transmits a sense of wonder for the stories, events and affective worlds that surround homespaces and the sensorial worlds of the village and Goa.

This dual dimension of domesticity and archival ability is fundamental to understand the relevance of the house and homespace as sight of oceanic reflexions in the Goan built heritage (Menezes, 2019). It is indeed extraordinary, as in other places deeply connected with migratory histories, this simultaneous rootedness in the local landscapes and wider seascapes, perceptible in dim lighted spaces, framed photographs or miniature ships, among many other objects, papers and tales. This translates the extremely diverse modes we can perceive the relevance of oceanic horizons in understanding Goa, its cultural heritages and social history that is deeply linked with centuries old histories of maritime trade, migration and circulation.

While may be seen as disparate elements, architectural decorative techniques and materials, old photographs or family stories are components of a maritime character that has informed, and changed, Goa's history and landscapes.

Understanding the maritime vocation of Goa through the lens of house aesthetics and homespaces enables their integration in much wider material histories of oceanic inclination (Meier, 2016; Pombo, 2018; Um, 2009). If Goan vernacular architecture has been studied mostly for its uniqueness, it deserves to be seen as integrating larger Indian Ocean artistic flows and connected histories and art histories (Singh, 2017; Subrahmanyam, 2005). Previous research I was able to carry out in several coastal and insular urban spaces as the old city of Diu (former Portuguese island off the Indian Western Coast), Zanzibar or Dar es Salaam (in Tanzania) allows me to analyse Goan homespaces as one more layer in an ensemble of traces of oceanic connections in architecture, visual and material cultures and family biographies and senses of belonging (Pombo, 2018).

In Goa, the powerfully centripetal element that is the family house, particularly for diaspora family stories, allows us to discover the ocean and its cultural worlds through affects that surround timeworn plastered walls, though objects gaining dust on side tables, porcelains from Macao supported by rusty nails or *carepa* windows that after many decades still illuminate inner rooms with a candid and soft light, waiting for someone to fully open them after the rains. These powerful evocations open Goan balconies to watery horizons and monsoonal geographies, and turn domestic objects and images in pieces of affective and historical museum.

Notes

1. For an overview of the catalogue, see: https://www.goa1556.in.
2. Goaphoto is a photography festival in Goa initiated by Lola Mac Dougall, that since 2015 has been intervening in public and private spaces across the state. The last two editions, in 2017 and 2019, took private domestic spaces as exhibition venues and as theme for the curated works exhibited. For the 2019 edition see: https://goaphoto.in/aldona-through-family-eyes/.
3. For the lyrics, see: https://www.mail-archive.com/gulf-goans@yahoogroups.com/msg01805.html.
4. For a comprehensive catalogue of her work, see her website: https://dayanitasingh.net.

References

Alpers, E., & Catlin-Jairazbhoy, A. (Eds.). (2008). *Sidis and scholars: Essays on African Indians*. Red Sea press.

Boswell, R. (2017). Sensuous stories in the Indian Ocean islands. *The Senses and Society, 12*(2), 193–208.

Carita, H., & Sapieha, N. (2002). *Palaces of Goa: Models and types of Indo-Portuguese architecture*. Cartago.

Carvalho, S. (2014). *A Railway runs through*. Cinammon Teal Publishing.

Carvalho, S. (2016). *Baker Butcher doctor diplomat. Goan pioneers of East Africa*. Selma Carvalho.

Carvalho, S. (2018). A house of many mansions. Retrieved July 30, 2021, from https://www.joaoroqueliteraryjournal.com/nonfiction-1/2018/3/16/nothing-new-on-the-portuguese?rq=africans

Dastkari Haat Samiti. (2019). Goa's windows: A heritage of shell work.

Fernandes, N. (2012). *Taj Mahal Foxtrot: The story of Bombay's Jazz*. Roli Books.

Ferrão, R. B. (2014). The other black ocean: Indo-Portuguese slavery and African-ness elsewhere in margaret Mascarenhas's skin. *Research in African Literatures, 45*(3), 27–47.

Frenz, M. (2013). Representing the Portuguese Empire: Goan consuls in British East Africa, c. 1910–1963. In *Imperial migrations. Colonial communities and diaspora in the Portuguese world* (pp. 193–212). Palgrave Macmillan.

Frenz, M. (2014). *Community, memory and migration in a globalizing world. The Goan experience, c. 1890–1980*. Oxford University Press.

Gupta, P. (2009). The disquieting of history: Portuguese (De)colonization and Goan migration in the Indian Ocean. *Journal of Asian and African Studies, 44*(1), 19–47.

Gupta, P. (2016). Visuality and Diasporic dynamism: Goans in Mozambique and Zanzibar. *African Studies, 75*(2), 257–277.

Henn, A. (2014). *Hindu-catholic encounters in Goa*. Indiana University Press.

Mascarenhas, M. (2011). *Skin*. Golden Heart Emporium.

Meier, P. (2013). At home in the world: Portrait photography and Swahili Mercantile aesthetics. In G. Salami & M. B. Visonà (Eds.), *A companion to modern African art* (pp. 96–112). John Wiley & Sons.

Meier, P. (2016). *Swahili port cities: The architecture of elsewhere*. Indiana University Press.

Menezes, V. (2019). The Saligao Blues: Goa's global village. Mint. https://www.livemint.com/mintlounge/indulge/the-saligao-blues-goa-s-global-village-1550777523403.html. Accessed on 29 April 2022.

Newman, R. (2001). *Of umbrellas, goddesses and dreams: Essays in Goan culture and society*. Other India Press.

Pereira, J. (2001). The evolution of the Goan hindu temple. In J. Pereira & P. Pal (Eds.), *India & Portugal—Cultural Interactions* (pp. 88–97). Marg.

Perez, R. M. (2012). *The Tulsi and the cross*. Orient BlackSwan.

Pinto, J. (1991). *Slavery in Portuguese India (1510–1842)*. Himalaya Publishing House.

Pombo, P. (2006). Shri Betal Prasannah. The temples of Betal of Priol Velinga and Querim. *Oriente, 16*, 115–128.

Pombo, P. (2018). Beyond the margins: Place, narratives and maritime circuits in Diu. *South Asian Studies, 34*(1), 33–46.

Pombo, P. (2019). Water cartographies of Goa: Khazans, sedimentation and dissolution of coastal cultural landscapes. *Journal of Heritage Management, 4*(2), 192–207.

Rodricks, W. (2012). *Moda Goa-history and style*. Harper Collins.

Rosales, M. (2009). *Objects, scents and tastes from a distant home: Goan life experiences in Africa. Dve domovini/Two Homelands* (Vol. 29).

Singh, K. (2017). Colonial, International, Global: Connecting and Disconnecting Art Histories. *Art in Translation 9*, no. sup1 (March 31): 34–47.

Silveira, A. (2008). *Lived heritage, shared space: The courtyard house of Goa*. Yoda Press.

Souza, T. R. de. (1989). French slave-trading in Portuguese Goa (1773–1791). In T. R. de Souza (Ed.), *Essays in Goan history* (pp. 119–132). Concept Publishing Company.

Subrahmanyam, S. (2005). *Explorations in connected history*. Oxford University Press.

8

Salt, Boats and Customs: Maritime Princely States in Western India, 1910–1932

Varsha Patel

Introduction

Indian Ocean history has for a long time battled euro-centrism. Scholars have unpacked the bias in previous scholarship that conceived of the Indian Ocean as a static zone that became active only with the arrival of colonialism. Ashin Das Gupta's pioneering work emphasized the importance and prominence of Indian merchants in the Gulf of Cambay in the early 1700s (Das Gupta, 1979). However, today Indian Ocean historians who highlight the agency of indigenous peoples (Ho, 2006; Martin, 2008; Palsetia, 2008) outnumber those with a Eurocentric bias. There is now a new India-centric bias that excludes Africa and most of East Asia from the field of Indian Ocean history (Mukherjee, 2013). Scholars have demonstrated that 'there is no simple correlation between the rise of European powers and the decline of Asian trade in Indian Ocean history' (Rajat Kanta Ray, 1995). However, the activities of small-scale merchants who did not always align with, or circumvented the networks of empire, and of the ports of Maritime Princely states of India that were indirectly under the British empire remain to be told.

V. Patel (✉)
Independent Researcher, New Delhi, India
e-mail: varpat2000@gmail.com

Princely states in India were outside the territory of British Empire but not outside its gaze: they comprised one third of the territory of India in 1912 and 1932 (Ramusack, 2004, p. xiv; Roy, 2006, p. 4). Most of the larger ports of colonial India were in the British territories of Bombay, Calcutta and Madras. The ports of Princely states that fringed the empire in India and were ruled by native Māhārājās (called Princes in British records) are relatively unknown. An examination of the routes of salt through the Princely states' ports and land routes into British India reveals the complex relationships of these states with each other and with the British empire between 1910 and 1932. Routes come under the broader idea of 'circulatory regimes' or 'the totality of circulations occurring in a given society and their outcomes' (Markovits et al., 2003, p. 3). Records of experiences of these maritime connections also throw light on the impact of colonialism on the residents of maritime Western India in the early 1900s.

The unfolding of colonialism and of Western science coincided with the charm of commerce and the lure of profits. Concurrently, laws and legislation began to increasingly govern, regulate and divide the ocean during this period. In the archival documents of the 1800s and 1900s, the Ocean becomes a resource, a source of profit and a passage for enabling the movements of goods and peoples. The field of Indian Ocean history continues to overemphasize trade and commerce. This essay too, builds upon archival records wherein the ocean emerges as a backdrop for contestations over revenue and power between the British empire and different Princely states. All the actors share a common interest in increasing profits by modernizing ports and enabling the circulation of salt and safeguarding their revenues. Other conceptions of the ocean, while they would have been present are missed in these formal documents from archives of British India and of two Princely states (Bhavnagar and Baroda).

This chapter focuses on the peninsula of Saurāshtra in Gujarat, Western India. This region includes the Gulf of Cambay and the Gulf of Kutchch, and it occupied a central nodal point in Indian Ocean trade until the 1900s. Saurāshtra has also been described as a 'miniature continent' with around two hundred different Princely states, estates and *jagirs* with a population of approximately three million in the year 1900 (Indian States Enquiry Committee, 1932, p. 78). Histories of these complex regions are often excluded from textbooks that focus on British ruled colonial India. Archival records documenting the past of these regions are also comparatively fewer.

The first section of this chapter introduces maritime Saurāshtra, a geographically unique space for salt production that encompassed the politically fragmented provinces that fringed the British empire in Western India in the early 1900s. The next section revolves around the land-based customs cordon that configured the routes of the salt trade. It presents an insight into the workings of the Princely states under the reign of the British empire that placed restrictions on Princely ports and barricaded the routes of trade, imposing strict customs barriers. The third section presents the case of sea routes of salt trading, discussing the case of a Princely state that requested a permit to export salt that might not have hindered British profits. Next, this chapter traces the role of *dhows* and country boats: these were likely to have availed of the livelihood opportunities of the illicit salt trade offered by the maritime provinces of Saurāshtra, with their unique geography.

Maritime Saurāshtra: Salt, Geography, Polity

Saurāshtra (Kāthiawār and Kattywar) is sandwiched between the Gulf of Khambhat (Cambay) and the Gulf of Kutchchh (Kutch and Cutch) in the Gujarati speaking region of Western India.[1] Gujarāt's ports were key nodes in Indian Ocean trade networks since the eighth century and Gujarātis were established in the ports of Oman and Aden well before the Portuguese entered maritime trade of the Indian Ocean (Grancho, 2015, p. 260; Machado, 2009, p. 55; Markovits, 2008, p. 11; Sheikh, 2014, p. 44). Peninsular Gujarāt or Saurāshtra has a seaboard of 600 miles and an area of around 23,445 square meters (Indian States Enquiry Committee, 1932, p. 78). The region of Kutchchh has a salty marshland, left behind by the receding sea, and this white salt desert covering 10,000 square miles guarded the Princely state of Kutchchh (Goswami, 2011, pp. 15–16). The coastal areas of Kāthiawār and Kutchchh have been described as a vast natural salt pan. The salt made in the salt pans or *āgars* covering 4953 km^2 within the Little Rann of Kutchchh is different from the natural sea salt that is made with sea water along the Saurāshtra littoral (Indian States Enquiry Committee, 1932).

The idea of a 'regime of circulation' highlights interconnections between modes of transport (Ahuja, 2004, p. 75). The arrival of the railways in Saurāshtra in the 1800s, which was resisted by the British rule, was ushered in by the Māhārājā of Bhāvnagar (1723–1948) in Saurāshtra. Henceforth, the ports of the maritime Princely states were perceived as competitors of British Indian ports and some of these ports could access markets in Delhi and central India more easily and faster than the British ports of Bombay and Karachi (Ramusack, 2004, pp. 202–203; Indian States Enquiry Committee, 1932, p. 98). During the 1800s, British officials invoked paramountcy to justify diverse policies, including the taking over of ports (Ramusack, 2004, pp. 97–98). Eight of the total fourteen ports under Princely states in India fell within the Gujarāti speaking region while all the other ports of the Indian subcontinent were within British Indian territory (McLeod, 1999, pp. 88–113). With the railways, there came a concurrent increase in the barriers restricting the movement of goods between the ports of Princely India (that facilitated the import of goods due to cheaper customs duties and geographic location) and their final destination in British India where most of the goods were consumed. These included smuggled silver bullion, cotton imported in Bhāvnagar for mills in Ahamedābād, and goods that interfered with the revenue sources and authority of British India.[2]

The British empire was the single largest exporter of salt to India. Imported salt arrived at the ports of India's eastern provinces, mainly Bengal, Bihar and Burma.[3] Salt duty comprised around five per cent of the Government of India's total revenue (Foundation of Indian Chambers of Commerce and Industry, henceforth FICCI, 1930). These maritime revenue sources for the British empire were still marginal in comparison to land revenue (McLeod, 1999; Roy, 1886). Indian salt manufacturers were hindered from competing in the salt market due to the regulations for safeguarding British revenue that interfered with the 'freedom, convenience and economy of trade'. They faced several restrictions, including a cap on the tonnage of vessels they could use, regulations regarding transport, high rates for steamer freight, higher rates for railway freight, the exclusion of salt transported by rail from the bonded warehouses (which received only salt imported by sea), and the absence of inland warehouses for salt transported by rail (Taxation Committee Report, Marginal note to para. 137: 168 In FICCI, 1930, pp. 173, 194, 199).

According to archival records and secondary literature, the terrestrial customs barriers were more effective than the legal instruments that hindered maritime routes and the patrolling of shoreline frontiers. The longest terrestrial customs cordon was a hedge (1843–1870) that ran through the breath of India stretching over 2300 miles in 1869 from the River Indus in North-West Indian to the extreme south of the Central provinces, reaching the banks of River Mahanadi in Madras (South East India). It was guarded by an army of 12,911 officers and men at an annual cost of Rs. [Rupees] 16,25,000' (Moxham, 2001). This larger land-based barrier was complimented by the Virāmgām customs cordon that guarded the British Indian territories immediately bordering the maritime Princely states in Saurāshtrā and Kutchchh. Over time, agreements, treaties and laws moved in tandem with each other and were thus transformed, as were portions of the customs cordons and relationships between British India and the many Princely states of Saurāshtrā.

The Government of India, like the provincial governments during the period of British expansion in India, faced 'great difficulty' in 'protecting salt revenue and ensuring its collection'. By 1880 an effective degree of British government control was established over salt sources throughout India' and agreements between 1883 and 1885 with salt producing princely states established a practical monopoly over the production and removed customs barriers previously maintained against the States (Indian States Enquiry Committee, 1932, pp. 74, 82). The salt monopoly which 'long sustained the policy of the government of India' involved a system of collecting excise duty at the sources of production, ensuring that the entire population of India, whether resident of British India or of the States (*with the exception of Kathiawar and Cutch* and certain areas supplied from Mandi), paid a salt tax at the rate from time to time in force in British India' (emphasis is mine) (Indian States Enquiry Committee, 1932, p. 76). Further, this taxation benefited the Princely states' rulers as the residents contributed to salt revenues on the same basis as the residents of British India (Indian States Enquiry Committee, 1932, p. 76).

The British Indian government acquired additional salt works from principal salt producing States (to meet an increased demand) and organized the suppression of all other states' salt works. Moreover, British India endeavoured to eliminate transit duties upon salt since many states taxed salt that passed through their territory on its way from production centres in other states or in British India towards final consumers residing in different places, including British India and the various Princely states. Regular excise upon salt in Bombay (the closest British Indian presidency to Saurāshtra) began as early as 1838, according to a report dated 30 July 1870 (FICCI, 1930, p. 104). Hence, the 'East India company found themselves compelled to establish customs lines for the collection of salt import duties at their frontier'. Echoing the importance of salt revenue for British India (1834–1854) is the work of G. H. Smith, (commissioner of customs) who expanded the customs line around Bengal over twenty years. He was instrumental in lowering taxes on 'lesser' items, including tobacco, in order to concentrate on the smuggling of salt (Kurlansky, 2002, p. 339).

The regions of Saurāshtra and Kutchchh were self-sufficient in salt farming, and their needs were not met by British Indian sources. Kutchchh was connected to the rest of India by caravan routes passing through the desert during dry season. According to British India, Kutchchh would be eager to access markets by rail and sea (Indian States Enquiry Committee, 1932, p. 80). The report of the Indian States Enquiry Committee introduces the salt rights of Kāthiawār and Cutch in the following words: 'they have no commercial agreements, they receive no compensation for the restrictions under which they carry on their salt trade; and they have had to submit to a policy dictated by one interest only. Although, other states were party to bilateral agreements under which they received valuable consideration for the rights which they had surrendered' (Indian States Enquiry Committee, 1932, p. 80).

Movement and sale of Saurāshtrā salt was unrestricted within Saurāshtra but it carried no duty payable to British India. The entire region of Saurāshtrā enjoyed immunity from having to pay salt revenue to British India. However, the maritime states that produced salt (Bhāvnagar, Janjirā [Jāffarābād], Junāgadh, Porbandar, Nawānagar, Barodā, Morvi, Dhāranghadrā, Kutchchh) collected and kept the salt revenue from their inland neighbours. Archival records note that the value of the immunity was 'debited' against the inland states, as their people did not contribute to British India's salt revenue. The value of immunity was calculated at Rs. 7,90,906 for Kāthiawār and Cutch while the total value of immunity that all states enjoyed was Rs. 38,15,151 (Indian States Enquiry Committee, 1932, p. 82). The next two sections of this chapter discuss barriers against the movement of salt. The next section discusses an urban customs cordon in Saurāshtra that restricted the trade of the Princely states' ports and contributed to the work of the longer customs cordon known as the Great Hedge of India that spanned British India, containing customs duties and profits for the British empire.

Land Routes of Salt: The Virāmgām Customs Cordon

The Virāmgām line or customs cordon over the land that separated the Princely states of Saurāshtra from British India was a check point in Virāmgām town that was an important infrastructure for curtailing smuggling. Here all passengers crossing the line were searched, and duties were levied on goods that crossed the border into British India (Indian States Enquiry Committee, 1932, pp. 99–101). This customs cordon was initially established in 1903 (or 1905 according to some sources), in order to curtail the circulation of smuggled silver from Muscat in Kāthiāwār (if allowed to enter British India, this silver threatened to bring down the price of the rupee, which the British at the time were compelled to purchase with the British gold coin). It was temporarily lifted between 1917 and 1927 (Mehta, 2009, p. 129). Eric Beverley (2013, p. 213) shows how local people used the liminal space of the frontier of British India bordering the Princely state of Hyderabad in South India as a resource for their livelihoods. Along the Saurāshtra frontier in Western India, too, archival records document the illicit movements of goods that often evaded the customs duties.

Reports of the Bhāvnagar Princely state and letters to the British Government of India register harassment that travellers and traders faced at the Virāmgām customs cordon (Bhavnagar Darbar 1917). Recurrent requests from the Princely states for the removal of this customs barrier are well documented. This customs cordon's interface with different social groups is likely to have been very different from those whose routes traversed the Great Hedge of India. In contrast to Moxham's (2001) 'great impenetrable hedge of thorny trees, evil plants, stones, ditches which no man or beast could pass without being searched', the Virāmgām customs cordon seems to have been a check point and a railway station. Perhaps it was also maintained partly by the staff of the Bhāvnagar Princely state, that had the status of a British Indian port.

Following new agreements with the Princely states, the restriction on the routes of trade between Kāthiawād, Kutch and British India, was temporarily lifted between 1917 and 1927. In 1917, the Princely states that unsealed their port infrastructure were under the impression that the customs cordon would not be re-imposed. Operating within the framework of agreements and treaties after the Virāmgām customs cordon was lifted in 1917, the Jām Sāhib of Nawānagar princely state modernized his port of Bedi, and his ports' customs duties increased from Rs. 30,00,000 in 1925–1926 to Rs. 78,00,000 in 1926–1927. Jām Sāhib's modern dockyard was making two per cent of the total customs revenue of British India in 1926–1927 (McLeod, 1999, pp. 95–96; Ramusack, 2004, p. 203). Finding their revenues threatened by the modern ports of Princely states, the British Indian government devised a certificate system under which the princes had to remit, to the Government of India, the duty collected on goods that were imported into British India from the Princely states. The British collected customs duties according to their established rate at each border crossing. Thus, goods entering the Indian subcontinent through the Princely states' ports had to pay twice, once on arrival while landing at Princely states' ports and then again when entering British India at the terrestrial customs cordon (Ramusack, 2004, p. 203). The report of the Indian States Enquiry Committee registers the 1927 protest of the states of Barodā, Nawānagar, Junāgadh, Morvi, and Porbandar, wherein they opposed the unjust termination of the 1917 Agreement following which their developed ports were labelled 'unhealthily developed', despite the fact that they had abided by all the treaties in place (Indian States Enquiry Committee, 1932, pp. 106–107). Port Bhāvnagar was an exception since it had the status of a British port due to a special treaty of 1860–1864. Such

'British ports' levied duty at British Indian rates that was in turn retained by the Princely state (Indian States Enquiry Committee, 1932, p. 98).

After a modification in the remittance system in 1929, the princes could keep duties collected below Rs. 2,00,000, and this could appease the smaller Princely states. However, Princely states of Nawānagar and Barodā with their large ports were still at a disadvantage. They 'objected to customs policies at the Round Table Conference in 1930 (and during subsequent negotiations over the federation') Jām Sāhib's Nawānagar, repeatedly appealed for a 'fair deal' and was consistently ignored (Ramusack, 2004, p. 203).

In 1927 after the Virāmgām line was re-imposed, the princes were eligible to hold only revenue on goods consumed within Saurāshtra. After the abolition of the inland customs line, portions of which coincided with the Great Hedge of India described earlier in this paper, the duty was uniformly reduced throughout British India. However, a certain section of the Salt Act 1882 conferred considerable power in the hands of the governor in council and 'provided infinite scope for mischief' (FICCI, 1930, p. 148). In contrast, the report of the Indian States Enquiry Commitee mentions that salt agreements were negotiated with States which, in the last resort, had the full right to reject the terms offered. Here too, the only exceptions were the 'special cases of Kathiawar and Cutch' where it was 'found necessary to invoke the principles of Paramountcy' and agreements were concluded between 1883 and 1885.[4] The Virāmgām customs cordon continued to exist even after the independence of India in 1947. It was finally removed in July 1948 (Mehta, 2009, p. 128).

During the late 1800s 'there was no proposal to station British customs officials in Kāthiāwār ports' (Mehta, 2009, p. 129).[5] In the early 1900s, British India proposed to install its own officers, who would manage the customs offices of the Princely states. The princes saw this as an intrusion on their and their state's sovereignty or, as they put it, *izzat* (honour, dignity, respect). While the princes largely refused these advances into their sovereign realms and refused the British proposal, at other points in time they conceded certain conditions. For instance, they agreed to send their trusted officials for training in British customs procedures after 1927 (when the Virāmgām customs cordon discussed earlier in this chapter was re-imposed). Therefore, the narratives of the land-based customs cordon and the legal framework with which it evolved depict the official route of duty paid salt as it made its way into maritime Saurāshtra. This section of the chapter has focused on land routes: the next section discusses sea routes, with a focus on the case of the curtailment of the legal salt trade on ships from the Princely state of Baroda to Calcutta in British India. This discussion of sea routes also includes country boats and *dhows* that are likely to have easily circumvented the formal regulations that rerouted ships.

Sea Routes of Salt: *Dhows* and Country Boats

According to Mehta, the colonial government passed the Sea Customs Acts in 1857, 1861, 1863, 1865 and 1886 compelling the Indian Princely states, including Travancore and Kochin in Southern India, to enter into agreements (Mehta, 2009, pp. 124–125). Some of the agreements that revolved around and involved customs barriers have already been discussed in the previous section: this section traces a specific maritime route of salt. Thereafter, it analyses excerpts from archival records and reads them 'in between the lines', in order to propose the existence of certain possible, if illicit and undocumented, circulations of salt.

The Okhāmandal region, with the port of Okhā, was surrounded by territory of the Nawānagar Princely state. But port Okhā came under the Princely state of Barodā, that governed it from the inland city of Barodā. In the remote Okhāmandal region, the westernmost tip of the Saurāshtra peninsula that juts out into the Arabian Sea, seawater washes the shore and the harsh sun turns it into salt. Sir T. Madhavrao (the Dewān of Barodā Darbār), in a letter (dated 11 June 1909) described Okhāmandal as follows:

> a small tract in the extremity of Peninsular Kathiawar... so isolated that it is very nearly an island in itself. In it salt is produced at numerous places and in considerable quantities. The production is the result of a natural process. The sea water flows inland, spreads over low ground, evaporates and leaves a crust of salt more or less thick according to the depth of brine. The salt is of excellent quality. It is hardly possible to find a region possessing greater natural facilities for the production of salt with regard to quality, quantity, cheapness or exportability.

A letter from the government of Bombay addressed to the Foreign Department, of the Government of India (dated April 1910) discusses the request of the Gāekwād of Barodā who wanted to export salt from Okhāmandal to British India for the payment of the usual British salt duty. In a letter dated 1880, a similar proposal by Barodā to export Okhāmandal salt to British India was refused by the British government of Bombay. Firstly, because the prevention of smuggling from Okhāmandal would require an elaborate establishment such as one that existed along the coastline of Broach and Surat districts (in Southern Gujarat) that came under British India. Secondly, only Khārāghodā salt (*baragraha* variety) was being consumed in Ahmedabad district (British India), which made it easy to detect natural salt smuggled from Kāthiawār that was of a different type (*ghassia*). Allowing imports of salt from Okhāmandal that cannot be distinguished in appearance from the natural salt of Kāthiawār (*ghassia*) to British India would spur smuggling over the land routes and disrupt the check on smuggling salt, since all smuggled natural salt could be presented as duty paid Okhāmandal salt.

A similar case is that of the salt works of Dhāranghadrā state in the Rann of Kutchchh. The salt works here were faced with the threat of closure because the salt quality was similar to the Khārāghodā salt that was manufactured and consumed in Ahmedābād in British India. From archival records, it appears that the best method for preventing salt smuggling was to detect it by the type of salt that consumers were allowed to access. Dhāranghadrā salt was indistinguishable in appearance and texture from the salt consumed in British India. Therefore, the salt manufacturers of Dhāranghadrā were compelled to collect duty at British Indian rates which were considerably higher than those prevalent in the rest of Saurāshtrā, under the Princely states. Dhāranghadrā's salt works could not survive in the Saurāshtrā market.

The salt that was identical in appearance, colour, texture to the salt produced and consumed in western territories of British India was not allowed to enter British India, because of the difficulties that it would pose for the detection of smuggling. The need for objectivity and ease of observation privileged the materiality of salt over trust in people, or even in documents that could attest to the payment status (customs duty) of salt. This finding slightly contrasts with the critical role of texts and documents for governance that scholars emphasize (Hull, 2012). The pursuit of objectivity introduces distance, leaving power in hands of experts (Porter, 1995). This role of the material properties of salt and its implications for governance offers a new understanding of the colonial encounter along the fringes of the British empire in maritime Western India.

The initial stance taken by the Bombay government was backed by the Political department of the government of India (letter dated 25 July 1910). The Governor General of India had, in a previous letter (dated 18 August 1880), contradicted the linkage between possible detection of smuggling and friction with the Barodā Darbār, by suggesting that the 'export of Okhāmandal salt' could be controlled by a British officer. Additionally, he advised officials of the Barodā government to adopt 'thoroughly efficient measures' to *prevent the export of salt by land or by sea* from Amreli Mahals (in Baroda) and *from Okhāmandal into any Kāthiawār state* (emphasis is mine). The only condition under which the government of Bombay 'deemed it safe to allow the import of Okhāmandal salt' was that salt be taxed by British staff which would, in turn, impose on the Darbār of the concerned Princely state an expenditure of nearly Rs. 600 per month on a shore office, and about Rs. 50 per month on a patrol boat. The Governor in Council was expecting this condition to be prohibitive. Much later, during discussions at Mount Abu in 1927, it was acknowledged that the Princely states saw such impositions as an encroachment on their sovereignty (Indian States Enquiry Committee, 1932: 101–103).

In contrast to the Bombay government, the Inspector General of Excise and Salt (Foreign department, Government of India, in letter addressed to the Bombay Government, dated 30 May 1910) emphasized that the chief market that Barodā Darbār desired permission to enter was Calcutta (on the East coast of India) and thus the danger of possible smuggling to the districts in Western India (Ahmedābād) did not exist. He requested the Bombay government to reconsider the question of allowing export of salt from Barodā exclusively to Calcutta, as the competition with British Indian salt manufacturers would also be immaterial. The bulk of the salt imported in Calcutta came from abroad and the exporters (and smugglers) were to abide by the Transport of Salt Act 1879. Consequently, salt would be transported under restrictions similar to, or more severe than, the ones applied to transport from British Indian ports: these restrictions included, among others, the weighing and supervision of shipments by a British officer, and a prohibition on calling at intermediate ports.

However, diverging from the expectations of British India, the Barodā Darbār decided to accept the prohibitive restrictions in order to export salt. In response, British India reiterated the need to refuse Barodā Darbār permission to export salt and emphasized that 'no conditions could sufficiently safeguard the imperial interests involved or justify a departure from the accepted policy of the British Indian Government (to protect revenue)'. Therefore, the Barodā Darbār was effectively obstructed from shipping salt to the British Indian port of Calcutta and also to intermediary ports in British India from 1909 until 1911. Hence, a possible maritime route for salt, involving the movement of salt from Okhāmandal on the westernmost tip of India to Bengal in eastern India, and, through Bengal, to consumers in central and northern India, remained, until 1911, only a possibility. Later, however, Saurāshtrā states applying for permission to export salt to Calcutta, without stopping at intermediary ports and observing the condition that salt be directly loaded onto steamers, were being permitted to send salt to Calcutta (Indian States Enquiry Committee, 1932, p. 73).

Concurrently, smaller quantities of salt were moved on country crafts along the coast of Western India. The coastal districts of Surat and Broach in British India were tightly patrolled, but it was impossible for the British empire to patrol the entire Indian coastline. Consequently, small boats and *dhows* were simply outlawed and barred from calling upon ports in British India. As the report quoted here testifies, the 'shipment (of salt) by sea to any Indian ports (British Indian) is forbidden and all salt traffic carried in *dhows* is made illegal'. This was 'because of the facilities for smuggling' that the 'innumerable creeks existing on the coast of Western India' offer (Indian States Enquiry Committee, 1932, p. 78).

Consulting archival records for colonial Orissa (East coast of India), Ahuja mentions the role of 'smaller coastal vessels' that operated from 'small village portlets' that were chronically silting and were unsuitable for large European ships but which carried 'salt, food grains, textiles, to ports in the Bay of Bengal' (Ahuja, 2004, p. 79). For timber, the coastal boats of the Gulf of Khambhat had been going till the Malabar coast (South West India) around 1750 (Das Gupta, 1979, p. 2). It is clear that small wooden sailing boats involved in coastal trade existed around this time (Hornell, 1920; Vaidya, 1945).

While the *dhow* trade was made illegal in order to prevent smuggling through the creeks by British India, to what extent were the coastal, small-scale, maritime routes of salt actually dissolved by this law? Ramusack writes, that 'smugglers tended to favour the more lightly patrolled ports in the Princely states' (Ramusack, 2004, pp. 202–203). Nordstorm complicates the ideas of legal and illegal while talking about the kind of business at sea that falls, in a way, outside the complex boundaries of terrestrial laws (Nordstorm, 2007).

Given the ubiquitous nature of salt and the sheer numbers of small sailing boats together with the affordances of the Saurāshtra littoral with its creeks, estuaries, small channels that enable small-scale coastal trade, it is likely that small boats that were too minor to be detected. These boats were likely involved in ferrying salt on a smaller scale between the coastal waters of Saurāshtra and British India (1900–1932) without denting British India's salt revenue on a large scale. Porous borders and the opportunities and possibilities that the liminal space of a frontier would offer are reflected in the narratives of both maritime and terrestrial movements of salt along the Saurāshtrā fringe of the British empire. Scholars show how the 'frontier zone' serves as a description of a 'borderland' while it brings out the 'ambivalent and productive character of particular spaces' and that the borders of

Princely states were used as resources that provided livelihood opportunities to residents of the border zones (Beverley, 2013, p. 243).

Kate Boehme (2015) shows how the British Indian actions for preventing smuggling in early nineteenth-century Western India were demonstrations of power and attempts to control the region. According to her, the quality and flow of actual goods smuggled were not that damaging to the British Indian economy, and the trade that was labelled 'smuggling' was a mere continuation of private trade from precolonial times into the colonial era. The porous line between private trade and smuggling is also reflected in the analysis of Western Indian Ocean trade for a previous period from 1720 till 1740 (Om Prakash, 2007).

Conclusion

The narrative of the circulation of salt between Saurāshtrā and British India between 1911 and 1932 focuses on a space and a time that has not been adequately examined by mainstream studies of colonial India. In studies of the Western Indian Ocean too, Saurāshtrā's Princely ports are eclipsed by the colonial ports of Bombay and Karachi. The histories of smaller Princely ports that were barred from modernizing and expanding their trade are simultaneously buried under the colonial narrative. Maritime trade in the Western Indian Ocean was closely linked with both private trade and smuggling and the geography of the coast with numerous creeks and estuaries facilitated small-scale coastal trade. This chapter has examined the infrastructure and impacts of land and the sea-based barriers that curtailed the movement of salt. It highlighted the position of the maritime Princely states in Western India that braced the British empire's severe restrictions on sea borne trade, and the heavy customs duties that colonialism and exploitation entailed. This contributes to postcolonial studies of the Indian Ocean by highlighting the agency and activities of the rulers, officials and the ports of maritime Princely states, underscoring their importance. Concurrently, the text has brought out the differences among officials of the British governments in colonial India in responding to the threat that maritime Princely states posed to British revenues. It complicates and contributes to an understanding of the British empire from its fringes and of the Indian Princely states in early 1900s Western Indian Ocean. It shows how, in the early 1900s, that Ocean had become a resource and a passage for goods for earning profits and safeguarding revenues. Science, the modernization of ports, railway infrastructure, and customs cordons and the law were all simultaneously deployed by both British India and the various Princely states (maritime and inland) in order to reap benefits for their populations and governments. This is a trend that continues today, with national governments claiming the Indian Ocean's heritage while aspiring for power, control and commercial opportunities.

Acknowledgments This research was supported by the Max Planck Institute for Social Anthropology between 2013 and 2016. Sincere thanks to Dr. John Eidson, Dr. David O'Kane, Dr. Stephen Reyna, Dr. James Carrier, Dr. Aida Alymbaeva, and to the participants in the colloquium 'Production of Imperial Space: Empires and Circulations', held at Sciences Po on 24 November 2017.

Notes

1. The emic name Saurāshtra is currently in use for Kāthiawār in the archival records. The same is true for Kutchch (Kutch and Cutch).
2. Printed indexes of the Commerce and Industry Department, Government of India between the years 1900–1940, National Archives of India, New Delhi.
3. Salt exports from Britain were valued at 63,20,957 in 1923–1924 and 80,15,288 in 1927–1918, as against the total salt imports of India valued at 47,06,910 in 1923–1924 and 9,468,996 in 1927–1928 (FICCI, 1930, pp. 133, 171, 198, 203).
4. Similarly, 'agreements of a commercial character embodying provision for compensation in respect of surrendered rights were entered into with about fifty states which owned, operated or had an interest in salt works, or were in a position to levy transit duties on an important scale' (Indian States Enquiry Committee, 1932, p. 74).
5. Mehta refers to the following record: Huzur English Office, Revenue Department No. 57 for 1895–1898 and Bombay Gazette 18 and 21 September 1895 (daily newspaper).

References

Ahuja, R. (2004). Opening up the country? Patterns of circulation and politics of communication in early colonial Orissa. *Studies in History, 20*(1), 73–130.

Beverley, E. L. (2013). Frontier as resource: Law, crime, and sovereignty on the margins of empire. *Comparative Studies in Society and History, 55*(2), 241–272.

Bhavnagar Darbar. (1917). *Report of administration of Bhāvnagar princely state 1915–1916*. Bhavnagar Government Press.

Boehme, K. (2015). Smuggling India: Deconstructing western India's illicit export trade, 1818–1870. *Journal of the Royal Asiatic Society, 25*(4), 685–705.
Das Gupta, A. (1979). *Indian merchants and the decline of Surat c. 1700–1750*. Franz Steiner Verlag.
FICCI. (1930). *Monograph on Common Salt*. Federation of Indian chambers of commerce and industry (FICCI).
Foreign Department. (1910, April). [unpublished manuscript. Foreign Department, Diary Salt and Co Diary no. 57-I-B. Foreign department, diary Salt and Co. dated 9 April 1910 received 14 April 1910]. *Index of commerce and Industry department*. (Customs, Pro July 9 1910, File 2 B). National Archives of India, New Delhi, India.
Goswamy, C. (2011). *The call of the sea Kutchchhi traders in Muscat and Zanzibar, c. 1800–1880*. Orient Blackswan.
Grancho, N. (2015). Diu as an interface of east and west: Comparative urban history in 'non-western'stories In Keller. S. & Pearson, M. (Eds.), *Port towns of Gujarāt*, (pp. 259–272). Primus Books.
Ho, E. (2006). *Graves of tarim: Geneology and mobility across the Indian Ocean*. University of California Press.
Hornell, J. (1920). *The origins and ethnological significance of Indian boat designs*. Asiatic Society.
Hull, M. S. (2012). *Government of paper: The materiality of bureaucracy in urban Pakistan*. University of California Press.
Indian States Enquiry Committee. (1932). *Report of the Indian states enquiry committee*. Government of India Press.
Kurlansky, M. (2002). *Salt: A world history*. Walker and Company.
Machado, P. (2009). Cloths of a new fashion: Indian Ocean networks of exchange and cloth zones of contact in Africa and India in the eighteenth and nineteenth centuries. In Riello, G., & Roy, T. *How India clothed the world: The world of South Asian textiles, 1500–1850* (pp. 53–85). Brill.
Markovits, C. (2008). *The global world of Indian merchants 1750–1947 traders of Sind from Bukhara to Panama (Cambridge studies in Indian history and society)*. Cambridge University Press.
Markovits, C., Pouchepadass, J., & Subrahmanyam, S. (2003). *Society and circulation: Mobile people and itinerant cultures in South Asia 1750–1950*. Anthem Press.
Martin, M. (2008). Hundi/hawala: The problem of definition. *Modern Asian Studies, 43*(4), 909–937.
McLeod, J. (1999). *Sovereignty, power, control: Politics in the state of western India, 1916–1947* (Brills Indological Library). Brill.
Mehta, M. (2009). *History of international trade and customs duties in Gujarāt*. Darshak Itihas Nidhi.
Moxham, R. (2001). *The great hedge of India*. Carroll and Graf Publishers.
Mukherjee, R. (2013). The Indian Ocean: Historians writing history. *Asian Review of World Histories, 1*(2), 295–307.

Nordstorm, C. (2007). *Global outlaws crime money and power in the contemporary world*. University of California Press.

Palsetia, J. S. (2008). The Parsis of India and the opium trade in China. *Contemporary Drug Problems, 35*(Winter), 647–687.

Porter, T. M. (1995). *Trust in numbers: The pursuit of objectivity in science and public life*. Princeton University Press.

Prakash, O. (2007). English private trade in the western Indian Ocean, 1720–1740. *Journal of the Economic and Social History of the Orient 50* (2/3 Spatial and Temporal Communities of Merchant Networks in South Asia and the Indian Ocean), 215–234.

Ramusack, B. (2004). *The Indian princes and their states (The Cambridge history of India III.)*. Cambridge University Press.

Ray, R. K. (1995). Asian capital in the age of European domination: The rise of the bazaar, 1800–1914. *Modern Asian Studies, 29*(3), 449–554.

Roy, A. L. (1886). English rule in India. *The North American Review, 142*(353), 356–370.

Roy, T. (2006). *Economic history of India 1857–1957*. Oxford University Press.

Sheikh, S. (2014). *Forging a region: Sultans, traders and pilgrims in Gujarat, 1200–1500*. Oxford University Press, Delhi.

Vaidya, K. B., (1945). *The sailing vessel traffic on the west coast of India and its future*. Popular Book Depot.

9

Navigating Environment, History, and Archaeology in Portsmouth Island, USA

Lynn Harris

An article in *Potters American Monthly—For the Higher Education of Women* (Marsh, 1878, p. 284) described a voyage from Charleston, South Carolina to New York when a ship sprang a serious leak off the coast of North Carolina. The crew fired a cannon to signal distress, and soon a sail of a small boat appeared on the horizon "floating over the water like swan with outstretched wings." As it drew closer, the passengers realized it was a pilot boat, crewed by a single black mariner, who informed them that they were at Ocracoke Inlet and offered to guide them safely across the treacherous inlet sand bar

---

Many thanks to the Program of Maritime Studies in the Department of History at East Carolina University for facilitating and supporting a variety of small field projects and a field school on Portsmouth Island from 2016 to 2019. Many M.A. students played a crucial role in data collection and site recordation, especially Ian Harrison, Kelsey Dwyer, and Ryan Marr. The project could not have been conducted without generous funding from the US National Park Service Preservation Training and Technology (PTT) Grant. Portsmouth island was one of several important case studies representing coastal heritage at risk on the North Carolina shoreline. Much appreciation to colleagues Thad Wasklewicz, Department of Geography, Planning and Environment and David Mallinson, Department of Geology, who played an important part in highlighting interdisciplinary issues, sharing expertise in new state-of-the art technologies as historic preservation tools, and introducing students to broader conceptual frameworks for collecting and analyzing data.

L. Harris (✉)
History Department, East Carolina University, Greenville, NC, USA
e-mail: harrisly@ecu.edu

or "swash" to the village of Portsmouth island. He noted in the African American coastal dialect of the time, "den you has safe anchoring, and den you get the leak prepared there. I'se been a pilot har for twenty years." As their ship approached Portsmouth through the swash passengers observed with apprehension a forest of shipwreck masts, serving as beacons to warn future mariners not to attempt this navigation challenge without a pilot. In contrast, the Portsmouth landscape was a pleasant surprise. The coastal terrain was covered in live oak, entwined vines, creeping plants, and wild perfumed flowers. The water at the anchorage was clear, teeming with all kinds of marine life.

A few days later while the leaking ship was undergoing repairs in Portsmouth, a passenger family—named Beaufort—had further contact with the talkative and engaging black mariner. They found out he was named Bob Blount and served as an enslaved waterman to a female descendent of the North Carolina trading empire of John Gray and Thomas Blount Merchants and their collaborative enterprise with John Wallace that dominated shipping at Portsmouth and Ocracoke islands in the early 1800s. It was miniature commercial metropolis connected to the larger southern seaboard port cities like Savannah, Wilmington, and Charleston. Bob, whom the Blount descendent hired out, offered to take them to visit the family castle noting "you see it am built on de rock, and it am called 'Shell Castle'. Many a storm de ole castle har stood. You see it are an ancien building." At the castle the guests were received by Mrs. Blount "…a very lady-like person of the old school; her manners were very polished." She entertained them with dinner, prepared by her servants, and overnight accommodations. The Beaufort's were astonished by her castle home, that they described as "An odd, non-descript sort of place rearing up as on strong foundation of oysters and other shells, which time and rush of water from the ocean had formed a compact body on which seaweed and other uncommon plants grew." But what surprised them most, was that a dignified woman could live in such an isolated place with no other white persons nearby. This unusual report, where the audience could "hear" the voice and dialect black southern mariner, aimed to share with readers the close experience of a wrecking and the service and value of these mariners to seaboard shipping.

Since the 1970s a growing number of scholarly works address the significant role of black watermen, both enslaved and free, along the eastern seaboard (Bolster, 1997; Cecelski, 2012; Dawson, 2006, 2018; Harris, 2014; Jarvis, 2002; Linebaugh & Rediker, 1990; Marin, 2007; Wood, 1996). Scholars identify black watermen as key agents of rebellious action and resistance to slavery. Local African American maritime culture of the eastern seaboard was intricately interwoven within the commercial fabric of the Atlantic World. For the enslaved population at Portsmouth, Ocracoke, and Shell Castle especially, living at an international crossroads meant a measure of freedom, access to information from northern ship crews, and possible opportunities for escape. Newspaper advertisements describe runaways during the years when Shell Castle flourished. North Carolina coastal enslaved African Americans, like others in South Carolina and Georgia, led highly cosmopolitan lives as divers, sailors, pilots, boatmen, fishermen, stevedores, and maritime tradesmen. By virtue of their work or their travels they were often hired out and closely connected with a network of African American watermen in ports up and down the eastern seaboard and in the Caribbean (Dawson, 2006, pp. 1327–1330; Harris, 2014, pp. 57–61, 75–77). Cecelski's analysis of black watermen and race in coastal maritime culture is one of the most valuable to North Carolina's vignette. It encompasses enslaved southern plantation boatmen, free blacks working in the fisheries, enslaved canal builders, and those heroic black watermen who aided other enslaved Africans in finding water routes to freedom (Cecelski, 2012, pp. xvi, 18–21, 47).

Historical Background of Portsmouth Island

Established in 1753, the Ocracoke Inlet town of Portsmouth functioned, not only as a service area for passing shipping, but as a lightering port where cargo from ocean-going vessels could be transferred to shallow-draft vessels capable of traversing the inner banks estuaries and tributaries that led to plantations and river port towns. Portsmouth grew to a peak population of 685 in 1860, with the first hundred years as the high point in commerce. The fortunes of the inhabitants were completely dependent on the relative navigability and stability of the inlet. These were the years of escalating development at Portsmouth when entrepreneurs, state, and federal government officials focused the most attention on this island. The population was comprised of one-third enslaved African Americans and a handful of free African Americans, later registered as mulattos or mixed race, in census records (Olson, 1982, pp. 60–61).

Though comparatively small, Portsmouth was situated at one of the most important points-of-entry, Ocracoke Inlet, along the Atlantic coast in post-Revolutionary America. Two merchants, Blount and Wallace, reclaimed and stabilized inlet sandbanks with oyster shells, building an essential port facility with a good anchorage and ability to repair and provision shipping. It was half a mile long and 60 feet wide boasting wharves and warehouses, a windmill, a gristmill, a store, a lumber yard, a notary public's office, a tavern, and a main building reported to be 300 feet long, all packed onto its tiny area. The 1789 development of the trans-shipment entrepot at Shell Castle, situated in the inlet between Portsmouth and Ocracoke islands, was not only an ingenious example of early land reclamation but a maritime marketing strategy based on navigation convenience. Portsmouth evolved as the base for lightering services, piloting, customs collection, and health care related to the inlet after Shell Castle's demise in 1818 (McGuinn, 2000, pp. 39–50; Olson, 1982, pp. 60–61) (Fig. 9.1).

Fig. 9.1 Ocracoke inlet North Carolina 1775. An Accurate Map of North and South Carolina with Their Indian Frontiers, Shewing in a distinct manner all the Mountains, Rivers, Swamps, Marshes, Bays, Creeks, Harbors, Sandbanks, and Soundings on the Coasts; with The Roads and Indian Paths; as well as The Boundary or Provincial Lines, The Several Townships and other divisions of the Land in Both the Provinces; the whole from Actual Surveys. By Henry Mouzon and Others Public Domain: https://commons.wikimedia.org/wiki/File:Ocracoke_inlet_north_carolina_1775.jpg

In 1806 Congress US Committee on Commerce and Manufactures recommended surveys of shoals to make the area safer for shipping noting that, "…it is supposed that there is no part of the American coast where vessels are more exposed to shipwreck than when they are passing along the shores of North Carolina…" (Herndon, 1972, p. 242). In 1846, two strong hurricanes cut Oregon Inlet and deepened the existing Hatteras Inlet to the northeast, making Ocracoke Inlet a less desirable shipping lane by comparison. The waters around Portsmouth's harbor also began to shoal up, speeding up its decline as a port. The Civil War (1861–1865) was yet another blow as many people fled to the mainland when Union soldiers came to occupy the Outer Banks. Many did not return after the war had ended and the Portsmouth village continued its decline, hastened along by the occasional storm or hurricane. The mammoth 1933 Atlantic hurricane season also served as a benchmark in the island's population decline, though more as a focal point of memory and a symbol of decline than the real cause of it. A further blow was the decommissioning of the US Life-Saving Station there in 1937 and closing of the post office in 1959. In 1967 Portsmouth Island and village had already been acquired by the National Park Service then incorporated into the new Cape Lookout National. Marian Gray Babb and Nora Dixon were the last two elderly Euro American residents who departed the island in 1971 after the death of African American Henry Pigott. Although Pigott was approximately the same age, he was essentially the caregiver of the two women.

The Demographics

Phillip McGuinn's (2000) research on Shell Castle includes significant detail on the lives of enslaved community who worked in and around Ocracoke Inlet during its years as a commercial center. Carteret County census schedules from 1790 to 1860 lists the enslaved people living at Portsmouth and at Shell Castle during its zenith. The Wallace family owned fifteen slaves who served at the Castle. Transactions in slave trade and correspondence reveal some information about the roles and treatment of their workforce. Perry was a "respected" boat pilot and Peter an "old and faithful" house servant. Perry was hired out for his lucrative services as a waterman. Living at a port enclave was also a liability for runaways. Four enslaved men hired by Wallace stole a boat in 1793 and escaped from bondage. Reports reveal runaways successfully reaching northern ports like Philadelphia or taking boats inland to river ports using these smaller towns as transit areas. There are numerous references to enslaved women and girls working in the Castle. A young girl named Kate became the house matriarch and had at least five children. Other transactions indicate that several other young girls were purchased for Shell Castle families (McGuinn, 2000, pp. 261–266).

Largely unanalysed data suggests that greater detail on enslaved African Americans here can be found in genealogies, wills, deed records, court records, and correspondence of key individuals like John Gray Blount (1752–1833) planter, merchant, shipper, land speculator, and politician, including many letters from his partner-brothers (Blount Papers 1706–1900). The association of larger scale enslaved African Americans with a few major families increases the prospect of learning about the lives of those at Portsmouth village, either via archival research or through archaeology around specific sites. Areas of focus are in and around the current town center, Doctor's Creek, and along the inlet-facing shoreline, where the Wallace lands and David Wallace, Sr. and David Wallace, Jr. home, as well as some of the Gaskill properties were situated—because they owned the largest number of enslaved people. Currently, this is one of the most vulnerable and at-risk shorelines of the area for erosion. The relative extent of enslaved African holdings among the Portsmouth white population waxed and waned over time, but overall concentration of enslaved African holding, was always relatively small, compared to the inland plantations—yet critical to maritime commerce of the 1800s.

Occupations

Analyzing 1820 census records for Portsmouth, Kenneth Burke found that most of the men were engaged in commercial activities focused on lightering, fishing, and navigation, though six worked in manufacturing. At that time, most of the ships calling Portsmouth their home port were smaller schooners not appropriate for seagoing travel and probably used as lighters. The size of ships based at Portsmouth increased from 1816 to 1839. In 1850, the population of 346 whites and 117 enslaved Africans included a majority of the adult white men employed as pilots, mariners, and boatmen (Burke, 1958, pp. 35–38, 52). Thus, during the century when thousands of ships going "to and from the most remarkable places" passed through Ocracoke Inlet, these black and white residents of Portsmouth were not fundamentally isolated or particularly provincial, but were instead deeply entangled with state, national, and international politics, and with the social and cultural worlds of other American ports north and south, the multinational ports of the Caribbean, and the inland towns that dotted the shores of North Carolina's huge estuaries and rivers.

In 1842 the Committee of Commerce of United States Congress reported, "Ocracoke Inlet is the outlet for all commerce of the State of North Carolina…and the whole extent of the country around them. One thousand four hundred sail of loaded vessels pass through the aforesaid inlet in the space of twelve months bound for various ports…[it is] not uncommon to see 30 to sixty sails of vessels at anchor in the roads at the same time" (Mallinson, 1998, pp. 10–11). As early as 1810, enslaved pilots began taking advantage of navigation and piloting opportunities with increased shipping at the inlet. It appears they had liberties to take on piloting tasks independently, but it remained a concern to merchants and white pilots. Joseph Bradley wrote to the Blount's at Shell Castle, "there is some stragling Negroes passing here piloting without branch or responsibility [unaffiliated to a European family or licensed as a pilot]. Captain Wallace and others are determined to prosecute them…." (Morgan, 1982, p. 136).

The volume of shipping at the inlet attracted various groups of people, including African Americans—some were freed slaves, others were slaves brought by their owners to work as sailors, lighter crews, and pilots. It became increasingly evident that for African Americans, enslaved or freed, one of the most prized trades was that of a pilot. The unlicensed black pilots that worked in the area quickly became problematic for the local licensed pilots. It became such a contentious and competitive issue that several of the licensed pilots joined together to petition Governor Josiah Martin and the General Assembly. The list of pilots included John Williams, George Bell, John Bragg, Adam Gaskins, Richard Wade, and Simon Hall.

The issue of black pilots at Ocracoke Inlet received little attention by the government, and complaints continued to occur through to the 1860s, even against black pilots licensed to pilot the inlet who were a source of commercial competition (Cecelski, 2012, pp. 48–49; Howard, 2016; State Records, 1790, p. 167).

Portsmouth and Ocracoke Fishing Industries

Of specific focus in this chapter are the Portsmouth Pamlico Sound and coastal fisheries, an important part of the economy where African Americans played a crucial role both in the 1700s and 1800s. In 1737 John Brickell, in *The Natural History of North-Carolina*, described an incident involving whaling and the Outer Banks:

> These Monsters [whales] are very numerous on the Coasts of North-Carolina, and the Bone and Oil would be a great Advantage to the Inhabitants that live on the Sand-Banks along the Ocean, if they were as dexterous and industrious in Fishing ….Inhabitants dwell upon the Banks near the Sea for that Intent, and the benefit of Wrecks of Vessels which are sometimes driven in upon these Coasts. Not many Years ago there were two Boats that came from the Northward to Ocacock Island, to fish, and carried away that Season Three Hundred and Forty Barrels of Oil, beside the Bone, but these Fishermen going away without paying the Tenths to the Governor, they never appeared to fish on these Coasts afterwards, or any other that I ever could hear of. (Brickell, 1737, p. 220)

Although shore-based whaling was always seasonal, and never amounted to more than a minor industry on the Outer Banks, whale oil was quite valuable as a lubricant and lamp fuel. Whale oil was especially prized for fuel in Outer Banks lighthouses. During the Proprietary period (prior to 1729), whale oil even became an official medium of currency. Sperm whales, humpback whales, blackfish (short-finned pilot whales), and others were sometimes chased, but the right whale (so named because it was the "right" whale to hunt) was the chief target. Right whales typically swam closer to shore, were more compliant, and floated after being killed. The typical Outer Banks whaleboat was 20–25 feet long, double-ended, high in the bow and stern, and constructed of lapped planks. Four men rowed the boat, a fifth served as steersman, while a sixth, often the captain, remained in the bow, prepared to throw the harpoon. The earliest harpoons employed in North Carolina were of the simple single-flue or two-flue variety. Later, "toggle-irons" (harpoons with a pivoting barbed head secured with a wooden shear pin) were used. After penetrating the whale's muscle, tension on the harpoon line broke the shear pin, turning the barbed point at a right angle making it difficult to dislodge (Bradley, 2015; Brimley, 1894; Howard, 2015; Simpson & Simpson, 1988, 1990).

In 1793, John Gray Blount and John Wallace the owners of Shell Castle initiated one of the earliest porpoise (bottle nose dolphin) fisheries, perhaps the earliest such operation. After the Civil War several other operations were established on the Outer Banks. A porpoise fishing enterprise included around fifteen to eighteen enslaved Africans in four small boats. The fishermen surrounded the marine mammals and captured them in heavy, large-mesh nets. Once trapped, thirty to forty porpoises at a time were surrounded by a smaller seine and hauled closer to shore. Enslaved African boatmen waded into the water and knifed the dolphins, thereafter they gaffed the animals and dragged them ashore. The next step was to cut off the flippers and dorsal fins and strip off the skin and blubber to process the oil in a fire on the beach. Porpoises were prized for their oil, and for their skins, which produced a supple, waterproof leather suitable for fishing boots. Porpoise meat was sometimes consumed locally, although its strong, oily flavor prevented it from being marketed commercially. A single right whale could yield more than 1000 gallons of high-quality oil for lamp fuel or lubricants. One dolphin might yield six to eight gallons of oil. A single large haul of one hundred or more dolphins could provide 750 gallons of oil, and thus a more reliable source of income (Cecelski, 2015, pp. 49–79; Huss, 2019; True, 1885, pp. 32–38).

Historians of the Outer Banks agree that commercial menhaden processing began soon after the Civil War when the Excelsior Oil and Guano Company of Rhode Island built a factory at Portsmouth Village. Greer's (1917) U. S. Commissioner of Fisheries report also says that a factory was established on Harker's Island in 1865. Menhaden oil, whose value was recognized later, was first extracted by rotting the fish in casks. However, a new way to extract menhaden oil (used in paint, soap, miners' lamps, and tanning) was discovered about 1850, setting the stage for later industrialization of the enterprise. By 1860, the first menhaden-cooking factories appeared, driven partly by the increasing scarcity of whale oil. Steam extraction brought a boost to the industry, first in land-based factories and then by oceangoing processors (Greer, 1917, p. 5).

George Brown Goode's brief retrospective account (1884, pp. 495–496) of the Portsmouth factory (and the menhaden industry in North Carolina) appears to be the most complete. For starters, Goode said the state lay "practically [at] the southern limit" of the Atlantic coast menhaden fisheries, and that no one had been able to turn a profit on it. They bought "modern apparatus," and hired "experienced northern seamen to handle the seines." But catches turned out to be small and of poor quality, and hot weather ruined the fish unless they were processed quickly—which they frequently could not be because of inlet conditions and frequent storms. The factory closed after about a year. An attempt by another Rhode Island company to establish a factory at Oregon Inlet failed after two years.

Over the long haul, however, the menhaden processing did survive and thrive. When the entire industry began to move south in the 1890s, much of it moved to North Carolina, where many of the jobs went to black workers. The state produced some 18,000,000 pounds of menhaden in 1902. By 1912, nearly 150 large steam and gasoline-powered menhaden vessels were serving forty-eight menhaden processing plants (employing more than 2000 people) on the Atlantic coast. North Carolina had twelve of them (Greer, 1917, p. 6).

The fish were processed in large screw presses. Much of the product was in the form of fish meal, used as an additive in poultry and livestock feed. Garrity-Blake (1994, p. 1) notes that the menhaden industry has evolved since the early nineteenth century, from an egalitarian organization composed of independent farmers and fishermen (especially in New England) to a hierarchical organization of capital-controlling manufacturers and wage laborers. Additionally, as it shifted geographically from New England to the southeast, the workforce, originally composed of native Yankees and immigrants, became a mix of rural southern whites and blacks. Although black history was affected by, and played itself out for decades within, newly formed autonomous black communities on the North Carolina coast, freed blacks also moved into existing and emerging coastal industries. Scholars have analyzed how class and race played out among menhaden fishermen in coastal North Carolina in the post-Civil War period, and for decades beyond, recent decades (Garrity-Blake, 1994; Greer, 1917, p. 6).

The most important fish that led to economic recovery on the Banks during the post-Civil War period, Fred Mallinson contends, was mullet. He quotes a Beaufort observer who in 1871 reported "enormous" numbers of mullet being harvested up to 500 barrels in a single haul and 12,000 barrels in a single September day of fishing. mullets were taken in vast numbers in nets—small dragnets in the sounds and much larger gill (or sweep) nets or seines in open water. The largest nets could be 12–18 feet deep and 900–1200 feet long. Sweep nets were 200–300 feet long and 4–6 feet deep. One or more small boats would tow the nets out to where lookouts had spotted a school, surround the fish with the net, beat on the boats to drive the fish into the net, then draw the fish-laden nets into the boat (Mallison, 1998, pp. 170–172; Stick, 2015, pp. 213–218) (Fig. 9.2).

The process for the largest nets was different. One end was attached to a rope on shore, while the other end was towed out to the school by boat, brought into a circle around the school, and then circled back to shore, where fifteen to twenty men—sometimes using "backing" seines behind the main one to pick up the overflow—were required to beach the catch. On shore at the temporary camps, men would stand at rough tables, slitting and gutting the fish before they were washed in sea water, salted, and packed in barrels. Since the fish bled into the salt, they would frequently be unpacked, washed, and repacked before sale. Fairly formalized "lay" systems were employed to determine how much each man was paid from the catch (Stick, 2015, p. 218).

Fig. 9.2 Fishermen on the Outer Banks land a mullet seine in this photograph, ca. 1884 (North Carolina Division of Archives and History)

Salted or smoked and packed in barrels, mullet was "savory and saleable." In 1880, a standard barrel brought $2.75–$3.50. A substantial portion of the catch was loaded on schooners, hauled across the sound, and traded with farmers for corn—five bushels of corn for a barrel of mullet. Some mullet fishermen were mainlanders who built seasonal camps on the coast and fished with the Bankers. Special conditions and methods at Portsmouth gave fish taken and packed there a special niche in the market. The foot-deep shoal waters of the sound allowed fishermen to surround the schooling fish, frighten them into the nets, and break their necks, leaving them in the nets until all had been killed before loading them into the boats. Onshore processing of mullet (removing the backbone, gutting, washing, and rubbing off the dark cavity lining) was a matter of great pride. On the market, their superior appearance and (many said) better taste put them in high demand (Smith, 1907, pp. 408–409).

In the Core Banks-Shackleford Banks area, mullet fishing thrived for about two decades, filling a demand from inside and outside the state for cheap fish. A report on the fishery industries of the United States for 1880 noted that shipments of salted mullet from [Carteret County] exceed the total shipments from all other portions of the Atlantic coast. In the late 1880s, when Carteret County was the center of mullet fishing in the United States, mullet fishing camps sprang up by the score along the sound-side banks from May to November, when the fish were running. If a half-dozen men in a small boat chasing a single whale with a harpoon and a drag defined one end of the spectrum of fishing techniques, mullet fishing was far out on the other end (Holland, 1968, pp. 20–21).

Tangible Heritage and Archaeological Surveys

Portsmouth is currently an uninhabited historic district maintained by the National Park Service as part of the Cape Lookout National Seashore. The village core has a total of 250 acres and 32 contributing buildings (18 main buildings; the rest outbuildings); the earliest dates to 1840, but the bulk of the buildings are from the 1920s and 1930s. This district was first listed in the National Register of Historic Places in 1978 (Brown, 1977). Currently, the author and colleagues are grant recipients tasked with compiling additional documentation to reflect current scholarship and new information about the history and material culture of the area, including a greatly expanded boundary to include related areas with ruins and archaeological potential for understanding this once-thriving area of North Carolina's Outer Banks.

Nine cemeteries and known grave sites are located throughout the Village area. Houses are separated from one another, and some are isolated by dense vegetation. The main vegetation on the island consists of dense, tall marsh grass, cedar trees, and pine trees. The island has sandy soil. The buildings in the village are connected by a network of narrow, dirt roads, including the main thoroughfares. The various historical structures and potential archaeological sites investigated are all within regions that are or will be directly impacted by sea-level rise and coastal erosion in the near future (<100 years). During the summers, people often visit the island and camp out overnight on the beach (camping is not allowed in the village). Facilities are very limited with a compost toilet near the Life-Saving Station and a restroom in the Salter house/visitors center, with no potable water, food, or electricity available (National Park Service, Cape Lookout NC: Visiting Portsmouth Village).

The Southeast Archeological Center (SEAC) is a branch of the National Park Service's Southeast Region. In assisting parks with their cultural resource management needs, SEAC facilitates long-term protection of archaeological resources and compiles and utilizes the archaeological information obtained from these resources. In addition to annually generating numerous archaeological reports, as mandated by federal law and park operations, SEAC is the repository for over six million artifacts that make up the Southeast Region's research collections and contribute to its cultural database. SEAC is staffed by professional NPS archaeologists and regularly employs archaeology students and other archaeological contractors throughout the Southeast (South Eastern Archeological Center, 2020). Below is a discussion of some tangible heritage examples pertinent to this topic.

Shell Castle

In 1999, a collaborative underwater archaeology project was undertaken on the site involving the NC Underwater Archaeology Branch and East Carolina University, Program in Maritime Studies. The National Park Service, US Coast Guard, and Ocracoke Historical Society provided further logistical and financial. The team conducted remote sensing operations, mapped the remaining submerged structure, and recovered historic ceramics. The main underwater feature on the seabed was the foundation of shell castle comprising stone, bricks, and mortar in an L-shape. A secondary feature was a thirty by 20-foot area concentration of loamy soil and shells containing a line of stakes. Archaeologists obtained wood samples from dock cribbing, beams foundations, planks, and trunnels. All were softwoods commonly used in building construction during that time. They compared the archaeological information with the illustration of the structures depicted on the pitcher, marketed to promote Shell Castle, which showed details of design. The conclusion was that it was a crib wharf found in other North Carolina, New York, and Boston ports. The dive team recovered large volumes of ceramic sherds from inside the cribbing. Comparisons of the Shell Castle owners probate records and the ceramic assemblage—teacups, saucers, tea pots, tureens, pitchers, slop bowls—reinforces the finding that Wallace was associated with upper tiers society. One discrepancy between the probate record and the archaeological data was the presence of annular ware bowls. During the first half of the nineteenth century, this type of bowl was most frequently associated with enslaved community cabin archaeological deposits, whereas the planters possessed more transfer printed wares (McGuinn, 2000, pp. 373–374, 391, 400, 419, 425) (Fig. 9.3).

Fishing Huts

In the 1880s R. Edward Earl (Goode et al., 1884, 564), a biologist studying North Carolina coastal fisheries, took photographs of fishing huts where black and white mullet fishermen lived and worked together in the Ocracoke and Portsmouth area. These distinctive, circular, thatched huts with conical or hemispherical roofs were featured in an article titled "Human Habitation" in *National Geographic* in 1908 by Collier Cobb, a North Carolina University geologist. He reported several more on islands of Carteret county like Cedar Island, Carrot Island, Core, and Shackleford banks. Scholars suggest these structures may represent a continuation of African hut design brought to North America and included features such as a central fireplace, thatching

Fig. 9.3 Shell Castle Pitcher made for John Wallace circa 1805–1810. Image bears the inscription "A North View of Governors Shell Castle and Harbour, North Carolina" (Image from NC Museum of History)

Fig. 9.4 Image of a typical black and white mullet gang at Brown Goode, ed. (1884). *The Fisheries and Fishery Industries of the Commission of Fish and Fisheries*, p. 562

and lathing building techniques prevalent for many years in parts of West Africa. There was a small hole at the apex of the roof to release smoke, a framework of red cedar or live oak limbs, with fire resistant salt marsh grass and black needle rush, bound together with bear grass (yucca). In cold weather the fishermen banked the outside of the hut with extra sand. The huts do not have similarities in European or American traditions of vernacular building. There are similar designs in Afro Caribbean villages established by maroons and other isolated communities (Cecelski, 1993, pp. 4, 9–10) (Fig. 9.4).

Other temporary structures included later plank-built fishing huts as well as houses. In the 1800s and 1900s these were rolled around and moved with high seas and storms. The locals built "mounds" and located the houses on hills to prevent flooding. The hammocks in the back-barrier marsh, where the individual homes were located, are the tops of former sand dunes formed on an old interior algal flat when sea level was some feet lower than it is today and have been buried by marsh peat in response to development of the wide series of beach ridges on the ocean side and rising sea level. These hammocks are slowly being buried by the ongoing rise of sea level—this ongoing drowning process historically forced some of the houses to be moved to higher hammocks. Portsmouth had three "storm houses" put together with wooden pegs and whalebones and, in a bad storm, all the islanders would gather in one or another of these temporary structures. The area of Portsmouth Island outside of the current historic district contains numerous archaeological resources. The vegetation cover and tree falls create challenging survey conditions. It is likely that every hammock outside the historic district had a site on it. Informants describe houses being moved and materials salvaged. Ada Styron Roberts recalled the use of whalebone to reinforce "storm houses" (Cecelski, 2008, p. 51).

Taverns and Anchorage

Several interviewees recalled their elders telling stories about "Washington Row," a section of taverns/brothels frequented by sailors during the 1800s (Cecelski, 2018). Further investigations of the locations or ruins of these structures could yield information about Portsmouth's identity as a small seaport with characteristics in common to other seaports. An 1806 map shows a cluster of three buildings is shown below Haulover Point labeled with the name "Watering Place" possibly the tavern and lightering anchorage for Portsmouth. In June 2019 the Program in Maritime studies at East Carolina University conducted a shoreline survey (NC State Site 31CR194**30) that encompassed a search area from the waterline to approximately 20 m inland and to 20 m out into the sound. The terrestrial team surveyed the inland half, while the water team waded and snorkeled through ~0–3 m water in the Pamlico Sound and up portions of Doctors Creek. Structures and sites in proximity to the artifact-rich search area were the houses inhabited by Frank Gaskill and Henry Piggot, as well as Haulover Point and Grey's Menhaden factory. The teams located, mapped, and identified a range of ceramic and glass fragments, marbles, a coin, and bricks relevant for adding to or corroborating the Portsmouth community's life (Harris et al., 2018). The field team located several submerged wooden barrels in a small bay in the vicinity of the Frank Gaskill house. Lids were missing, and the barrels filled with sediment. These artifacts have potential to contribute toward understanding waterfront maritime activities related to features shown on historic maps such as a windmill, taverns, and a gristmill. One possibility, according to other

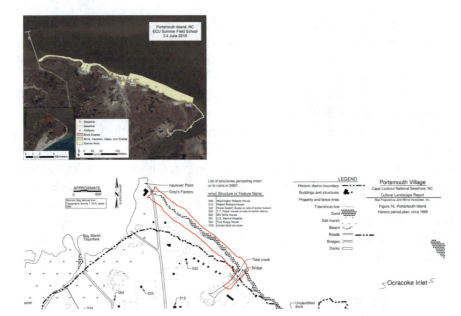

Fig. 9.5 Survey Area 2019 at Portsmouth Island near Haulover Point and up Doctors Creek. Program in Maritime Studies, East Carolina University image

archaeological case studies, is that these are barrel wells. The technique used wooden barrels stacked end to end lining a hole down to ground water. Barrel wells usually were constructed by stacking two (or more) wooden barrels on top of one another with the lower barrel filled with clean sand to serve as a filter (Austin et al., 2011) (Figs. 9.5 and 9.6).

Cemeteries

There are ten documented cemeteries on Portsmouth Island with preservation status varying with age and proximity to the historic village. Island stewards, with Portsmouth island family descendants still living in North Carolina, take care of certain cemeteries. The Keeler-Styron cemetery, associated with African American individuals, includes ten grave sites. Marble and granite headstones and a few footstones mark seven graves dating between 1836 and 1932. One grave is outlined in a single row of mortared brick, slowly disintegrating. Some of the headstones are weathering and eroding.

Like the better-preserved cemeteries closer to town, it is likely that it was previously enclosed by a fence to prevent livestock from accessing the area. Rose Pickett (with the next generation the name changed to Piggott), an enslaved African woman and her free African American children, Isaac, and

Fig. 9.6 Water survey team aligned 2 m apart on a tape, searching for submerged artifacts and structures on an eroding bank. Program in Maritime Studies, East Carolina University image

Leah, have graves in this cemetery. Leah was Henry Pigott's mother. The illustrious Samuel Tolson is buried in this cemetery in proximity to her grave. Park Service researchers note Tolson was born on Portsmouth Island on November 7, 1840. He worked as a mariner and lived on the island until he passed away in 1929. Uncle Sam, a Euro American, as he was fondly known, is remembered by fellow villagers for his long-stemmed clay pipe, stylish dress suit, and his expert dancing. Off the island, he was better known for his resemblance to the infamous assassin of President Abraham Lincoln, John Wilkes Booth. This resulted in a false arrest and imprisonment until residents of Portsmouth island vouched for his identity (Sam Tolson, National Park Service Website). Interview transcripts from National Park Service interviews Connie Mason conducted, indicate that Sam Tolson may have been Henry Pigott's father. He and all his siblings are listed as mulatto in the census records. Leah Pigott was born in June 1867 in Portsmouth. Her father was African American, Benjamin Willis was 19 and her mother, Rosetta (Rose), was 30 (Coastal Voices, Connie Mason Interview with Chester Lynn, # 10-009, 2011, p. 12). Henry Piggott's grave, along with his sister Elizabeth "Lizzie" Piggott's grave, is situated in the Babb Dixon cemetery along with the last surviving residents of the now deserted island—Lillian Babb, Arthur Dixon, Nora Dixon, and two parakeets named Pete and Dick (Fig. 9.7).

Fig. 9.7 Headstones of enslaved grandmother Rose Pickett (1836–1909) and her free African American grandson Henry Pigott (1896–1971). Program in Maritime Studies, East Carolina University image

Intangible Heritage

"Coastal Voices: Linking Generations," an oral history collection, is now available for free online viewing. The collection includes audio recordings and transcripts of 30 oral history interviews conducted by Outer Banks residents between 1978 and 2003 for the Southern Oral History Project and the National Park Service Cape Hatteras National Seashore Ethnohistory Project (Coastal Voices). The collection is housed in the Core Sound Waterfowl Museum and Heritage Center's online oral history archive.

Oral Histories and Chanties

Several interviews are pertinent to the experiences of African American fishermen. Although the informants did not reside on Portsmouth island, African American crews practiced the same fishing techniques there in the 1900s. Ernest Davis, from Beaufort, worked from sunup to long past sundown as a chantey man. At 74 years old, he was one of a few chantey men still living in the Cape Lookout area. He began working in 1955, when he was 16 years old. He noted, "It would be so cold, your hands would freeze up. It was tough back then. A workday would often stretch into the wee hours, especially when the crew had a big roundup." "They were just hardworkin' fish....Sometimes it would take up there to the middle of the night to get'em. You gotta stay there and wrestle with'em. You can't turn'em loose." In the thick of the fish fight, he and other men sang. Every crew member was expected to lift his voice while lifting the net. They had a finely tuned system. A rhythm. "And you had to sing it together, too, or you wouldn't be there the next day,"

Davis says. This singing was part of the job description—keep quiet, and you wouldn't keep your job. "Yeah, they'd pull you off."

As the bows of the two boats came closer together, the net closed, or pursed, at the bottom to keep the menhaden from escaping. Then came the hard part: raising the net, bulging with tens of thousands of fish, literally tons of fish, so they could be densely packed for bailing onto the mother boat. This was backbreaking, muscle-aching, hand-callusing labor that demanded these men work together—pull together—in unison. To keep everyone synchronized, the men sang songs that were known as chanteys. The workers, away from home for days, weeks, or months at a time, came up with the words themselves. They were bluesy songs, suggestive of homesickness and longing. One chantey printed in Barbara Garrity-Blake (1994) work goes as follows,

> *The fish factory:*
> *I left my baby*
> *Standing in the back door crying.*
> *She said, "Daddy, don't go,*
> *Lord, Lord, Daddy, don't go.*

Some chanteys had a stick-it-to-the-man tone, expressing the solidarity the crew members shared with one another. For example:

> *Captain, if you fire me, fire me, fire me*
> *Captain, if you fire me,*
> *You got to fire my buddies, too.*

James "Poppy" Frazier of Harlowe describes pulling the purse seine net by hand before the mechanized hydraulic power block was introduced in the early 1960s. He describes the jobs of a captain, a bunt puller, and a fish bailer. Primrose Jones was a cook on a menhaden vessel for many years. He talked about the job of a cook: what he cooked, buying groceries, his help, his favorite meals. Local companies in Carteret County provided generations of workers—boat captains, boat pilots, engine runners, ring setters, fish bailers, factory foremen, shore engineers. The last factory, Beaufort Fisheries on Front Street, closed in 2005 and was razed a few years later. Although nothing remains of the structure, an old net reel is on display (Fig. 9.8).

Euro-American's interviews dominate the oral history portal for Portsmouth Island. Frances' mother, Clara Salter Gaskins, was born in Portsmouth. Her grandparents, Theodore and Annie Salter were central to life in Portsmouth. Theodore was a store owner and Annie was the postmaster. Frances' mother was interviewed extensively in the 1970s and 1980s. Chester's Lynn, now a resident of Ocracoke island, provided details about

Fig. 9.8 Menhaden Netting Operation (NC State Archives)

personalities on Portsmouth island including the last remaining African American family. The oral historian, Connie Mason, asked questions of his knowledge of interracial relationships and paternity of Henry Piggott. He favored other discussion topics especially the Portsmouth landscape and, as a keen gardener, he was knowledgeable about the fauna on the island. His other interest was finding archaeological sites and antiques on the island.

Homecoming Festivals and Events

A working committee known as Friends of Portsmouth Island was formed in late 1989 under the sponsorship of the Carteret County Historical Society. The statement of purpose consisted of three areas: to promote and encourage the preservation of the historic structures, furnishings, and sites of Portsmouth Island; to collect and preserve artifacts, photographs, documents, and manuscripts of Portsmouth Island for deposit in the Carteret County Museum of History and Art; and to foster and promote public knowledge of and interest in Portsmouth Island's past, present and future. The organization sponsored its first homecoming event on April 25, 1992. Seven people who were born in Portsmouth were attended. Homecomings were sponsored by FPI every second year starting in 1994. The 2020 Homecoming was canceled due to the COVID-19 crisis, and the next will be held in 2022. Traditionally, homecomings in the south are for the members of a particular church or family, but the Portsmouth Homecoming welcomes everyone. In fact, about half of the participants at homecoming are first-time visitors to the village. The homecoming service, in the past held in the church, but now held on the grounds in front of the church, includes singing, special music, recognition of guests and other visitors, a short devotional, and good fellowship. Dress for the day is very informal (Friends of Portsmouth Island, 1996; NPS Portsmouth Homecoming, 2020).

Race Relations Narratives

According to Howard (2011b) among Ocracoke Inlet communities, "Interbreeding added to the confusion of race relations on the coast. Masters and crews of sailing vessels occasionally made alliances with black women in port, or on board their ships. Their mulatto offspring kept race relations on shore fluid and ambiguous." In the late 1800s after the Civil War, not many of Portsmouth's former black residents returned to the island. The primary narrative oral history white Portsmouth descendants and National Park Service websites and brochures offer to the public regarding race relations in the town is sentimental, affectionate, and often vague. They are remembered and cast as part of the island's narrative by the white residents who were children at the time, or repeated stories from their parents and grandparents. The popular "miniature" story of postwar Portsmouth centers around the family of Henry Pigott, its last black resident, always described as unfailingly obliging and devoted. Pigott (1896–1971), a descendent of enslaved Africans, was black and poor but nevertheless "a friend to all." Race relations in the town and on the island were harmonious and unproblematic. "Family" was the preferred terminology for describing those relations (Taylor, 2009, pp. 38–39). Enslaved Rose's children bear the surnames Bragg, Ireland, Abbot or Abbott, Pickett, Pigott, and Willis, the names of the white families of Portsmouth and several are listed as mulatto in the 1910 census, the first year that categories of "color of race" were expanded. Rose's great-granddaughter believes "there was an invasion into her [Rose's] life from the men on the island" (National Park Service, 2015, p. 30).

The more sentimental narrative continues to be contextualized by the heritage stewards as the centerpiece story of Henry Pigott and his relatives, who were content and more comfortable with the old ante bellum paternalistic system. However, Pigott and his sister Lizzie were illiterate as they were not allowed to attend the only all-white school on Portsmouth island, later relying on others to write letters for them to order supplies from the mainland. "There were never any segregation rules," writes Ellen Fulcher Cloud in *Portsmouth: The Way It Was*, "except what the blacks imposed upon themselves" (Cloud, 2006, p. 97). After Pigott's death in 1971, the Park Service produced a brochure about Henry Pigott (Pigott, 2000). It delineated what was to become the standard story: Pigott was descended from enslaved Africans (Fig. 9.9). His ancestors stayed in Portsmouth after most former enslaved Africans left. His grandmother Rosa Abbot was a jack-of-all-trades (midwife, doctor and nurse, gristmill worker) who also fished for her living. In interviews Salter and Willis state that they don't remember Henry and Leah's father:

Fig. 9.9 NPS Signage on Portsmouth Island about Tom Piggott (Photo by Lynn Harris)

They never talked about him, nor did they mention who he was or where he was. We never asked. . ... [Henry] was not dark in color, he was like an Indian in appearance. . . . We never heard about a color barrier in those days. There was no need; we were all in the work together. . . . The Pigotts attended the Methodist Church that we did. They visited with us and lived among us. . . . [A] finer man I never knew. . . . Neither Henry nor Lizzie ever married. (Willis & Salter, 2004, pp. 38–39)

Environmental Threats to Portsmouth

Like most North Carolina Inlets, Ocracoke Inlet is a dynamic part of the coastline and more vulnerable to transformation than others. Portsmouth Town is situated on the northern end of North Core Banks (also known as Portsmouth Banks or Portsmouth Island). Core Banks is a narrow barrier island with low elevation, dominated by inlet and wash over processes (Heron et al., 1984). Highest dunes (ca. 7 m) occur in the Portsmouth Town area. The upland regions, consisting of dunes, are separated by tidal creeks which drain a tidal flat (algal flat) region situated between the highest areas (where the town resides) and the ocean shoreline. The island is situated on the downdrift side of Ocracoke Inlet—thus, the shoreline is highly dynamic, with erosion or accretion rates sometimes exceeding tens of meters during a single storm event. The position of Ocracoke Inlet has remained relatively stable during historical times, likely because of the control exerted by a relict river channel (paleo-Pamlico Creek) situated beneath the inlet (Mallinson et al., 2010).

Sea-Level Rise and Shoreline Erosion

Long-term rates of sea-level rise in this area were measured using peat records from near Cedar Island (Kemp et al., 2017), ca. 10 km southwest of Portsmouth Town. The tide gauge at Beaufort, NC approximately 60 km to the southwest, provides data since 1953. Both methods indicate a rate of rise of ca. 3 mmy. Shoreline erosion rates vary considerably seasonally and annually because of the position adjacent to Ocracoke Inlet. Portsmouth Town is situated on low land, on a dynamic barrier island.

However, the adjacent inlet shoreline has remained somewhat stable over the last 25 years. Furthermore, the tidal flat region, and the seaward side of the uplands are accreting sand, and marsh vegetation has replaced open algal flats in the backshore on the eastern extent of the island. This suggests that sedimentation may help to maintain this island even in the face of rising sea level.

Vulnerability Assessment

The historic sites at Portsmouth Island occur at low elevations on a dynamic barrier island. Many of the sites occur within 0–1.5 m above sea level, meaning they are likely to be inundated by rising seas over the next century (Fig. 9.10). Thieler and Hammer-Klose (1999) place it in their high to very high-risk category. Sediment erosion, transport, and deposition in the form of inlet and spit migration, and shoreline erosion and overwash will likely shift the present ocean shoreline to the north, likely further filling the algal flats, but eroding the sand hills. It is unclear what may happen to the inlet shoreline as it depends on unpredictable storm energy and sediment transport around the inlet, affecting the channel position. If the trend of the last 30 years continues, Shell Castle is slowly eroding away and remaining portions will soon be completely submerged.

Heritage at Risk Survey

To facilitate a national need in cultural resource management, researchers from different disciplines at ECU (Geography, Geology, Coastal Resource Management, and Maritime History) work together to adapt and develop several technologies and techniques on cultural heritage examples that are unstable, eroding, or deteriorating situated both on land and underwater.

132   L. Harris

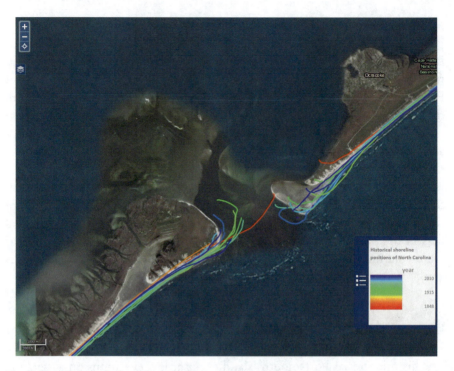

Fig. 9.10 Historical shoreline positions since 1848 for the Portsmouth Town to Ocracoke (Harris et al., 2018; Thieler & Hammer-Klose, 1999)

The Portsmouth Historic District structures: Life-Saving Station, Methodist Church, and Henry Pigott's house, were case study sites selected based upon criteria such as state or national historic significance, conservation management challenges, and probability to serve as an intellectual platform to segue between preservation of a historic icon and research questions to develop student theses and dissertation topics (Harris et al., 2018).

The Terrain Analysis Laboratory in the Department of Geography, Planning, and Environment used three Leica HDS terrestrial laser scanners. A terrestrial laser scanner is a tripod-mounted instrument and emits laser light at a specified vertical and horizontal spacing. Any surface reflecting the laser light back to the scanner will be recorded as a point at every 6.3 mm across the entire surface. Each point records the location of the surface, its height, the surface intensity (amount of energy reflected from each object), and rgb values (from a photo taken with the laser scanner). This information can be used to measure features, assess structural changes over time, assist in restoration of the sites by identifying features that are decaying, preserve the current condition of the site, develop educational and research-educational experiences, and visualize the features by a variety of media including 3D models printed from 3D printers (See 3D scans links in endnote).

Discussion and Future Directions

There are significant avenues for further or more in-depth research on Ocracoke Inlet communities encompassing both tangible and intangible heritage. For instance, investigation into similarities associated with African coastal traditions, like the fishing huts (Cecelski, 1993). A cross-cultural study of fishing techniques with small boats and nets from beaches or in estuaries is rich with possibilities. Historian Emmanuel Akyeampong (2000, 2001, 2007), who studies the African Diaspora, suggests there are several unexplored areas of historical exchange especially in technology. Drawing upon his article indigenous "Knowledge and Maritime Fishing in West Africa: The Case of Ghana" there was a long tradition use of a variety of fishing nets cast from the shore or from boats. Along the West African coast with its heavy surf, European ship captains had often recruited Mina and Kru fishermen with their canoes and used them in the loading and offloading of goods. Oral traditions recounted how the Fanti had introduced them to canoes, nets, and the cultural knowledge of the marine world in the second half of the eighteenth century.

African and African American gender roles in fishing traditions may also add further insights. Women often oversee fish distribution, preservation, and preparation once the catch comes to shore. They may also be involved with net manufacture and repair. Oral histories of Portsmouth island note, very peripherally, that women were involved in crabbing, oystering, clamming, making, and repairing fishing nets. Interestingly, after the Depression years netmakers on the island participated in a stimulus initiative that contracted fishing net makers to make hammocks and basketball nets (Coastal Voices, 10-001, 2022, 17). Research might focus on how technology and skills sets associated with fishing translated and gave economic agency to black and white islanders of both genders.

Exposed shipwrecks and timbers periodically wash up at various locations around Portsmouth island. Systematic monitoring may contribute toward greater understanding of local coastal change and historic site formation processes. Maritime cultural resource managers around the world are increasingly aware that these sites and fragments are a neglected database for archaeological research and potential for public outreach and engagement. Portsmouth was a significant colonial maritime entrepôt and has high potential for periodically exposed beach shipwrecks to contribute toward the historical narrative (Jones, 2017).

With specialist analysis and detailed documentation and analysis, timbers washed up on a beach may contribute information on the date of wreck, vessel type and function, craftsmanship, building techniques or characteristics, wood identification and place of construction, and even ethnicity of the builders. Along the south eastern seaboard enslaved African Americans built and used boats bringing traditions from their countries of origins and merging them with new traditions. For example, the most traditional boat at Ocracoke Inlet, was called a *cooner or kunner*. It was a distinct kind of regional dug-out workboat made from two logs joined by a keel log and then decked and fitted with masts. Descendants of boatbuilders clearly recalled their family cooners. Long before Down East boat builders created what are now considered the region's classic work boats early in the twentieth century, the cooner and its larger cousin, the *periauger*, were *the* classic workboats in North Carolina. Currently one such example resides in the NC Maritime Museum in Beaufort known as the *Bonnerton* boat built in the 1870s (Harris, 2014).

House sites and ruins need further investigation on potential to yield information about architecture, diet, foodways, historic landscaping, and social status. During the antebellum period, African Americans were the property, mistresses, and mulatto children of Portsmouth Euro American residents. Census records detail slave holders on the island up until 1860 and include the number of enslaved Africans owned, ages, and gender. Future research could integrate historical, ethnographic, and archaeological data with an in-depth review of slave owners' personal records, census data, their property locations, and interviews with slave owner and enslaved descendants, bring together both black and white perspectives and memories.

Other than archaeological subsurface testing or excavation, it might be productive to compile and analyze surface archaeological features like ruins or artifact scatters, combining geospatial overlays of historic images of structures. In addition, landscape data sets like aerial imagery, first-hand accounts regarding the location of archaeological deposits, land management history, GPR and LiDAR data showing subtle variation in surface elevation related to archaeological features such as unmarked graves or building foundations. New uses of drones for up-to-date flyover aerial imagery and state-of-the art 3D animation mapping purposes could also be considered with special permissions from NPS (Harris et al., 2018).

Geologists note that tide washing into Doctors Creek or pounding the shore is creating erosion and artifact dispersal along the shoreline where East Carolina University conducted 2019 field school mapping operations. Hundreds of surface artifacts scattered in the marsh and on sandy coves have potential as material culture sources to develop research questions that contribute toward a deeper understanding about life on the island like forms of recreation (marbles), wealth and economy (coins), social status and foodways (ceramics), medical practices and social settings (glassware and toiletries), and industrial activities (machinery, ships timbers, slips, runways, and hardware).

The descendants of Portsmouth residents suggest the historic landscape in the 1800s and 1900s "died of rapacity," with the community allowing livestock to devour every leaf and shrub of the sand-binding vegetation, causing the large sand dunes to fly away with the wind and fill up the harbor. Their biannual homecoming gatherings are filled with nostalgia. As one person noted:

> I carry so may memories of Portsmouth in my heart, it would be impossible for me to tell "in mournful tradition" all I feel about the once thriving seaport with its harbor, warehouses, academy, and shipyards, all of which had declined and left only a remote, close-knit village in my day. Gradually, with no children, no pastor, no cats, no dogs, no ponies or cattle, no noise except birds, sea, and wind, it wasted away and became an old man dreaming in the sun, "sans teeth, sans eyes, sans taste, sans everything. (Howard, 2011a)

While nurturing the memories of the descendants with heritage tourism initiatives and continuing preservation of the historic structures on Portsmouth island, the National Park Service is actively seeking solutions to management challenges on North Carolina barrier islands on two fronts. Firstly, addressing the need to highlight tangible and intangible African American history more inclusively and authentically. Secondly, dealing with the current and predicted coastal change. National seashores have high investments in assets at risk from sea-level rise over the next 100–150 years according to National Park Service studies (Peek et al., 2015). One of the most valuable assets at risk at Cape Lookout is the Portsmouth Village historic landscape. Like Shell Castle, much of this historic landscape may be a submerged landscape and will likely become a management asset, or problem, for underwater archaeologists in the not-too-distant future.

References

Akyeampong, E. (2000). Africans in the diaspora: The diaspora and Africa. *African Affairs, 99*(395), 183–215. Available at: https://www.jstor.org/stable/723808. Accessed June 2021.

Akyeampong, E. (2001). *Between the sea and the lagoon: An eco-social history of the Anlo of Southeastern Ghana, c.1850 to recent times*. James Currey Press.

Akyeampong, E. (2007). Indigenous knowledge and maritime fishing in West Africa: The case of Ghana. *Tribes and Tribals, 1*, 173–182.

Austin, R., Hendryx, G., Worthington, B., & Wells, D. (2011). Excavation of a mid-nineteenth century barrel well and associated features at Fort Brooke, Tampa, Florida. *Florida Anthropologist, 64*, 163–185.

Bolster, W. (1997). *Black Jacks: African American seamen in the age of sail*. Harvard University Press.

Bradley, R. (2015). *Where were the whalers? An investigation of the archaeological, historical, and cultural influences of North Carolina whaling* (Master's thesis). East Carolina University. Available at: http://hdl.handle.net/10342/5089. Accessed June 2021.

Brickell, J. (1737). *The natural history of North Carolina*. Johnson Publishing Company.

Brimley, H. (1894). Whale fishing in North Carolina. *Bulletin NC Department of Agriculture, 14*(7), 4–8.

Brown, L (1977). "Portsmouth Village Historic District". *National register of historic places—Nomination and inventory*. North Carolina State Historic Preservation Office. Available at: https://files.nc.gov/ncdcr/nr/CR0007.pdf. Accessed June 2021.

Burke, K. (1958). *The history of Portsmouth, North Carolina, from its founding in 1753 to its evacuation in the face of federal forces in 1861*. Insta-Print Inc.

Cape Lookout National Seashore. (2011). *Audio tour Portsmouth village*. Dobbs Ferry, RBH Multimedia. Available at: https://www.nps.gov/calo/learn/photosmultimedia/upload/CALO-AS-1564-narration-script.pdf. Accessed June 2021.

Cecelski, D. (1993). The hidden world of Mullet camps: African American architecture on the North Carolina Coast. *The North Carolina Historical Review, 70*(1), 1–13.

Cecelski, D. (2008). Playing Croquet until dark: Voices of Portsmouth islanders. *North Carolina Folklore Journal, 51*, 43–53.

Cecelski, D. (2012). *The Waterman's song: Enslaved Africans and freedom in maritime North Carolina*. University of North Carolina.

Cecelski, D. (2015). Of time and the sea: Nye's clock oil and the bottlenose Dolphin Fishery at Hatteras Island, North Carolina, in the early twentieth century. *The North Carolina Historical Review, 92*(1), 49–79. Available at: http://www.jstor.org/stable/44113249. Accessed June 2021

Cecelski, D. (2018). *Remembering Portsmouth Island*. Available at: https://davidcecelski.com/2018/05/18/remembering-portsmouth-island/. Accessed June 2021.

Cloud, E. (2006). *Portsmouth the way it was: Island history* (Vol. III). Heritage Books.

Coastal Voices. (2022). *Carolina Coastal Voices: An oral history of the outer banks and down east NC*. Available at: https://www.carolinacoastalvoices.com/. Accessed April 2022.

Dawson, K. (2006). Enslaved swimmers and divers in the Atlantic world. *The Journal of American History, 92*(4), 1327–1355.

Dawson, K. (2018). *Undercurrents of power: Aquatic culture in the African diaspora*. University of Pennsylvania Press.

Friends of Portsmouth Island. (1996). Available at: https://friendsofportsmouthisland.org/fopi/. Accessed April 2022.

Garrity-Blake, B. (1994). *The fish factory: Work and meaning for black and white fishermen of the American menhaden industry*. University of Tennessee Press.

Goode, G. B., Allen, J. A., Elliott, H. W., True, F. W., Ingersoll, E., Ryder, J. A., Rathbun, R., Earll, R. E., Wilcox, W. A., Clark, A. H., Mather, F., Collins J. W., McDonald. M., Stearns, S., Jordan, D. S., Bean, T. H., Kumlien, L., Scudder, N. P., Gilbert, C. H., ... Swan, J. G. (1884). *The fisheries and fishery industries of the United States*. Prepared through the co-operation of the commissioner of fisheries and the superintendent of the tenth census. Govt. Print. Off.

Greer, R. (1917). *The Menhaden Industry of the Atlantic Coast: Appendix III to the Report of the U.S. Commissioner of Fisheries for 1914*. Bureau of Fisheries.

Harris, L. (2014). African canoe to plantation crew: Tracing African memory and legacy. *Coriolis: Interdisciplinary Journal of Maritime Studies, 4*(2), 35–52. Available at: http://ijms.nmdl.org/article/view/13011. Accessed June 2021.

Harris, L., Wasklewicz T., & Mallison, D. (2018). *Coastal heritage at risk*. NPS Preservation Training and Technology Grant Number: P16AP00372/DUNS. Piggot House. https://sketchfab.com/3d-models/henry-piggot-house-2f972bb7fe9e42a0ac8afaaca35220f9

Herndon, G. (1972). The 1806 survey of the North Carolina Coast, Cape Hatteras to Cape Fear. *The North Carolina Historical Review, 49*(3), 242–253.

Heron, S., Moslow, T. F., Berelson, W. M., Herbert, J. R., Steele, G. A., III., & Susman, K. R. (1984). Holocene sedimentation of a wave-dominated barrier-island shoreline: Cape lookout, North Carolina. *Marine Geology, 60*, 413–434.

Holland, F. (1968). *A survey history of Cape Lookout National Seashore*. Division of History, Office of Archaeology and Historic Preservation, National Park Service, US Department of the Interior.

Howard, P. (2011a). *Wedding on Portsmouth Island*. Available at: https://www.villagecraftsmen.com/tag/m-mason-daniels/. Accessed June 2021.

Howard, P. (2011b). *Slavery on Ocracoke*. Available at: https://www.villagecraftsmen.com/slavery-on-ocracoke/. Accessed June 2021.

Howard, P. (2015). *Whale and porpoise fishing on the outer banks*. Available at: https://www.villagecraftsmen.com/tag/try-yard/. Accessed June 2021.

Howard, B. (2016). *Navigating historical waters: A study of the pilots and original settlers of Ocracoke Island* (Master's thesis). East Carolina University. Available at: http://hdl.handle.net/10342/5346. Accessed June 2021.

Huss, G. (2019). *Of blood, salt, and oil: An archaeological, geographical, and historical study of North Carolina's Dolphin Fishery* (Master's thesis). East Carolina University. Available at http://hdl.handle.net/10342/7245. Accessed June 2021.

Jarvis, M. (2002). Maritime masters and seafaring slaves in Bermuda, 1680–1783. *The William and Mary Quarterly, 59*(3), 585–622.

Jones, J. (2017). *On a sea of sand: A comparative analysis of the challenges to beached wreck site stability and management* (Doctoral dissertation). East Carolina University. Available at: http://hdl.handle.net/10342/6210. Accessed June 2021.

Kemp, A. C., Kegel, J. J., Culver, S. J., Barber, D. C., Mallinson, D., Leorri, E., Bernhardt, C. E., Cahill, N., Riggs, S. R., Woodson, A. L., & Horton, B. P. (2017). An extended late Holocene relative sea-level history for North Carolina, USA. *Quaternary Science Reviews, 160*, 13–30.

Linebaugh, P., & Rediker, M. (1990). The many-headed hydra: Sailors, slaves, and the Atlantic working class in the eighteenth century. *Journal of Historical Sociology, 3*(3), 225–252.

Mallison, F. (1998). *The Civil War on the Outer Banks: A history of the late rebellion along the Coast of North Carolina from Carteret to Currituck, with comments on prewar conditions and an account of postwar recovery*. McFarland and Company.

Mallinson, D., Culver, S., Riggs, S., Thieler, E. R., Foster, D., Wehmiller, J., Farrell, K., & Pierson, J. (2010). Regional seismic stratigraphy and controls on the Quaternary evolution of the Cape Hatteras region of the Atlantic passive margin; USA. *Marine Geology, 268*, 16–33.

Marin, C. (2007). *Coercion, cooperation, and conflict along the Charleston waterfront, 1739–1785: Navigating the social waters of an Atlantic port city*. University of Pittsburgh.

Marsh, R. (1878). A storm off Ocracoke Inlet-Rock Castle. *Potters American Monthly–For the Higher Education of Women, 10*, 283–284. John E. Potter & Co., 1875–1882. Available at: https://babel.hathitrust.org/cgi/pt?id=umn.31951002793381d&view=1up&seq=7&q1=ocracoke. Accessed June 2021.

McGuinn, P. (2000). *Shell Castle, a North Carolina entrepot, 1789–1820: A historical and archaeological investigation* (Master's thesis). East Carolina University. Available at: http://hdl.handle.net/10342/6635. Accessed June 2021.

Morgan, D. (1982). *John Grey Blount Papers, 1803–1833* (Vol. 5). Historical Publications Section.

National Park Service, Cape Lookout NC. (2015). *Henry Piggott House, Portsmouth village*. Historic Structure Report. Available at: http://npshistory.com/publications/calo/hsr-henry-pigott-house.pdf. Accessed June 2021.

NPS Portsmouth Homecoming. (2020). Available at: https://friendsofportsmouthisland.org/fopi/category/homecoming-2018/. Accessed April 2022.

Olson, S. (1982). *Portsmouth village historic resource study*. National Park Service Report.

Peek, K., Young, R. & Beavers, R., Hoffman, C., D., B. & Norton, S. (2015). *Adapting to climate change in Coastal Parks: Estimating the exposure of park assets to 1 m of sea level rise.* NPS Report. Available at: https://doi.org/10.13140/RG.2.1.3659.9122. Accessed June 2021.

Pigott, H. (2000). *Visitor brochure.* Cape Lookout National Seashore. Available at: http://www.nps.gov/calo/planyourvisit/upload/Henry2000.pdf. Accessed June 2021.

Portsmouth Church. https://sketchfab.com/3d-models/church-portsmouth-eda4ad03170a4f91a8cc1ef0ca53b1e1

Portsmouth Life Saving Station. https://sketchfab.com/3d-models/portsmouth-life-saving-station-b0bca79537c34f3e8f2a5620bec47a87

Sam Tolson. (2018). National Park Service Website. Available at: https://www.nps.gov/calo/learn/historyculture/tolson.htm. Accessed June 2021)

Simpson, M., & Simpson, S. (1988). The pursuit of Leviathan: A history of whaling on the North Carolina coast. *The North Carolina Historical Review, 65*(1), 1–51.

Simpson, M. & Simpson, S. (1990). *Whaling on the North Carolina Coast.* North Carolina Division of Archives.

Smith, H. M. (1907). *The fishes of North Carolina* (Vol. 2). EM Uzzell & Company, State Printers and Binders.

South Eastern Archeological Center. (2020). Available at: https://www.nps.gov/orgs/1539/index.htm. Accessed June 2021.

State Records. (1790). *Records of North Carolina, 1709–1790.* North Carolina Division of Archives and History.

Stick, D. (2015). *The Outer Banks of North Carolina, 1584–1958.* UNC Press Books.

Taylor, S. (2009). *A home transformed: Narratives of home, loss, longing and the miniature from Portsmouth Island, North Carolina* (Master thesis). University of North Carolina. https://doi.org/10.17615/e5f3-vc19. Accessed June 2021.

Thieler, E., & Hammer-Klose, E. (1999). *National assessment of coastal vulnerability to sea-level rise: Preliminary results for the U.S. Atlantic Coast.* U.S. Geological Survey Open File Report, 99-593. Available at: https://pubs.usgs.gov/of/1999/of99-593/. Accessed June 2021.

True, F. (1885). The porpoise fishery of Cape Hatteras. *Transactions of the American Fisheries Society, 14*(1), 32–38.

Willis, D. S., & Salter B. (2004). *Portsmouth Island: Short stories, history.* Montville Publications.

Wood, P. (1996). *Black majority: Negroes in colonial South Carolina from 1670 through the Stono rebellion.* WW Norton & Company.

10

Multi-Spirited Waters in Lynton Burger's She Down There

Confidence Joseph

In this chapter, I ask: What happens when we de-centre land in Southern African narratives and allow for a turn towards watery spaces? To this end, I think with Isabel Hofmeyr's notion of multi-spirited waters which suggests alternative modes of perception where water is animated with ancestral, celestial, and diabolical beings (2019). The multi-species entanglements between humans, sea creatures, and spiritual beings create the inspirited waters of *She Down There* (2020). I consider Lynton Burger's deployment of the mythical water goddess—Sedna—to blur the lines between the human and the non-human, the natural and the supernatural in his novel. I argue that the creolized waters of *She Down There* (through the intermingling of indigenous Canadian Haidan, South African Khoisan, and the Mozambican Bitonga water myths) provides a global approach to reading for water in Southern African literature. I further explore the ways in which the novel extends our

---

This chapter is part of my PhD thesis titled 'Of Water and Water Spirits in Southern African Literature'. I extend my gratitude to the Andrew. W Mellon Trust Foundation for the grant that made the research possible.

C. Joseph (✉)
Witwatersrand University, Johannesburg, South Africa
e-mail: coejoseph14@gmail.com

understanding of amphibian aesthetics as the blended myths mirror intermingling elements of land and water. The novel introduces the reader to a vast watery world that extends to the driest areas of the Karoo in South Africa.

Synopsis of the Novel

In *She Down There*, the author invites us to think with sea creatures, the oceans, and spiritual beings to problematize the line between human and non-human. There is also a farm dam and deep pools that are captured on rock paintings of the Karoo that are significant to the plot. The cyclical narrative mode of the novel displaces linear temporalities as past, present, and future conflate. Burger underscores the plight of the world's oceans through an underwater perspective that personifies the ocean and its creatures. While this anthropomorphism advances the cause of animal/environment preservation, other critics question its unwitting projection of human desires onto animals and the environment. As Peter Rutherford (cited in Wilson, 2019) suggests, "the staging of intimacy between human and non-human animals is not only a way of encouraging people to care for wild animals and nature but also a way of disciplining them" (p. 720). The other misgiving about the novel is that Burger's passion for environmental activism burdens it with a moralistic and didactic tenor. Despite these misgivings, the author insists that he only wanted to tell a story and share the wonder of the ocean and its creatures.

Burger invokes "compassion, affect, sentiment and cognition" (Villanueva, 2019, p. 189) through emotive diction and imagery. His representation of the non-human draws from his experience as a marine biologist, professional diver, and underwater photographer. A photograph of him swimming with a dugong captured in Bazaruto Island of Mozambique serves to authenticate his narrative. Burger describes himself as a "writer who delight[s] in exploring the beauty and the essence of our oceans, and the relationships people have with nature and each other when nature is at the centre."[1] He goes on to state that "[a]s an underwater photographer, I combine my love affair with the aesthetics of the underwater world with my keen interest in abstract and contemporary art" (ibid.). This fascination with the underwater aesthetics and photography manifests in *She Down There* as the reader is drawn to the depths of the ocean to meet various sea creatures.

Burger's intimate relationship with the ocean informs the depiction of the protagonist Claire, a marine biologist who "feels as comfortable underwater as on land" (p. 20). Claire and her fiancée Todd "share the desire to understand how it all works down [there and to] …use this knowledge to protect what is left" (p. 22). Outside her desire to understand and preserve the ecology of the oceans, Claire has deep ancestral links that tie her to water. Her great grandfather was a whaler while her grandmother Naan requests to have her remains spread on both "…the land…[and] the sea" (p. 14). Elsewhere in the novel, Naan reminds Claire, "[y]ou know who you are…We are sea people" (p. 21). She further sustains their connection to the sea when she intimates that "[y]ou my Claire, remind me of Sedna. You want to live down there and serve all those sea creatures, don't you just?" (p. 22). The passion to protect marine life which begins in the waters of the Pacific Ocean later takes her to the Indian Ocean following the tragic death of her fiancée Todd during an underwater expedition.

The bulk of the action of the novel takes place in Praia do Tofo in Mozambique. Tofo is a small coastal village in the Southern part of Mozambique. It is a popular tourist destination because of its warm waters, manta rays, and whale sharks. In the novel, the inspirited waters of Tofo become a major setting that facilitates an interspecies entanglement. The title of the novel, *She Down There*, pulls the reader under the waters of the North Pacific and the Indian Ocean to meet the "She" who inhabits the depths of the oceans. Though we later learn that the water goddess's name is Sedna, Burger symbolically identifies her as "She down there" and so creates a narrative space that encompasses the diverse forms and names that this creature assumes all over the world.

The story begins in 1768 in the Commander Islands of the North Pacific Ocean where we are introduced to "She Down There" who is also known as Sedna or the Half-Away Woman in Haida Gwaai, British Colombia. Her mission is to save sea creatures from poachers. Sedna is the voice, ruler, and protector of the sea and all its creatures. In the words of Naan she is the:

> Half-Away Woman. Her name is her destiny: half woman, half sea creature. Down with the octopus she dives. She swims out beyond the waves with the sea lions and the orcas. She rolls with the sea otters in the kelp. She rests in the intertidal-that place which is half sea, half land…she sees all in the sea. She feels all. And she is forever destined to be the voice of the sea creatures, the one the Shamans have to dive down to each spring to appease, to ask to release the animals for the summer hunt. [26]

The narrative moves from 1768, North Pacific Ocean where we witness Sedna trying to save the endangered Steller sea cows to 1991 in Southern Haida Gwaii Archipelago where we meet one of the protagonists Claire. Haida Gwaii is a collection of islands surrounded by large bodies of water which shape Haida culture. According to the Marine Plan Partnership for the North Pacific Coast, "the intimate relationship with the marine environment is reflected in Haida traditional knowledge, a living body of knowledge that is built upon past and current Haida experiences throughout the archipelago" (http: mappocean.org/). From Haida Gwaii, the action of the novel shifts to the Boesmansvlakte Farm, which is in the Karoo region of South Africa. The year is 1980. Karoo means the "land of thirst" in the language of the Khoisan, its earliest inhabitants. Although the region was once submerged underwater (see Day et al., 2015), nowadays it lacks surface water. In response to this lack, residents drill boreholes and construct dams. To detect water that is trapped underground, local people use the forked-stick method which involves "locating groundwater by walking the surface of a property while holding a forked stick…that responds when the person moves above a location that will yield an adequate flow of water to a drilled well" (King, n.d). The Karoo is also, surprising for a semi-desert region, a land of mermaid myths. Wendy Hardie (2008) states that "the mermaid has been an integral part of the Karoo folklore for years…There are mysterious San rock paintings of creatures with the upper body of a human, and a fish tale, on the walls of caves high up in the Little Karoo Mountains" (n.p).

The Karoo setting and local beliefs inform Burger's depiction of Klaas Afrikaner the second protagonist whom he refers to as a "bush merman." When we first meet him, Klaas just like his father is a labourer at Boesmansvlakte Farm. Here, "[t]he windmill draws from the vast, brackish sea trapped deep below this dry, weathered land. Once a swamp inhabited by dinosaurs, but hardly vegetated now" (p. 58). Just as in the arid Manyene Tribal Trust Lands in Mungoshi's *Waiting for the Rain* where underground water dominates, the idea of a sea trapped underground in *She Down There* recognizes the presence of water in its seeming absence. As would be shown later in the chapter, it is in this semi-desert place that Klaas' ironic connection to water manifests. He later enlists as one of the first Coloured divers in the South African navy. He wants to be "…like Jacques Cousteau…to be with the sea animals. The navy will teach [him] how" (p. 73). After completing his training, he is deployed in a clandestine mission to destroy ANC structures in Maputo. Klaas sinks into a dilemma as he is forced to choose between duty and Gwen his childhood lover who happens to be one of the members of the ANC hiding in Maputo. He chooses Gwen. Unable to go back home, he resettles in Tofo, Mozambique where he leads a quiet life as a dive master and later meets Claire with whom he has a daughter, Alice. Burger's protagonists have inextricable ties with water and the way they relate to sea animals muddles the line between the human and non-human.

She Down There can be read as both an ecological thriller and a romantic adventure. On one level, it details the love between the two protagonists, Claire Lutrisque, born of a French-Canadian father and a Haidan mother and Klaas Afrikaner, a coloured South African of Khoisan origin. The deep love for water and its creatures takes the two protagonists to the Tofo Islands of Mozambique towards the end of the novel. On another level, the novel foregrounds the friction between humans and the non-human, in this case the ocean and its creatures. In thinking with Villanueva (2019, p. 188), I read the ocean "as an affective space in which animals communicate their suffering and also display their individuality." I also draw on Danson Mwangi's conceptualization of a multi-species landscape where several organisms including humans interact with non-humans to shape, create, and form an integral part of one another's environment (2019). This multi-species landscape is captured in the final scene of the novel when Alice shares her dream with her mother Claire:

> A dream woke me up Mummy. It felt so real. I was sitting in a clearing in a forest. It was nighttime and there was a fire, and all around the fire were animals. Sea animals. I woke up with my heart bumping. It didn't make sense…There were all staring at me, Mummy. Like they wanted me to do something. Then an old lady was there. She was smiling, and she said, 'You can do it.' But what, Mummy? I just know that I want to be with the sea animals. To help them… [233–234]

A multi-species lens recognizes non-humans as beings with legitimate interests and rights, hence the sea animals pleading with the little girl to save them. In the novel, we are presented with sea creatures who in the words of Eduardo Kohn albeit in a different context, "engage with the world and with each other as selves — that is, as beings that have a point of view" (2007, 4). The use of 'who' instead of 'it' in reference to sea creatures in the novel, seeks to place them on the same level with humans. This is in line with anthropomorphism[2] that advances the cause of the non-human but is crippled with the risk of projecting human ideas on the same. As Joanne Mierek (2010, p. 46) explains, "[b]ecause we are unable to separate ourselves from our own subjectivity, we cannot objectively approach another's." However, since no better means of allowing animals to speak for themselves exists, Burger's attempt deserves credit. In his novel, so to speak, sea creatures "speak," and the onus is on humans to "listen." In this instance, to speak and to listen transcends our conventional comprehension. Mierek (ibid., p. 7) reminds us that:

[s]haring thoughts, intentions, and feelings with animals does not depend on language; that is, although the ability to talk about the relationship does rely on language, the ability to have it doesn't. With animals, 'thoughts' can be understood as the focus of attention through vocalizing or eye contact, as when a dog checks in or glances at the door or the leash or the food dish.

Mierek's intimation informs Burger's depiction of the human–animal relationship that I consider in greater detail later in this chapter.

Multi-Spirited Waters in *She Down There*

In the novel, the multi-spirited waters comprise mythical beings such as Sedna and ancestral spirits like Naan that reside in the ocean. In addition, we have living human beings like Claire and Klaas who feel at home underwater as on land, and spirit sea animals like turtles, orcas, mantas, sea horses, whales, and dolphins. In the words of Naan "the sea is inhabited by many supernatural creatures, the underwater people or ChaaGaan XaadaGay. Creatures from a time when the veil between humans and other animals was thin" (pp. 30–31). The ocean becomes a space that dissolves interspecies hierarchizations popular in land-based narratives. All these lifeforms intermingle and transform the ocean into inspirited waters.

Burger draws from Canadian Haidan, South African Khoisan and the Mozambican Bitonga beliefs on water myths and in the process creolizes the oceans. Through the stories that his characters share, Burger imbues the oceans with a host of water spirits who though languaged differently share some similar traits. To begin with, Burger draws on one of the oldest Haidan myths of the sea. It is the tale of Sedna that has been passed down from generation to generation. The myth explains how different sea creatures came to be and how Sedna became their ruler. The legend has it that Sedna, the most beautiful girl in the village shunned all men and fell for a raven, the spirit bird who disguised himself as a whaler from another village. The two went on to settle on a faraway island. Her departure riled the elders who ordered her father to get her back. On his way back from fetching his daughter:

> Before he could reach the village, a raven came flapping low across the water. And as he drew near, he transformed and filled the sky above them. The great spirit bird flapped around the canoe, screeching and beating the surface of the water in a fury, so that the sea rose up, becoming one with the mist. Her father was terrified. [24]

The reader is introduced to the great spirit bird which expands the spiritual world of the novel that now brings water, earth, and sky together. Moreover, the passage sustains the mythical atmosphere of the novel as the three elements unite. Frightened and eager to appease the fury of the raven, Sedna's father threw her into the ocean. As she struggled to hold on to the canoe, the father, "leaned forward and hacked at her knuckles…He reached down to where her still warm fingers lay and hastily threw them overboard" (p. 25).

> Sedna and her severed fingers drifted down into the cold depths. One by one the fingers turned into sea creatures. A sea lion, a seal, an otter. One thumb became a rockfish, the other a grey whale. One finger turned into a whole school of herring, while another became an orca. They all circled around Sedna, whose braids were unravelling, the dark hair swirling in the current. They were joined by the other animals of the ocean who heard the orcas and the whales calling out her name. All the creatures swam with her in a mighty vortex, and from then onwards, she was Sedna. Ruler of the sea creatures. [p. 25]

Through the fantastic mode, Sedna's death, which at first suggests an ending is transformed into a beginning, a multi-species turn that explains the interconnectedness of diverse creatures in the novel. The legend becomes part of Haidan creation myths and Sedna transforms into a mystical being—half-woman, half-fish. From her chopped hands emerges the different creatures that populate the ocean. It is no surprise then that Sedna feels the pain of even the smallest sea creature and desires to protect them. They are part of each other. Furthermore, Sedna's hybridity—half-fish, half-woman—impacts the representation of Claire and Klaas. Just like Sedna who is a hybrid being, Claire is half-Haidan, half-European. The similarities between the two extend to the work both perform for the ocean. In a podcast discussing his novel, Burger states that "through her conservation work Claire becomes Sedna as a protector of the sea creatures."[3] It is therefore possible to read Claire as a contemporary water spirit that continues in the tradition of her Haidan ancestors who as "indigenous peoples regard the inland waters, rivers, wetlands, sea, islands, reefs, sandbars and sea grass beds as an inseparable part of their estate" (Altman & Jackson, 2008, p. 169). In the instructive words of Naan to Claire, "we are sea people. Through me you are a descendent of the first people who came from the breasts of SGuuluu Jaad, Foam Woman, from between the tides at Xd' gi, the sacred reef near SGaang Gwaay…" (p. 21). For that reason, Claire makes it her life mission to save the marine creatures, the same way Sedna has always done. In as much as Sedna makes the depths of the ocean her home, Claire's profession as a marine biologist allows her access to the same depths of the oceans.

From the cold waters of the North Pacific Ocean Sedna moves to the warm waters of the Southern Indian Ocean. In this way, the author notes that "Sedna seems to be the unifying thread of these continental waters" (ibid.). She operates in both the Pacific and the Indian Oceans. In addition, the waters in the novel are not only connected through Sedna but also through the characters—Clair, Klaas, and Benito—who seem to flow from one body of water to the next, bringing with them their embodied knowledges about the ocean and the life within it. Their stories foreground the sacredness of these waters.

This sacredness echoes in Putuma's poem *Water,* that perceives the ocean as a spiritual space where the living can connect with their ancestors, and where rituals of various kinds take place. The belief that the ocean is inhabited by numerous spirits (dead, alive, human, animal, and plant) is sustained in Burger's novel when Claire spills her grandmother's ashes according to her will (p. 17). Submerging Naan's ashes in the ocean can be read as a ritual that reconnects her with her origins. Earlier in the novel, Naan informs Claire that she is a direct descendant of the Foam Woman, She Down There. This ancestral connection to a half-human, half-fish creature helps to explain why Claire continues to sense Sedna's presence even in the Indian Ocean. Similarly, she still senses her grandmother "like she's still around, goes with me wherever I go. Especially underwater" (p. 113).

The ocean space in the novel is also populated by spirit sea animals. Writing in the context of Zimbabwean Shona beliefs, Clemence Makamure and Vengesai Chimininge (2015) argue that totems are "seen as personifications of spiritual beings" (p. 11). Claire's totem is the Sea Otter which is believed to be one with the ocean. Her grandmother persuades her to "flow like Otter flows" (p. 30) and this further explains Claire's intimate connection with the ocean. Her grandmother feels obliged to share the stories of her people and the ocean. Their totem is the Turtle, the messenger. Makamure and Chimininge go on to argue that:

> Such a conception of the relationship between the human society and the environment foster sound environmental ethics that does not only take into account the well-being of the individual and his community, but also that of the environment. The environment should be construed as an end in itself in a similar way that human beings are perceived, rather than viewing it as a mere means to some human ends. [p. 8]

In the case of Klaas, he is connected to the manta. He encounters one as "she hangs, hardly moving her wide wings, maintaining eye contact with him. Her eye bores into him and time ceases. *My spirit manta*" (p. 185).

Mashudu Mashige suggests that "[b]eliefs regarding totems can vary, from merely adopting one as a whim, to adopting an animal that a person sees representing favourable traits reflected in their own behaviour or appearance. Other clans believe their totem literally acts as a spirit guide" (2021, p. 16). In the novel, the sea creatures act as spirit guides and seem to materialize when needed the most. These spirit animals share profound and intimate moments with the human characters. Totems are therefore used in the novel to underscore the relational connection between the human and the other-than-human world —he ocean and its creatures.

As I have shown, the oceans in this novel are populated by various kinds of spirits; they can also be accessed in various ways. One such means is diving which according to Melody Jue allows people to move in ways that they cannot move on land, that is three-dimensional movement.[4] In the documentary film "My Octopus Teacher" [Dirs: Pippa Ehrlich & James Reed, 2020], Craig Foster echoes that "[w]hat is so amazing about this environment is you are in a three-dimensional forest, and you can jump off the top and go wherever you want. You are flying basically. You might as well be on another planet. You naturally just get more relaxed in the water." The enhanced underwater movement in turn enables perspectival shifts in terms of perceptions of the ocean and conservation movements that the novel grapples with. Foster goes on to explain that plunging under the ocean implies "stepping into this completely different world. You go into that water and it is extremely liberating. All your worries and problems and life drama just dissolve. You slowly start to care about all the animals, even the tiniest little animals. You realize that everyone is very important" (ibid.). Apart from diving, the animated oceans of the novel can also be accessed through dreams where "… there is no veil between the spirits of humans and the spirits of other creatures. This is the realm of the supernatural creatures, the ChaaGan XaadiGay" (pp. 10–11), according to a Haidan Shamaan. That in the depths of the ocean there is no veil between human and animal spirits points to the belief in existence of a universal consciousness that unites all life and non-life forms. The practice of accessing water bodies through dreams, recalls the initiation rituals of *sangomas* in some parts of Southern Africa. The initiates might not physically submerge underwater but can access these water depths and the spirit realm through dreams (Mutwa, 2003).

The foregoing resonates with Khoisan access to the underwater world through dream-like trances in *She Down There*. Divers also experience a trance-like sensation underwater because of the difference in pressure levels. Foster (2020) narrates in "My Octopus Teacher" that ocean water "upgrades the brain because you are getting this flood of chemicals every time you immerse in that cold water. Your whole body becomes alive. And then as your body adapts it just becomes easier and easier." Elsewhere in a podcast interview with Nancy Richards, Burger observes that "being underwater is an altered state, it brings about a dream state because of the pressure and floating in three dimensions, like being in space where you can move in any direction you want" (Burger, 2020, n.p). It is through similar dream-like trances that the reader experiences the underwater world of the Great Karoo in South Africa. The Kalahari Meerkat Project article titled "Kalahari Bushmen" suggests that because the Khoi-San people "lived in a very dry area, water to them has a very magical power that could revive them." In the semi-arid region of the Karoo, it is not surprising that we are introduced to the underwater people through cave rock paintings. Khoisan artists can access and paint underwater spirits in a trance state as they "become one with the stars and the wind and the living, dry land of… ancestors" (2021, p. 11). In the midst of this dryness however, there is water that is trapped underground, and in the deep pools. As Herbert Aschwanden explains, "[p]ools, springs and swamps sustain the life of the rivers and they give the vital water. Therefore, such places are regarded as an origin of the fertility of nature. Also, they safeguard human life and are thus to be especially respected as sacred places" (1933, p. 9).

These sacred pools seem to be the home of the strange looking beings as shown by the Khoisan rock painting captured in *She Down There* as quoted below:

> When he is ready, he dips a supple twig into the ochre paste in the ostrich shell he holds and lifts it reverently to apply the pigment to the rock. Stroke by stroke, the mystical abhumans emerge. A reminder to be left for his grandchildren, and theirs, that life is not always as it seems. And perhaps one day one of his people will find a way to go where he could not. Down there, into that dark-blue, mysterious sea. [pp. 11–12]

The painting represents the Khoisan links to water and water spirits. The novel's depiction of the sea creatures which draws on ancient rock paintings shows that the mermaid myth has been around for a long time (see Gary Varner, 2010). The rupestrian abhumans help narrow the gap between human beings and non-human creatures emphasizing the intermingling of worlds. The prophetic utterance from the Khoisan artist, "perhaps one day one of his people will find a way to go where he could not. Down there, into that dark blue, mysterious sea" (ibid.) manifests years later through Klaas. He experiences what his forefathers had frozen on the rocks. This materializes the moment Klaas falls from a windmill into the farm dam and experiences "what it's like underwater" (p. 60) for the first time. He finds himself in "a cool, all-encompassing wetness. It's strangely familiar, comforting...." (p. 60). The ironical comforting familiarity resonates with his bush merman status: a water spirit in a waterless space. As the impact of the fall sets in, he loses consciousness, and has visions of:

> strange human like creatures drifting in a forest of seaweed. These creatures are dancing and drumming as well as swimming in a sky that is water. It's as if the world of the land has met the world of underwater...They have the heads, arms, and torsos of human beings, but from the waist down they are like dolphins. [p. 60]

This moment leads to a series of dreams where Klaas is underwater, a world that is strangely familiar for someone who had spent his whole life in semi-desert Karoo. He "dreams of the strange swimming people who appeared in his watery vision. And during the day he looks toward the dam, thinks of the real creatures that live there. Water now contains magic for him. The sea becomes a dream" (p. 63). Through this encounter, the dam becomes a sacred space that links Klaas to his ancestors and sustains the mythical atmosphere of the novel. These experiences cultivate his intimate relationship with the ocean later at Tofo in Mozambique where he works as a diving master. He is aware of the other beings that are found in the depths of the ocean which his ancestors captured on the rock paintings and he approaches the ocean with some reverence and demands the same from his tourist clients who visit the Tofo Islands for recreation. He urges them to respect every sea creature seen during the diving expeditions.

As the novel ends, the Indian Ocean becomes a melting pot that brings together different characters who seem to share similar knowledge and attitudes to the ocean. In addition to the Haidian and South African water myths, the character Benito brings the beliefs of the Mozambican Bitonga people similar to those of East African communities. The unifying thread of these beliefs is that the ocean is populated with supernatural beings and spirits. These beings communicate with humans in different ways including the sounds dolphins make which reminds one of the Khoisan click language. Again, these animals communicate with humans through eye contact, movement, and thoughts. This ability to cross-communicate bridges the gap between humans and the non-human.

While the above considers diverse indigenous water myths that feed into the notion of inspirited waters, the next section examines the representation of sea creatures in the novel which constitutes Burger's mission to give voice and agency to voiceless and apparently passive oceanic beings.

Thinking with Underwater Animals

In *She Down There*, we are invited to sink and think with the sea animals, to appreciate them as beings who possess individual feelings, interests, and needs that are separate from human desires. Kira Gee (2019) cites Rachel Carson (1937) who poses the provocative question: Who knows the ocean? Gee goes on to suggest that our "earth bound senses" limit our knowledge of the ocean. We cannot grasp:

> the foam and surge of the tide that beats over the crab hiding under the seaweed of his tide pool home; or the lilt of the long, slow swells of mid-ocean, where shoals of wandering fish prey and are preyed upon. [2019, p. 23]

To better engage this aquatic world, we need to "shed our human perceptions of length and breadth and time and place and enter vicariously into a universe of all-pervading water" (ibid.). The sea creatures that we encounter in *She Down There* are imbued with human-like behaviours and feelings: they are individual characters who love, mother, play, and feel loss and pain. Burger's representation resonates with Lori Gruen's concept of entangled empathy. She defines the concept as:

> a type of caring perception focused on attending to another's experience of well-being. It is an experiential process involving a blend of emotion and cognition in which we recognise we are in relationships with others and are called upon to be responsive and responsible in these relationships by attending to another's needs, interests, desires, vulnerabilities, hopes and sensitivities. [2015, p. 227]

While some have regarded this humanizing of animals as a human egocentric gesture, others maintain that it helps in the protection of non-human life. The argument is that thinking with and understanding "nonhuman animals and their agency can inform how humans empathise, relate, and interact with animals" (Villanueva, 2019, p. 192). In Burger's words, "I needed to go beyond the facts and figures that amaze us about underwater. I wanted to bring an emotional and human story to that wonder and explore what lies between the facts and figures and explore the interaction with sea creatures."[5] It is this experience that he shares with the reader to bridge the gap between two seemingly different worlds.

One way to bridge this gulf is to stop conceiving sea creatures as one whole mass but to individualize them through name designations. National Geographic has documentaries where they follow named individual sea creatures to comprehend them within their particular worlds. In this way, they cease to be mere sea creatures but become distinct sentient beings with agency. This however is not without its problems. For instance, Wilson argues that in attempting to represent the other, "questions about translation, decipherability, and the dangers of misunderstanding" (2019, p. 713) may arise. In the novel the first sea creatures we meet are the Steller sea cows, the last of their kind due to poaching. In opening the novel with a close encounter between Steller sea cows and the poachers, Burger frames his narrative as the fight between men and the sea creatures. The sea cows seem content in their world. Like humans, they love, mate, feel, and communicate. This foregrounds the author's intention of rendering subjectivity and individuality to these sea creatures. The novel opens with a beautiful mating scene that underscores their passion-driven relations. However, this erotic scene contrasts sharply with the pain and death of the sea cows that comes immediately after. The somber words used to capture the ordeal highlight the callousness of it all. The male sea cow shudders in pain as he is impaled by the poachers and helplessly calls out to his partner for help. She is sadly oblivious to the danger they are in. The narrator writes:

> But all she can see in that deep-set eye is a complete lack of fear- a being who has never known a foe. The sea cow is confused. Her mate is calling out to her as if she isn't near. A strange floating thing has come between them. Below her, through the clear water, the comforting sounds of the reef come rushing up to her. Her eyes close involuntarily as a blinding pain arcs across her back. [p. 9]

Human beings pose the greatest threat to the survival of marine life. The underwater world without the intrusion of men provides shelter, captured in the phrase "the comforting sounds of the reef." Left to its own devices, the marine world has a self-regulating mechanism that seems to ensure the survival of the different sea animal species.

We also meet "a large yellow-eye rockfish, close to a metre in length, the fish is gravid, her belly visibly swollen with live young…*precious, pregnant life*" (p. 52). In this description, focus is extended to the life that is about to be brought forth. This perception of sea creatures as mothers is also depicted in the encounter between Klaas and a pregnant manta. Klaas' "eyes go to its belly and narrow with deep appreciation when he notices the swelling.

A smile spreads on his face. *You're pregnant!* He arches away. Swings around gracefully in the water to meet the mother's eye once more" (p. 184). In the same manner that Claire's heart swells when she encounters the pregnant rockfish, Klaas is mesmerized by the pregnant manta. He considers the sea creature not just as a manta but a proud mother to be. We are told that "she is grace, she is poise, she is one with the sea. As is he. He communicates, through his movements, through his flow, all of his loneliness. All of his joy at her pup, which must be coming soon" (pp. 184–185). Thinking of these creatures as mothers and families obliges one to protect them, for their sake and that of their offspring like the manner one thinks of protecting human life. The sea creatures are portrayed as sentient beings, living, loving, preying, and being preyed upon, precarious but seemingly content in their world. While Burger's representation of these sea creatures sometimes slides into idealism, for him the desperate desire to protect these animals from human-induced extinction justifies the means.

Another example is the representation of sharks as amiable beings which is captured below:

> She turns. He is in awe. She is magnificent. All powerful. The most beautiful creature he's ever seen. He can't tell if her round, black, expressionless eye is on him or not. As the light plays on her graceful lines, her body eases through the water like it's a part of it. Without hesitation, Klaas swims out, straight towards the four-metre great white. The shark completely ignores him, but instinct tells him not to venture closer—that the shark is well aware of him. He hangs in the water, frozen. She is magnificent. Massive. Sleek. She glides away from him, and he loses sight of her black eye. [p. 75]

The encounter departs from the monster image that most people are familiar with. Diction such as magnificent, beautiful, graceful, powerful, sleek, and massive is used to describe the shark Klaas encounters. These adjectives do not depict a dangerous animal but one that is regal and majestic. Instead of attacking Klaas, the shark targets a seal. Molly Edmonds (2020) relates those human beings are not part of the shark diet. She further notes the irony that even though sharks are mostly the victims of human poaching, they are labelled as dangerous predators who will kill anything in sight. The misapprehension contradicts scientific evidence demonstrating that sharks rarely attack humans. Of the few attacks, most are of mistaken identity where humans are misconstrued for large sea creatures like seals or dolphins. Normally sharks prefer fish and invertebrates like squid and clams to humans (https://animals.howstuufworks.com/fish/sharks/most-dangerous-sharks.htm). Burger seems to corroborate the view that sharks are harmless to humans. In place of the supposed monster, Klaas identifies with the shark and has an epiphanic moment: "He knows why he loves it down there. It's because he feels totally safe" (p. 76).

Another encounter involves Klaas and dolphins:

> the four dolphins surround him, and he makes eye contact with those glistening hyper-intelligent eyes before they veer off playfully, clicking and whistling all the while. He can't help himself. He 'talks' to them as they twirl around him, in what he believes is some version of Bushman language. [p. 150]

Klaas perceives that the dolphins talk back to him, inviting him to "Come! Ride this wave with us!" (ibid.). When he surfaces for air, Klaas is "filled with pure child-like joy of having played with kindred souls" (ibid.). Again, the term "kindred souls" suggests a deep relationship and attachment between this human and the dolphins—interspecies connection. The pure child-like joy he experiences underwater captures Klaas' connection to the sea creatures and to water. The deep reverence is also manifest in the way that Claire formulates her relationship with the sea creatures. As she intimates to Benito towards the end of the novel, "… my own inner journey is moving toward a practice of connecting personally with sea creatures. As beings. Individuals. Sometimes it honestly feels like I'm talking to them. Hearing them" (p. 166). This illustrative human–sea animal relationship resonates with Foster's connection to an octopus in "My Octopus Teacher." Foster narrates that "a lot of people say that an octopus is like an alien. But the strange thing is that as you get closer them, you realize that we are very similar in a lot of ways…" Through compelling cinematic techniques, the film visualizes a deep relationship that grows between Foster and the octopus. Just like Klaas and Clair, Foster feels welcome and at home with sea creatures.

Benito (who heads the Tofo marine project in Mozambique) links this reverential attitude towards sea beings to Credo Mutwa's philosophy from Zulu traditional African knowledge, dreams, and practices. Mutwa, a well-known Sangoma and spiritualist believes that there is a hidden lake in the spiritual world where the knowledge of everything resides and is accessible—past, present, and future—in the form of small silver fishes (Mutwa, 2003). It is in this vein that sea creatures such as whales and dolphins are held in great reverence.

Building on the idea that people will always care about what they know and understand, Burger paints an arresting image of the ocean world (sea creatures, human beings, and its flora) as part of his environmental activism. In Claire's words, "[h]ow do we get people who can't dive down here to appreciate, and to protect, this miracle of life" (p. 123). This miracle of life encompasses both marine creatures and the coral reefs that sustain them. In another illustrative instance, we read that:

> Claire's hand comes up and moves involuntarily in a slow, regal wave, acknowledging the subtle shades: the greens, oranges, yellows and faded browns that indicate healthy coral growth. And in between, the soft corals, the sponges and tunicates, in an array of colours. Bunches, branches, fans. All of these permanently attached invertebrates are filtering the seawater soup for nutrients. Feeding, cycling, recycling. And with light angling down through the blue, catching the throngs of fish, she becomes aware of a gentle warmth spreading throughout her body as she breathes easefully. She feels into the reef's vulnerability. She knows that if the sun heats up just a few degrees, the polyps will spit out the algae and go into a kind of hibernation. And if the warm water persists, even for a few days, the polyps will starve and die. And the wonderland she is floating over will turn white overnight. Dead, bleached limestone sculptures. Millions of fish without food. [p. 123]

The world depicted in the above passage captures a living, self-sufficient marine bionetwork that sustains, cycles, and recycles. It is also a beautiful, multi-coloured world that nurtures a vast underwater ecosystem. Through evocative diction, Burger depicts a world which is enchanting but under threat. The underwater world is not just beautiful but also fragile and susceptible to human intrusion and climate change. Claire's anxieties for the reef are echoed in the Coral Reef Alliance Report (2005) that states that:

> rising sea temperatures and sea levels and increasing frequency of storms will increase coral mortality and seriously endanger coral reefs, especially those already under stress. These climatic changes could become the proverbial straw that breaks the camel's back for reefs facing stresses such as poor water quality, destructive fishing and tourism impacts. (n.p)

The common image of the ocean that most people are familiar with derives from television programmes, nature magazines, and paintings. In these different multimedia representations, the ocean space is always a mysterious place outside of our lived experiences. The perpetuation of "otherness" is responsible for how the ocean and its creatures have been objectified and exploited. In departure, Burger not only shares the beauty of the oceans but also alerts us to the fragility of these spaces in hopes of changing the humans' relationship with the seas and its creatures for the better.

Conclusion

In this chapter I have shown how *She Down There* draws on water-based indigenous beliefs to represent oceanic heritages. The author draws on the mythical water goddess Sedna and diverse sea creatures to blur the lines between the human and the non-human, the natural and the supernatural, the land and the ocean, the past and the present, the rational and otherwise. In reconciling the human to the ocean, Burger does not restrict us to the materiality of the ocean but instead unites us with the ocean and its creatures. By immersing the readers underwater, it becomes possible to think and feel with sea animals.

Notes

1. https://www.lyntonburger.com/about.
2. For more on anthropomorphism see Fisher Fisher, John Andrew (1996) "The Myth of Anthropomorphism." In *Readings in Animal Cognition*, edited by Marc Bekoff and Dale Jamieson, 3–16. Cambridge, Mass: MIT Press and Lorraine Daston and Gregg Mitman (2005) Thinking with Animals: New Perspectives on Anthropomorphism, Columbia, Columbia University Press.
3. Richards, Nancy. "Nancy Richards speaks to author Lynton Burger", Episode 17, Country Life, 8 June 2020, https://anchorfm/country-life/episodes/Country-Life-Podcast.
4. (Hannabach Cathy. "Melody Jue on Thinking Through Seawater", Imagine Otherwise, Episode 97, 9 October 2019, https://ideasonfire.net/97-melody-jue/).
5. (Nancy Richards. "Nancy Richards speaks to author Lynton Burger", Episode 17, Country Life, 8 June 2020, https://anchorfm/country-life/episodes/Country-Life-Podcast).

References

Altman, J., & Jackson, S. (2008). Indigenous rights and water policy: Perspectives from tropical Northern Australia. *Australian Indigenous Law Review, 13*(1), 27–48.
Aschwanden, H. (1933). *1989 Karanga Mythology*. Mambo Press.
Burger, L. (2020). *She down there*. Penguin Random House.
Day, M.O., Ramezani, J., Bowring, S., Sadler, P., Erwin, D., Fernando A, F. and Rubidge, B. (2015). When and how did the terrestrial mid-Permian mass extinction occur. Evidence from the tetrapod record of the Karoo Basin, South Africa.

Proceedings of the Royal Society B: Biological Sciences, Vol. 282, No. 1811, https://doi.org/10.1098/rspb.2015.0834

Edmonds M. (2020). Shark facts vs. shark myths. https://www.worldwildlife.org/stories/shark-facts-vs-shark-myths, viewed 16 February 2021.

Ehrlich, P., and J. Reed [Dirs.]. (2020). *My Octopus Teacher*. Netflix

Fisher, J A. (1996). The myth of anthropomorphism. In Colin Allen and Dale Jamieson (Eds.), *Readings in Animal Cognition* (3–16). MIT Press and Lorraine

Gee, K. (2019). The Ocean perspective: Past, present, future. In Zaucha, Jacek and Kira Gee (Eds.), *Maritime Spatial Planning: Past, Present, Future*, 23–45. Palgrave Macmillan.

Gruen, L. (2015). *Entangled empathy: An alternative ethic for our relationships with animals*. Lantern Publishing & Media.

Hannabach, C. (2019, October 9). Melody Jue on thinking through seawater. Imagine otherwise, podcast. Accessed 14 January 2020, https://ideasonfire.net/97-melody-jue/

Hardie, W. & Morgan M. (2008). *Searching for mermaids in the Karoo*. Wendy Hardie productions and South African broadcasting cooperation. https://searchworks.stanford.edu/view/10572460

Hofmeyr, I. (2019). Provisional notes on hydrocolonialism. *English Language Notes*, 57(1), 11–20.

King, H. Dowsing. (n.d.). As a method of finding underground water. https://geology.com/articles/water-dowsing/. Accessed 6 July 2020.

Kohn, E. (2013). *How forests think*. University of California Press.

Makamure, C., & Chimininge, V. (2015). Totem, Taboos and sacred places: An analysis of Karanga people's environmental conservation and management practices. *International Journal of Humanities and Social Science Invention*, 4(11), 7–12.

Mashige, M. (2011). Essences of presence in the construction of identity. *Southern African Journal for Folklore Studies*, 21(1), 13–27.

Mierek, J. (2010). *Interrelating with animals: Nonhuman Selves in the literary imagination*. Graduate College of the University of Illinois.

Mungoshi, C. (1975). *Waiting for the rain*. Heinemann Educational.

Mutwa, C. (2003). *Zulu Shaman: Dreams, prophecies, and mysteries (Song of the stars)*. Destiny Books.

Mwangi, D. (2019). *Multi-species entanglement: Human-baboon interactions in Nthongoni, eastern Kenya*. Durham University.

Putuma, K. (2017). *Collective Amnesia*. Uhlanga.

Richards, N. (2020, June 8). Nancy Richards speaks to author Lynton Burger. podcast, https://anchorfm/country-life/episodes/Country-Life-Podcasts. Accessed 4 September 2020.

Varner, G. (2010). *Ghosts, spirits & the afterlife in native American Folklore and religion*, Lulu.com, Kindle Edition.

Villanueva, G. (2019). Animals are their best advocates: Interspecies relations, embodied actions, and entangled activism. *Animal Studies Journal, 8*(1), 190–217.

Who are the Bushmen?—The Kalahari Meerkat Project. (2013). https://www.yumpu.com/en/document/view/5541722/1-who-are-the-bushmen-the-kalahari-meerkat-project. Accessed 4 December 2020.

Wilson, H. (2019). Contact zones: Multispecies scholarship through imperial eyes. *Environment and Planning: Nature and Space, 2*(4), 712–731.

11

Women Rising, Women Diving in Vanuatu and Australia

Cobi Calyx

We sweat and cry salt water, so we know that the ocean is really in our blood. (Teresia Teaiwa 1998 in Katerina Teaiwa 2020)

This chapter focuses on women's experiences in histories of diving and shellfish use and trading, while noting that in many societies, people of both genders and beyond gender binaries have participated in such marine activities. There is value in exploring women's histories and realities explicitly given their past erasure and devaluation (Waring & Steinem, 1988; Hooks, 2000; Harris, 2009; Haraway, 2016; Pfeiffer & Butz, 2005), and the fact that their sensory experiences are overlooked (Boswell, 2019). This chapter offers some insight into First Nations voices in the ocean studies space, noting that some First Nation peoples have not felt safe to publicly articulate their voices and values, and that for the most part, they remain largely ignored as narrators of humanity's ocean stories (Lee, 2018; Teaiwa, 2020; Underhill-Sem, 2020).

The research presented in this chapter is supported by the University of New South Wales (UNSW) Career Advancement Fund for Female Academics Returning from Maternity Leave.

C. Calyx (✉)
UNSW Sydney, Sydney, NSW, Australia
e-mail: cobi.calyx@unsw.edu.au

A further issue is that many First Nations peoples, do not feel "…fully legible to colonialism's eye and [feel that they] cannot be defined by its sciences nor described through its grammar of power" (Paperson, 2010). Dominant frames presented for oceans management and sustainability can also conceal the knowledges of indigenous peoples including First Nation peoples (Hau'ofa, 1993). The social categories created in hegemonic (Western) epistemologies can disregard gender diversity, as well as relations that cross the species divide (DeLoughrey, 2017). As Haraway notes, the latter can dissolve and transform relationships (Haraway, 2015), shaping new perspectives and understandings.

This chapter also picks up on the residual effects of conflict on Australian frontiers, specifically the ways in which First Nations women's traditional marine hunting in coastal areas has been ignored and marginalised (Stronach et al., 2019). As I argue, such marginalisation did not only appear in text it was experienced in an embodied manner (Bignall, 2010) and had visceral effects (Boswell, 2019). In this regard women divers in the southern oceans were not and are not always 'free' divers. They are affected by colonial legacies and its patriarchal, gendered conceptualisation of health (Shildrick, 1997; Ussher, 2006). It is important therefore, in considering global 'blue heritage' that one understands how women's contributions to ocean sciences and to human life with water, have been obliterated in and marginalised by historical records (Harding, 2016).

Interdisciplinary perspectives in ocean sciences are equally important. These allow a shifting of perspectives from masculinised transoceanic crossings and naval ambitions to oceanic submersions, embodied and gendered experiences of water and materiality in the oceans space (DeLoughrey, 2017). Interdisciplinary approaches also help to 'reground' ocean studies by emphasising the colonial and often hierarchical nature of interaction in certain settings, such as the Pacific (DeLoughrey, 2019) foregrounding the fleshliness of colonial encounters. These thoughts came together for me when I encountered the following objectionable claim about women divers, their physical capabilities and skill as divers:

> The female population and its eventual susceptibility to diving accidents poses a problem that requires study, because many publications have shown a higher incidence of decompression accidents in women. (Seyer, 2012, p. 183)

The quotation is far removed from more nuanced and deep analyses of women engaging with and across (sea) waters. Faris (2019) notes in 'Sisters of Ocean and Ice' that globally a rich, multi-layered, gendered, south-south activism is emerging that foregrounds the place of women in addressing issues of climate change and the participation of women across ocean domains. Faris and Tuhiwai-Smith (2015) state that in a still patriarchal world, there is opposition to such activist women and in using a diving idiom, such women may be 'rising too quickly' for the comfort of those used to male presence and dominance.

Island Women with Water and Land

If water is life and land is our first teacher (Tuhiwai-Smith et al., 2018), then divers like explorers, undertake journeys underwater for lifelong education. In the following I offer a partly analytical, partly self-reflexive account of how island women's lives are enmeshed with water and land in Vanuatu and surrounding islands. I show that while dominant frames of diving, and being a diver prevail, women divers—often marginalised in the male-dominated diving community, craft their own narrative of diving, that is embodied and intergenerational. In this regard, the narrative I offer aligns with comments made by ecofeminist Neimanis (2012), who argues that the ocean is archive, source of meaning and matter. It is not merely a medium for transoceanic voyages or ambitions. Reflecting on the situation of women divers, I propose in the following paragraphs, that for such women, the ocean is a palpable substance (i.e. matter) which they engage with. It is also a source of meaning and identity, both localised and intergenerational. Finally, the ocean is also archive, for it provides a means to capture the histories and stories of women across nation states and patriarchal narratives. I also argue that some women divers in the southern oceans of the Pacific are not 'yet' at the stage where they can dispense with gendered binaries and perceive themselves as independent beings who can freely dive and be 'part' of water.

A Kaurna elder known by various names in the landscape (Amery & Yambo Williams, 2002; Yambo Kartanya, 2002) and her kin shaped my learnings on land, which settlers from several islands like me interpret partially. Nudges from a playful baby dolphin near stromatolites by the Indian Ocean, as well as sensing the smells of the Patawalonga where dolphins are rarer, impacted land learnings too. As a diver familiar with various diving idioms, I perceive the learning process and efforts at gendered activism in the ocean sciences space to sharing breath (Batacharya & Wong, 2018). Divers sometimes share oxygen to stay alive. To endure and remain lucid in a potentially hostile space, they need to breathe out slowly more than in. The objective is to release accumulated toxins, and in the diving sense, to release carbon dioxide, remaining aware of one's purpose and placement at a particular point in the aqueous medium. Similarly, as a diver outside of water, I have begun to appreciate the importance of sharing my perspective as a woman diver across communities in the southern oceans of the Pacific.

Through reading, I have learned that 'Aboriginal women were the expert swimmers and divers in their communities, responsible for collecting abalone, mussels and crayfish…' (Cadzow, 2016). Palawa women of the island now called Tasmania, were documented in a range of sources as hunters, swimmers and divers (Stronach & Adair, 2020) on whom encroachers/poachers depended for survival (Cadzow, 2016). And, before the colony of South Australia, Ngarrindjeri women on the Coorong dived for abalone from rafts they made (Bogna, 1991). During colonisation permission to even ask to dive was controlled and fishing equipment granted remained the property of the Crown (Fowler et al., 2016). Development of fisheries legislation in Australia granting fishing licences further erased women from diving industries. A dispute between de facto spouses in South Australia under the Fisheries Act 1971, *Kelly vs Kelly*, led to recognition of the value of abalone licenses as transferable property (Bogna, 1991). Abalone diving became a licensed occupation dominated by men.

These southern oceans stories can be contrasted with accounts from Asia where women divers remain prominent and sometimes celebrated, such as *haenyŏ* in Korea (Gwon, 2005) and *ama* in Japan (Nukada, 1965). However, these celebrations are not being sustained as tourism promotions suggest (Kato, 2019; Ko, 2013). *Haenyŏ* diving has declined in recent years with the development of capitalism and a reordered gender division of labour (Gwon, 2005). Fortunately, a gendered activism in the ocean sciences space means that the voices of women such as *haenyŏ* and *ama* are beginning to be heard first hand, as initiatives including the SDGs link oceanic and human health outcomes to cultures and gender equity.

I encountered the quote about women's susceptibility to diving accidents while looking for histories of women divers in the Southern and Indian oceans, given family connections to local diving technology and the abalone industry. How these connections unfolded meant instead of speaking with my grandmother about diving around Mauritius, an archipelago her ancestors came from, we spoke about Vanuatu and about how women there dived and what their skills were. The subject came to us naturally, as I learned to dive in Vanuatu with a self-contained underwater breathing apparatus (SCUBA).

Thinking through the issue of women with water and land, it also occurred to me that my own experience as a diver brings new forms of knowledge to the interdisciplinary space of ocean sciences. It offers the frame of decontamination and detoxification (Calyx & Jessup, 2019) as well as the frame of first responders (Calyx, 2020) to the decolonisation literature. As a woman diver, publicly, I remain subject to a male gaze (Prasad, 2003). But the more I dive, and learn about indigenous and First Nations' diving practices and skills, the more I see that knowledge regarding their skills and the possibility of sharing indigenous knowledge with a wider audience, can decontaminate and detoxify knowledge held about First Nations and indigenous peoples. Diving is therefore yet another (watery) space which has been colonised and requires a decolonial 'lens'. Thus, and as I offer here, there are possibilities for activism in the watery space of the sea rather than outside of it, on land from the perspective perhaps of a territorialised, oceanic humanities.

Becoming a mother twice has made such activism more pressing, as intergenerational legacies are becoming more important. Part of my work as a diver, therefore, seeks to break the cycle of oppression and to offer another perspective on women divers. I was born and raised on peninsulas in between Mauritius and Vanuatu islands, living on the traditional lands of the Kaurna people (Yambo Kartanya, 2002), but I am not a Kaurna person. I pay my respects to elders, past and present, especially Ivaritji. She had no children; her story matters because in the context of the islands, great aunties matter as much as mothers (Amery & Yambo Williams, 2002). Connection with the past and specifically other, island pasts became possible through my grandmother, who, being from a pigmentocracy, often contrasted my fair skin with hers, sharing old photos and relating how being Mauritian Creole shaped her grandmother's perspectives and interactions with her. Touching skin and reflecting on her words during my first pregnancy offered an additional layer of embodied self-reflexivity (Pagis, 2009). It came to me that I am not merely me, as those who do not know me, nor am I merely a diver from these islands. I am a composite of island women who have lived with and near water. Thus writing self-reflexively about my experience as a diver is part of a 'fleshy becoming' (Huopalainen & Satama, 2020). I now realise how my body, now being a mother, binds me, netting me to the lineage of foremothers and aunties.

Like some island women, I have also been a migratory species, breaking the confines of 'the island' to circle the Earth, then hearing my body clock ticking, returned to my place of birth (Australia) to raise my own children. I have predominantly European ancestry, with Mauritian Creole ancestry predating the federation of English colonies on this continent now called Australia. If tried to go back where I came from (Mauritius), it would be too difficult to do so, not only because I only have Australian citizenship but because I now occupy a nebulous space as a person of mixed ancestry in a place (Australia) where assimilation is key (Bellino, 2020; Bignall, 2010; Fowler et al., 2016; Harkin, 2015) and I have 'come' from another place (Mauritius), where particularism, or identifying with a specific ethnic group is important.

My ancestors' mobility across transoceanic surfaces (DeLoughrey, 2017) however, does not touch on the issue of oceanic submersions, specifically the role and value of abalone diving.

Discussions with people from and related to local and surrounding First Nations about histories of diving and abalone (Cruse et al., 2005), raised further questions. Narratives of abalone diving are globally and publicly associated with a lucrative male-dominated modern industry. Time spent working with Indigenous peoples in Asia however, revealed an ongoing tradition of women divers there, more celebrated than those in southern Australia. Indigenous women divers in Asia are documented as diving as deep as twenty meters and staying underwater for several minutes without breathing technology (Ko, 2013). Even though knowledge of diving practices further south is less well known, there remains evidence from other seas and oceans of the abilities of women divers and one would hope that in the near future, these should inform texts on hyperbaric medicine.

Recognising the Contribution of Women Divers

Leading up to the quote, the medical handbook referred to research estimating the number of recreational divers based on the world market of diving apparatus. Calculating who counts as divers based on a commercial market brought to mind Marilyn Waring's iconic work *If Women Counted: A New Feminist Economics*. Given the problematisation of women in the introduction, one anticipates further discussion. However the only paragraph mentioning it that followed was about how women in the United States had a higher rate of accidents than those in France, explained as due to 'the fact that in France women dive in difficult conditions less often than men' (Seyer, 2012, p. 184). Discussion is limited to women from nations that have shaped Pacific colonialism (Teaiwa, 2020). Later, the author discussed underreporting of accidents, using male pronouns to describe people who neglect, ignore or hide their symptoms. Thus, the epidemiology of diving accidents, are made murky by what men are more likely to hide, yet women are cast as the ones rising too fast.

Some historical medical sources considered regarding gender in diving did not help much either. For instance the history of spearfishing and SCUBA diving in Australia (Byron, 2014) states 'underwater activities began in Australia in 1917', a patently false claim when one considers the long history of diving in Australia and its nearby territories. Such claims are now being contested, as women divers resist such erasure in historical and contemporary texts, 'we are not drowning, we are fighting' (Teaiwa, 2020).

Women's leadership in shellfish knowledge and trading globally is under-appreciated (Williams et al., 2012). Arguably women's role in peacekeeping in southern waters is better recognised. 'Most often, Pacific resistance movements are led by women, keepers of the land and peace, who historically have been less compromised by offers of wealth in exchange for land and power directed by traders, missionaries and colonial officials to male leaders' (Teaiwa, 2020). Historical records and scholarly journals include 'Welsh matrons on the Llanrhydian sands, Nguni foragers of the Transkei coast, Tasmanian and Maori abalone divers, Fuegian limpet bashers…' (Anderson, 1983) as well as celebrated communities in Asia, but rarely were the voices of these women included first hand. Women in southern oceans have traditionally been less counted as marine workers but have also been regarded as more reliable providers than men (Chapman, 1987; Cruse et al., 2005). Chilean women fishers have traditionally been unregistered direct producers (Gallardo-Fernández & Saunders, 2018). Ongoing traditions of women in northern Australia hunting shellfish have been documented (Meehan, 1975, 1977).

In South Africa, Indigenous coastal women in Transkei were reluctant to tell anthropologists the extent of their shellfish trading (Lasiak, 1993). The women's own voices were absent in this research about them. These women were reasonable in their reluctance to share the extent of their knowledge for anthropological research, in which their activity was problematised. Their wariness reflects experiences of Indigenous people internationally, such as the Gitxaała people of Canada who agreed to share their abalone knowledge with researchers, after which this culturally significant local species was decimated (Menzies, 2004).

In the 1830s, well before the federation of the Australian nation, missionaries were surprised to discover some Aboriginal women from islands now called Tasmania could speak French (Russell, 2012, p. 129). Some of these women were participants in multicultural marine industries, valued for their diving and hunting expertise. They sailed as far as Rodrigues and Mauritius (Russell, 2012, p. 131), with experience of both the Southern and Indian Oceans as well as their traditional island waters. This history, submerged until recent decades, offers evidence of how women divers were involved in southern intercontinental oceanic encounters long before their descendants were counted as citizens or people with the right to own property (Moreton-Robinson, 2015).

Women are divers and transoceanic crossers are also survivors. Mobility across islands from Tasmania to Tromelin attests to this. Women were not bound to their homes. For example there are historical accounts of Tasmanian Aboriginal women who sailed to Mauritius (Russell, 2012). Furthermore, women of unrecorded origins trafficked illegally as slaves on a French ship from Madagascar to Mauritius were among 160 shipwrecked near Tromelin Island; fifteen years later only eight women and a child remained there. Issur surmises that the women left the island (Issur, 2020).

SDGs, Ocean Users and Women's Identity

What do such histories (Russell, 2005) mean for how we progress towards SDGs? My view is that we need careful analysis of hidden or concealed histories and narratives. The UN should also be careful that in attempting to measure and count progress towards the SDGs, they do not fall into the trap of presenting all women as marginalised and potentially incapable of self-liberation. It is also key that there is no further erasure of the lived experiences of women and women divers and their coastal histories. Harm is also possible when we document women's knowledge and histories simply because it is deemed necessary or a last resort for such knowledge to be respected (Brodie & Gale, 2002; Kartinyeri & Anderson, 2008; Rigney & Hemming, 2014). Last, we need to consider that there may be irreconcilable ontologies (Russell, 2005) but that this should not lead to ignoring or erasing that which is not easily legible elsewhere.

Abalone diving in particular highlights licensed ocean users versus poachers—a dominant narrative in modern scholarship about abalone in South Africa (Raemaekers et al., 2011). When considering the illicit abalone trade (Steinberg, 2005) there are also hidden histories. For people descended from those who were first slaves then free people, what opportunities existed that were not illicit (Allen, 2015; Sivasundaram, 2021)? What about people forcibly relocated from sites where they were able to dive and fish after slavery supposedly ceased (Allen, 2015; Edmonds, 2017; Peabody, 2017)? Why do indigenous people have to suffer and not others (Ojanuga, 1993)? Furthermore, commercialised and illegal fishing continues in the contemporary waters of Aotearoa/New Zealand, Indonesia, Papua New Guinea and

South Africa (Tickler et al., 2018), among other places. Indigenous peoples, long dispossessed, appear to have little say about these uses of the oceans and coasts. Regarding my study of abalone diving, it is not clear yet, how many women are shifting from traditional, 'cultural' diving to ostensibly illegal collection of abalone and what this might mean for the complex process of achieving SDG 5, while respecting SDG 14 and efforts to conserve marine biodiversity. Clearly however, this is a research question worth pursuing.

Given the difficulty of speaking directly about women's situation and positioning in strong patriarchal societies such as Australia, metaphors and storying can be a more appropriate (Le Heron et al., 2020) means of sharing knowledge. 'If reality comes to us filtered through words, it can be re-written, in a constant *chassé-croisé* between fiction and reality' (Williams-Wanquet, 2007, p. 62).

In the 1800s on islands off the southern coast of Australia, anarchy prevailed (Merry et al., 2000); around several island seals were hunted to the verge of local extinction (Anderson, 2018; Ling, 2002; Taylor, 2000). Sealers, many from the northern hemisphere who arrived via the Indian Ocean, plundered southern oceans, including seals and women (Merry et al., 2000; Taylor, 2000). Diving women of southern islands had reason to hide from sailors, though not all did (Russell, 2005). It can be challenging to think of women migrating like other mammals (Ryan, 1989) or being illicitly taken. Radical myths (Ussher, 2006) of women in sealskin (Ellis, 2002) transcending species boundaries (Haraway, 2015), travelled with sailors from the north familiar with seal industries there (Christensen, 2016; Sellheim, 2018).

In showing my small son a video of a seal in waves for the first time, I was surprised by cuts in the film, to nets in which men catch baby seals to cut plastic from around their necks—unexpected gore. I had wanted to show him live seals surfing like dolphins, but it's hard to get out there; I have not surfed since his birth. At least he has seen dolphins in the flesh. Maybe they're not always dolphins, but that's what he calls out with excitement at spotting figures with fins in the waves. My board remains at another science-trained mother's house on the esplanade, gathering dust, our collaborative intention to share childcare on the beach unrealised yet. What do sons learn from their mothers being bound to the shore? He and I speak in French sometimes, we half-joke about the genetics of rejecting marriage to an African cousin. We will get out there, back in the surf, when our children are bigger. I swear. I buy coffee on the esplanade near a creek mouth, handing over my cup of recycled plastic as my baby waves to the barista. We relate her Peruvian and my Mauritian heritage as I recall where beans have been and we discuss which milk is of what ilk, all of it frothing white. I'm still breastfeeding. On my wall is a table runner from Peru, from a market near a surf break that was once Bolivia. It's full of coloured birds. Blue as well as red, natural pigments? The dunes and hooded plovers by the esplanade, trying to survive (Dennis & Masters, 2006), do not respect the boundaries of the council's delineations of where to drive or walk. Are their ripples and eggs illicit? Living into wrinkles means sharing culture nurturing my baby daughter as she carries her store of eggs in revolutions around the sun.

Diving in Different Locales

Diving in the diverse geographies of tropical Vanuatu and temperate South Australian gulfs in winter, a difference is clear. Whereas in tropical waters, wetsuits protect against reef cuts or stings, in southern waters they serve as a second skin against the cold. Freediving the Spencer Gulf in winter to see cuttlefish mating in rainbow display (Mills, 2018) is an exercise in cold endurance which for some, myself included, would be untenable without a wetsuit. The cold is not the only oppressive force against women divers in southern oceans. Wetsuits can serve other protective mechanisms, such as being barriers to sexploitation (Franklin & Carpenter, 2018). Jokes about being selkies can lighten the cognitive load (Haring, 2012) for women experiencing these gendered power dynamics, which for some may be greater disincentive to diving than environmental conditions.

The *chassé-croisé* dance of retelling and oral storytelling written about in a Mauritian context (Williams-Wanquet, 2007) echoes Aboriginal Australian yarning practices (Geia et al., 2013; Walker et al., 2014). In the century past, Indigenous people in Tasmania were falsely declared extinct; now however, as dominant paradigms have been challenged, more identify as Indigenous on the census than ever (Scott, 2017). Narratives of change in Tasmania are beginning to be reflected on smaller islands (Stronach et al., 2019; Taylor, 2008). This reflects changing social norms, or perhaps the passage of time, about the acceptability of discussing colonial frontier relationships that were deemed illicit (Taylor, 2008). The idea that writers can only write certain parts of the story is also falling way, thus Morill and Tuck's comment, 'I am interested in only telling certain parts, untelling certain parts, keeping the bodies and the parts from becoming a settlement' (Morrill & Tuck, 2016) is becoming redundant. Different types of knowledge are becoming more recognised in science (Harding, 2016), as part of community responses to negative development and unmitigated disasters (Kartinyeri & Anderson, 2008; Teaiwa, 2020; Underhill-Sem, 2020) including climate change (Milton Faris, 2019).

Women have persevered in professions where they were unwelcome. Accessing and interpreting records of such women today gives us pause to reflect on how what we do today—what we record and how we attribute it (Forsyth, 2012a, 2012b; Janke, 2021)—might shape future generations' understanding and reflections. In northern Vanuatu in 2016 I witnessed and listened to women's water music, a non-literary but tangible cultural heritage that is now receiving more scholarly attention which it did not do in earlier times (Dick, 2014, 2015; Grant, 2019).

The experience of hearing water music differs from experiences of touching water in making sounds or diving in, which involves collaboration and being in concert to manifest as music. First witnessing and hearing water music, then reading scholarly text about it, builds different neuronal associations than the reverse order.

> when we listen to music, we must refuse the idea that music happens only when the musician enters and picks up an instrument; music is also the anticipation of the performance and the noises of appreciation it generates and the speaking that happens through and around it, making it and loving it, being in it while listening. (Moten & Harney, 2013, p. 9)

In a way diving into water is like making and hearing water music. This is an embodied heritage that is perhaps not yet recognised, in the same way that Vanuatu women's water music was not recognised for centuries. 'Voices' and tones can be heard when entering the water and while in it and the roar of the waves can be like voices in a concert that are not only speaking or singing but also screaming (Youngblood, 2019). I heard Jetñil-Kijiner (2018) recorded, speaking of being aware of everything around the strands while weaving: 'what has happened before, what has happened after—she says I need to be like the women warriors from our past. Vigilant'. Her voice, consciously or not, renders audible the voices of women passed. In the same way, I feel that being in the water as a woman diver is a sacred heritage experience. While Brodie and Gale argue that it is 'the water around that island that is sacred', I contend that it is the experience of it that is sacred and special.

Ancestral Paths

Despite the paths of my ancestors I have swum but not sailed in the Indian Ocean. There are spectres there (Baker, 2018; Harkin, 2014; Issur, 2020). Living in the Pacific has reoriented my thoughts on the home (Mauritius) that is in the south. Travelling east led me to reflect on identities in the west and how a gendered positionality provides a different narrative of trajectories than those traditionally presented in academic scholarship (Underhill-Sem, 2020; Wehi et al., 2021). Which way do we rise? Thinking about archipelagos (Hamilton Faris, 2019) reiterates my questioning why we still call islands near to where I live, Nuyts Archipelago. Apparently is because this is where seals and sea lions have kept returning in the centuries since Nuyts visit (Shaughnessy et al., 2005). The Neptune Islands, closer still, remain named after a Roman god. What of the goddesses and the seven sisters (Andrews, 2004; Hamacher, 2015)? Listening to elders of First Nations since returning south from Vanuatu has reinforced how far we are as a species from understanding what is needed for human flourishing (Underhill-Sem, 2020; Wehi et al., 2021). It was hearing a prominent man (Regenvanu, 1999, 2008) speaking in a shared dominant colonial language about the value of traditional knowledge for resilience that led me to prioritise living in Vanuatu and pursuing the watery life of the diver. Before meeting this man face-to-face in Vanuatu, I learnt from *fa'afafine* in Samoa how people can embody social movements (Kanemasu & Liki, 2020). From this I came to realise that humans do not just merely embody social movements, they also live in embodied ways in the different locales and mediums in which they find themselves. Thus and as I said earlier, experiencing the water as a diver is a fully embodied and sensory experience. We need to share this knowledge of embodied human experience with the oceans and coasts so that we may more broadly, encourage others (including policymakers and government leaders) to understand 'material, corporeal, grounded, and symbolic readings across disciplinary boundaries, geographic areas, and temporal contexts' (Teaiwa, 2014, p. 119).

Differences in perceptions of taboo or appropriateness of diffracting sand talk have been understood from experiences in countries now known as Vanuatu and Australia (Dick, 2015; Forsyth, 2012a, 2012b; Yunkaporta, 2019), where archipelagos are more numerous than naming might suggest. These understandings are enhanced by scholarly work about what is taboo or sacred (Tuck & Yang, 2014). Relating diversely documented knowledge and its diffractions (Barad, 2014) involves treading carefully, respecting footprints covered over by tidal waters or human hands (Forsyth, 2012a, 2012b), taking care in how and when to share (Janke, 2021). Tides still rise as we care.

Conclusion

Framing women as marine hunters (Meehan, 1977) rather than gatherers is a lens through which women's work can be revised in understanding human relationships with oceans. Women's increasing agency in first hand relating of knowledge in the sciences can be analogous to changing from a foraging to hunting frame. We are not wandering to gather whatever opportunities present, but rather seeking out specific sources of nourishment for us, our families and extended kin. In giving gratitude to beings we seek out, we can better regulate our rising and submersion, better understand our agency in articulating our positions. This chapter, although abstracted for scholarly reflection and discussion offers an embodied and partly self-reflexive study of a woman diver's experience with the ocean and coast. It was written while hearing waves crash on the shore and contemplating seafood for lunch—mundane activities one might argue, in a context where serious scholarly reflection is required. However, and as suggested in this chapter, women and ordinary (often indigenous peoples') engagement with the sea has been perceived and treated as mundane, unimportant, non-visceral and marginal. By contrast, I have tried to show that for women in the pacific, diving is profoundly visceral, socio-political (gender liberating), historical (ancestral and diasporic) and epistemological (revealing little known knowledge foundations). You are invited, reader, to place yourself in the sea and to experience this fully sensory and historicised, human-marine environment and to realise that you are a part of this environment, that you are not apart. For those already deeply immersed in these worlds who need no invitation, thank you for immersing yourself.

References

Allen, R. B. (2015). *European slave trading in the Indian Ocean, 1500–1850*. Ohio University Press.

Amery, R., & Yambo Williams, G. (2002). Reclaiming through renaming: The reinstatement of Kaurna toponyms in Adelaide and the Adelaide Plains. In *The land is a map*. Pandanus Books.

Anderson, A. (1983). Anbarra shellfishing: Shell bed to shell midden by Betty Meehan. *Australian Archaeology, 16*(1), 157–159.

Anderson, R. (2018). The role of sealers, whalers and aboriginal people in the exploration of western Australia's southern ocean frontier. *The Great Circle, 40*(2), 1–27.

Andrews, M. (2004). *The seven sisters of the Pleiades: Stories from around the world*. Spinifex Press.

Baker, A. G. (2018). Camping in the shadow of the racist text. *Artlink, 38*(2), 14–21.

Barad, K. (2014). Diffracting diffraction: Cutting together-apart. *Parallax, 20*(3), 168–187.

Batacharya, S., & Wong, Y. L. R. (Eds.). (2018). *Sharing breath: Embodied learning and decolonization*. Athabasca University Press.

Bellino, E. (2020). Married women's nationality and the white Australia policy, 1920–1948. *law&history, 7*, 166.

Bignall, S. (2010). Affective assemblages: Ethics beyond enjoyment. *Deleuze and the postcolonial* (pp. 78–102). Routledge.

Bogna, R. (1991). *Fisheries regulation in South Australia (1836–1990): A short history*. Accessible online: https://www.academia.edu/download/56837316/Fisheries_Regulation_in_South_Australia_1836-1990.pdf

Boswell, R. (2019). Desensitized pasts and sensational futures in Mauritius and Zanzibar. *Journal of the Indian Ocean Region, 15*(1), 23–39.

Brodie, V., & Gale, M. A. (2002). *My side of the bridge: The life story of Veronica Brodie as told to Mary-Anne Gale*. Wakefield Press.

Byron, T. (2014). *History of spearfishing and scuba diving in Australia: The first 80 years 1917 to 1997*. Xlibris Corporation.

Cadzow, A. (2016). Guided by her: Aboriginal women's participation in Australian expeditions. *Brokers and Boundaries, 85*.

Calyx, C. (2020). Sustaining citizen science beyond an emergency. *Sustainability, 12*(11), 4522.

Calyx, C., & Jessup, B. (2019). Nuclear citizens jury: From local deliberations to transboundary and transgenerational legal dilemmas. *Environmental Communication, 13*(4), 491–504.

Chapman, M. D. (1987). Women's fishing in Oceania. *Human Ecology, 15*(3), 267–288.

Christensen, A (2016). Exploring Faroese and Icelandic relationships to nature in folktales: Kópakonan and Selshamurinn.

Cruse, B. Stewart, L., & Norman, S. (2005). *Mutton fish: The surviving culture of Aboriginal people and abalone on the south coast of New South Wales*. Aboriginal Studies Press.

DeLoughrey, E. (2017). Submarine futures of the Anthropocene. *Comparative Literature, 69*(1), 32–44.

DeLoughrey, E. M. (2019). *Allegories of the anthropocene* (p. 280). Duke University Press.

Dennis, T. E., & Masters, P. (2006). Long-term trends in the Hooded Plover thinornis rubricollis population on Kangaroo Island. *South Australia. South Australian Ornithologist, 34*(7/8), 258.

Dick, T. (2014). Vanuatu water music and the Mwerlap diaspora: Music, migration, tradition, and tourism. *AlterNative: An International Journal of Indigenous Peoples, 10*(4), 392–407.

Dick, T. (2015). Choreographing the Vanuatu aquapelago. *Shima: The International Journal of Research into Island Cultures, 9*(2).

Edmonds, P. (2017). Emancipation acts on the oceanic frontier? Intimacy, diplomacy, colonial invasion and the legal traces of 'protection' in the bass strait world, 1832. *Law & History, 4*(2), 20–44.

Ellis, P. (2002). Radical myths: Eliza Keary's little seal-skin and other poems (1874). *Victorian Poetry, 40*(4), 387–408.

Faris, J. H. (2019). Sisters of ocean and ice: On the hydro-feminism of Kathy Jetñil-Kijiner and Aka Niviâna's rise: From one island to another. *Shima, 13*(2).

Forsyth, M. (2012a). Lifting the lid on "the community": Who has the right to control access to traditional knowledge and expressions of culture? *International Journal of Cultural Property, 19*(1), 1–31.

Forsyth, M. (2012b). Do you want it gift wrapped? Protecting traditional knowledge in the Pacific Island countries. *Indigenous Peoples' Innovation: Intellectual Property Pathways to Development*, 189–214.

Fowler, M., Roberts, A., & Rigney, L. I. (2016). The 'very stillness of things': Object biographies of sailcloth and fishing net from the Point Pearce Aboriginal Mission (Burgiyana) colonial archive, South Australia. *World Archaeology, 48*(2), 210–225.

Franklin, R., & Carpenter, L. (2018). Surfing, sponsorship and sexploitation: The reality of being a female professional surfer. In *Surfing, sex, genders and sexualities* (pp. 50–70). Routledge.

Gallardo-Fernández, G. L., & Saunders, F. (2018). "Before we asked for permission, now we only give notice": Women's entrance into artisanal fisheries in Chile. *Maritime Studies, 17*(2), 177–188.

Geia, L. K., Hayes, B., & Usher, K. (2013). Yarning/Aboriginal storytelling: Towards an understanding of an Indigenous perspective and its implications for research practice. *Contemporary Nurse, 46*(1), 13–17.

Grant, C. (2019). Climate justice and cultural sustainability: The case of Etëtung (Vanuatu Women's Water Music). *The Asia Pacific Journal of Anthropology, 20*(1), 42–56.

Gwon, G. (2005). Changing labor processes of women's work: The Haenyo of Jeju Island. *Korean Studies, 29*, 114–136.

Hamacher, D. (2015). Identifying seasonal stars in Kaurna astronomical traditions. *Journal of Astronomical History and Heritage, 18*(1), 39.

Haraway, D. (2015). Anthropocene, capitalocene, plantationocene, chthulucene: Making kin. *Environmental Humanities, 6*(1), 159–165.

Haraway, D. J. (2016). *Staying with the trouble*. Duke University Press.

Harding, S. (2016). *Whose science? Whose knowledge?* Cornell University Press.

Haring, L. (2012). The Elusive Presence. *Western Folklore*, 239–256.

Harkin, N. (2014). The poetics of (re) mapping archives: Memory in the blood. *Journal of the Association for the Study of Australian Literature, 14*(3).

Harkin, N. (2015). *Dirty words*. Cordite Press.

Harris, L. M. (2009). Gender and emergent water governance: Comparative overview of neoliberalized natures and gender dimensions of privatization, devolution and marketization. *Gender, Place and Culture, 16*(4), 387–408.

Hau'ofa, E. (1993). Our sea of islands: A beginning. In E. Hau'ofa, V. Naidu & E. Waddell (Eds.), *A new oceania: Rediscovering our sea of islands* (pp. 2–19). School of Social and Economic Development, The University of the South Pacific in association with Beake House.

Hooks, b. (2000). *Feminist theory: From margin to center*. Pluto Press.

Huopalainen, A., & Satama, S. (2020). Writing birthing bodies: Exploring the entanglements between flesh and materiality in childbirth. *Culture and Organization, 26*(4), 333–354.

Issur, K. (2020). Mapping ocean-state Mauritius and its unlaid ghosts: Hydropolitics and literature in the Indian Ocean. *Cultural Dynamics, 32*(1–2), 117–131.

Janke, T. (2021). *True tracks: Respecting indigenous knowledge and culture*. NewSouth Publishing.

Jetñil-Kijiner, K. (2018, November 25). *Lorro: Of wings and seas 2018/spoken word performance*. Gallery of Modern Art, Brisbane. Accessible online: https://www.qagoma.qld.gov.au/whats-on/exhibitions/past-exhibitions/the-9th-asia-pacific-triennial-of-contemporary-art-apt9/artists/jetnil-kijiner

Kanemasu, Y., & Liki, A. (2020). 'Let fa'afafine shine like diamonds': Balancing accommodation, negotiation and resistance in gender-nonconforming Samoans' counter-hegemony. *Journal of Sociology*, 1440783320964538.

Kartinyeri, D., & Anderson, S. (2008). *Doreen Kartinyeri: My Ngarrindjeri calling*. Aboriginal Studies Press.

Kato, K. (2019). Gender and sustainability—Exploring ways of knowing—An ecohumanities perspective. *Journal of Sustainable Tourism, 27*(7), 939–956.

Ko, C. H. (2013). A new look at Korean gender roles: Jeju (Cheju) women as a world cultural heritage. *World Environment and Island Studies, 3*(1), 55–72.

Lasiak, T. (1993). The shellfish-gathering practices of indigenous coastal people in Transkei: Patterns, preferences and perceptions. *South African Journal of Ethnology, 16*(4), 115–120.

Le Heron, E., Le Heron, R., Logie, J., Greenaway, A., Allen, W., Blackett, P., Davies, K., Glavovic, B., & Hikuroa, D. (2020). Participatory processes as twenty-first-century social knowledge technology. *Sustaining seas: Oceanic space and the politics of care* (p. 155).

Lee, E. (2018). Black female cultural safety in Tebrakunna country. *Tourism and wellness: Travel for the good of all* (pp. 1–20).

Ling, J. K. (2002). Impact Of colonial sealing on seal stocks AroundAustralia, New Zealand and Subantarctic Islands between 150 And 170 degrees east. *Australian Mammalogy, 24*(1), 117–126.

Meehan, B. F. (1975). *Shell bed to shell midden*. PhD thesis, Australian National University. https://doi.org/10.25911/5d74e186644fc

Meehan, B. (1977). Hunters by the seashore. *Journal of Human Evolution*, *6*(4), 363–370.

Menzies, C. R. (2004). Putting words into action: Negotiating collaborative research in Gitxaala. *Canadian Journal of Native Education*, *28*(1/2), 15–32.

Merry, K., Murray-smith, S., & Stuart, I. (2000). The cross-cultural relationships between the sealers and the Tasmanian Aboriginal women at Bass Strait and Kangaroo Island in the early nineteenth century. *Journal of Australian Studies*, *66*, 73–84.

Mills, J. (2018). Swimming with aliens. *Overland*, *230*, 20–25.

Moreton-Robinson, A. (2015). *The white possessive: Property, power, and indigenous sovereignty*. University of Minnesota Press.

Morrill, A., & Tuck, E. (2016). Before dispossession, or surviving it. *Liminalities*, *12*(1), 1.

Moten, F., & Harney, S. (2013). *The undercommons: Fugitive planning & black study*. Minor Compositions.

Neimanis, A. (2012). Hydrofeminism: Or, on becoming a body of water. In C. Nigianni & F. Söderbäck (Eds.), *Undutiful daughters: Mobilizing future concepts, bodies and subjectivities in feminist thought and practice*. Palgrave Macmillan.

Nukada, M. (1965). Historical development of the ama's diving activities. *Physiology of breath-hold diving and the ama of Japan* (pp. 25–40).

Ojanuga, D. (1993). The medical ethics of the 'father of gynaecology', Dr J Marion Sims. *Journal of Medical Ethics*, *19*(1), 28–31.

Pagis, M. (2009). Embodied self-reflexivity. *Social Psychology Quarterly*, *72*(3), 265–283.

Paperson, L. (2010). The postcolonial ghetto: Seeing her shape and his hand. *Berkeley Review of Education*, *1*(1).

Peabody, S. (2017). *Madeleine's children: Family, freedom, secrets, and lies in France's Indian ocean colonies*. Oxford University Press.

Pfeiffer, J. M., & Butz, R. J. (2005). Assessing cultural and ecological variation in ethnobiological research: The importance of gender. *Journal of Ethnobiology*, *25*(2), 240–278.

Prasad, A. (2003). The gaze of the other: Postcolonial theory and organizational analysis. In *Postcolonial theory and organizational analysis: A critical engagement* (pp. 3–43). Palgrave Macmillan.

Raemaekers, S., Hauck, M., Bürgener, M., Mackenzie, A., Maharaj, G., Plagányi, É. E., & Britz, P. J. (2011). Review of the causes of the rise of the illegal South African abalone fishery and consequent closure of the rights-based fishery. *Ocean & Coastal Management*, *54*(6), 433–445.

Regenvanu, R. (1999). Afterword: Vanuatu perspectives on research. *Oceania*, *70*(1), 98–100.

Regenvanu, R. (2008). Issues with land reform in Vanuatu. *Journal of South Pacific Law*, *12*(1), 63–67.

Rigney, D., & Hemming, S. (2014). Is 'closing the gap' enough? Ngarrindjeri ontologies, reconciliation and caring for country. *Educational Philosophy and Theory, 46*(5), 536–545.

Russell, L. (2005). Indigenous knowledge and archives: Accessing hidden history and understandings. *Australian Academic & Research Libraries, 36*(2), 161–171.

Russell, L. (2012). *Roving mariners: Australian Aboriginal whalers and sealers in the Southern Oceans, 1790–1870*. SUNY Press.

Ryan, L. (1989). Patterns of migration in Tasmania: The Aboriginal experience. *Bulletin of the Centre for Tasmanian Historical Studies, 2*(2), 4–14.

Scott, P. (2017). People who were not there but are now! Aboriginality in Tasmania. In *Globalization and marginality in geographical space* (pp. 248–266). Routledge.

Sellheim, N. (2018). *The sealhunt in the global community*. BRILL.

Seyer, J. (2012). Epidemiology of decompression accidents during recreational diving. *Handbook on hyperbaric medicine* (p. 183). Springer.

Shaughnessy, P. D., Dennis, T. E., & Seager, P. G. (2005). Status of Australian sea lions, Neophoca cinerea, and New Zealand fur seals, Arctocephalus forsteri, on Eyre Peninsula and the far west coast of South Australia. *Wildlife Research, 32*(1), 85–101.

Shildrick, M. (1997). *Leaky bodies and boundaries: Feminism, postmodernism and (bio)ethics*. Routledge.

Sivasundaram, S. (2021). *Waves across the south*. University of Chicago Press.

Steinberg, J. (2005). The illicit abalone trade in South Africa. *Institute for Security Studies Papers, 2005*(105), 16.

Stronach, M., & Adair, D. (2020). Swimming for their lives: Palawa women of Lutruwita (Van Diemen's Land). *Sporting Traditions, 37*(2), 47–70.

Stronach, M., Adair, D., & Maxwell, H. (2019). 'Djabooly-djabooly: Why don't they swim?' The ebb and flow of water in the lives of Australian Aboriginal women. *Annals of Leisure Research, 22*(3), 286–304.

Taylor, R. (2000). Savages or Saviours?—The Australian sealers and Aboriginal Tasmanian survival. *Journal of Australian Studies, 24*(66), 73–84.

Taylor, R. (2008). *Unearthed: The Aboriginal Tasmanians of Kangaroo Island*. Wakefield Press.

Teaiwa, K. (2020). On decoloniality: A view from Oceania. *Postcolonial Studies, 23*(4), 601–603.

Teaiwa, K. M. (2014). *Consuming ocean island: Stories of people and phosphate from Banaba*. University of Indiana Press.

Tickler, D., Meeuwig, J. J., Bryant, K., David, F., Forrest, J. A., Gordon, E., Larsen, J. J., Oh, B., Pauly, D., Sumaila, U. R., & Zeller, D. (2018). Modern slavery and the race to fish. *Nature Communications, 9*(1), 1–9.

Tuck, E., & Yang, K. W. (2014). R-words: Refusing research. *Humanizing Research: Decolonizing Qualitative Inquiry with Youth and Communities, 223*, 248.

Tuhiwai-Smith, L. (2015). *Decolonizing methodologies: Research and indigenous peoples*. Zed Books.

Tuhiwai-Smith, L., Tuck, E., & Yang, K. W. (Eds.). (2018). *Indigenous and decolonizing studies in education: Mapping the long view*. Routledge.

Underhill-Sem, Y. T. R. R. O. T. (2020). The audacity of the ocean: Gendered politics of positionality in the Pacific. *Singapore Journal of Tropical Geography, 41*(3), 314–328.

Ussher, J. M. (2006). *Managing the monstrous feminine: Regulating the reproductive body*. Routledge.

Walker, M., Fredericks, B., Mills, K., & Anderson, D. (2014). "Yarning" as a method for community-based health research with indigenous women: The indigenous women's wellness research program. *Health Care for Women International, 35*(10), 1216–1226.

Waring, M., & Steinem, G. (1988). *If women counted: A new feminist economics*. Harper & Row.

Wehi, P. M., van Uitregt, V., Scott, N. J., Gillies, T., Beckwith, J., Rodgers, R. P., & Watene, K. (2021). Transforming Antarctic management and policy with wehi an Indigenous Māori lens. *Nature Ecology & Evolution, 5*(8), 1055–1059.

Williams, M. J., Porter, M., Choo, P. S., Kusakabe, K., Vuki, V., Gopal, N., & Reantaso, B. (2012). *Guest editorial: Gender in aquaculture and fisheries-moving the agenda forward.*

Williams-Wanquet, E. (2007). Lindsey Collen's 'the rape of Sita': Re-writing as ethics. *Commonwealth Essays and Studies, 29*(2), 55.

Yambo Kartanya, G. W. (2002) *Sustainable cultures and creating new cultures for sustainability.* Sustaining our Communities International Local Agenda 21 Conference, March 3–6, Kaurna Country

Youngblood, F. (2019). On un-silencing voices: Tarantismo and the gendered heritage of Apulia. *Folk Life, 57*(1), 42–55.

Yunkaporta, T. (2019). *Sand talk: How Indigenous thinking can save the world*. Text Publishing.

12

Speaking with the Sea: Divination and Identity in South Africa

Dominique Santos and Rev Thebe Shale

Introduction

How does the sea speak? What does it tell us when we listen? About the entangled histories that have played out over the waves and under them, transporting and transforming the unwanted, the captured, the powerful? What might it tell us about its rage? When it rises in anger and consumes boats like matchsticks, scattering lives and reordering destinies? How do we listen? In the land now known as the Eastern Cape, South Africa, but which has had many names, there is a long tradition of engagement with the Ocean as a spiritual entity. Using traditional technologies of communication to focus on the spiritual dimension of this stretch of Indian Ocean coast via the proliferation of slave shipwrecks of the fifteenth-seventeenth centuries, through divination enquiry with traditional healer and co-author Rev Thebe Shale,

---

The research presented in this chapter is funded by the African Studies Centre, Rhodes University, Makhanda, South Africa.

D. Santos · R. T. Shale (✉)
Rhodes University, Makhanda, South Africa
e-mail: thebe.mabandla@gmail.com

D. Santos
e-mail: d.santos@ru.ac.za

we encounter stories from the margins of archival accounts of shipwrecks, the place of shipwreck victims in local vernaculars of engagement with the ocean, and the ecological consequences of global settler colonialism, slavery and the ideology of unfettered economic growth.

Drawing on frames of divination methodology, subaltern studies and radical participation, this paper begins a provisional process of what such engagements might look like as complementary research methodologies that locate indigenous technology not as object of study, but lead investigator to explore more fulsome representations of shipwrecks, shipwreck survivors and their aftermath on land. We ask how the cosmology of the ocean from a local perspective might inform interpretations of shipwrecks and their aftermaths, interwoven with the voices accessed via divination of Antonio/Mangabome, an enslaved person from Malabar, and Diogo, a black Portuguese sailor, who both became rainmakers following shipwreck and whose stories we encountered briefly in written accounts of shipwreck survivors. We consider the intersection of this cosmological frame and divinatory insight with critique of the ecological consequences of the extractive world which has been created in the past few hundred years. This is revealing of the entanglements of ocean trade as "Blue Economy" with indigenous conceptions of the water, and ongoing concerns with social justice (Sharfman et al., 2012).

Divination Methodology

Divination as a practice which links the seen and unseen worlds through the deployment of various technologies of enquiry ranging from the reading of inorganic objects and biological remains, through trance and the channeling of entities, is globally practiced as a way of "coming to know" (Tedlock, 2001). It is ubiquitous in a plethora of African social contexts spanning the continent, taking a dizzying array of forms. While this may be the basic ground on which divination practices rest, as Peek (2013) calls it, "With such an array of divination systems, clearly one should be very cautious of any generalizations." Indeed, divination has been likened to anthropology (though anthropology has a significantly reduced sphere of influence in comparison!), and has been a feature of anthropological enquiry from the discipline's beginning (Callaway, 1872) "… both pursuits attempt to sort out foreign cultures, to determine meaningful utterances from background noise, and to translate from one system to another" (Benneta Jules-Rosette, 1975 in Peek, 2013).

12 Speaking with the Sea: Divination and Identity in South Africa

Divination is a key area in which knowledge is generated and legitimized. Yet, despite the centrality of these systems to the epistemological worlds they shape, they receive relatively little attention in comparison with other areas of African religious and political life (Peek, 2013, p. 14); taken less seriously than other forms of social practice by scholars, or rationalized according to criteria more relevant to the researcher than the divinatory practice, acting to, "…rationalise divination after the fact, totally removing it from the realm of intentionally effective action" (Tedlock, 2001, p. 195). This is not surprising when one considers that the Western academy has tended to focus on areas of study which mirror the social institutions and practices it recognizes as legitimate.

In their critique of the over-emphasis on textual analysis when considering divinatory practice, De Boeck and Devisch emphasize that divination, "…constitutes a space in which cognitive structures are transformed and new relations are generated in and between the fields of the human body, the social body and the cosmos" (1994, p. 100). It is precisely this "intentionally effective action" (Tedlock, 2001, p. 195) which makes divination a compelling tool for engaging with archives, produced within colonial ideological frames, but peopled by many who are on the outskirts of, and the Other to (Said, 1978), the making of the utopia of consumer capitalism. In this sense, while scholarly interest in divination has focused on the dialogic between diviner and client, the presence of entities, in addition to the diviner and client/community, is as key an actor in the affective quality of divinatory insight for the purposes of this article, and in considering how knowledge elicited through divination, via the diviner as interpreter, but originating from the participating entity who finds voice through the diviner, and invites the possibility for new relations with historical events, shadow archival figures and social body as an act of imagination derived from what is known, "a performative generation of relations and involvement in the life-world" (De Boeck & Devisch, 1994, p. 103). Extending Turner's analysis of the divination toolkits and process as symbolic restorative ritual for social cohesion, "Divination is a hermeneutic of disclosing, a worldmaking in perpetual emergence, compared to weaving: it posits a world of becoming, withering away and re-emergence, of death, gestation and rebirth" (1994, p. 109).

This is divination methodology, what Saidiya Hartman in her work to call forth the stories that lie in the gaps and silences of the archive around the Trans-Atlantic slave trade and African American history calls "critical fabulation" (2008). This method explores the potential of stories that are speculated on by those of us still living, stories which refuse the imposition of a provable subject with a coherent narrative. Those whose shadow forms we find in the archives and histories cannot be reclaimed, we can't hope to know the entirety of how they may have represented their own story, or even if the notion of story would have framed their experience. But, as ancestors, they offer us the possibility of examining our own story to generate other future possibilities. Working with the dead is about the living, crafting possibilities for imagining ourselves differently out of the silences in the official versions of history.

Subaltern Studies, Radical Participation and Dreaming Concerns

The theoretical lineages informing this work are drawn from subaltern studies, poised around Gayatri Spivak's question: "Can the subaltern speak?" (Spivak, 1988). These lineages carry cautionary voices as attempts to reconstitute subjects hidden from view in the colonial archive risk reproducing the norms of a particular kind of narrative rooted in a rational western tradition—coherent, processual, rational. Rosalind O'Hanlon, in her essay about the risks of attempting to recover "lost" and "hidden" historical voices, asks what, with all the risks of reproducing the power relations we are hopelessly entangled in, is the point of doing this salvage work to, "…attack historiography's dominant discourses… find a resistant presence which has not been completely emptied or extinguished by the hegemonic" (O'Hanlon, 1988, p. 175). For O'Hanlon, the work is less about recovering lost stories, and more about unraveling histories of domination which have provided us with only partial maps for showing us who we are, who are ancestors were, and what our descendants might become. Challenging dominant historical narratives permits us to, "…envision a realm of freedom in which we ourselves might speak. This is not to say that our project becomes thereby a private and merely selfish one: it is precisely on the predication of such a realm that we can think of our practice as a provider of insight and clarification" (O'Hanlon, 1988, p. 175).

Anthropologist Francis Nyamnjoh, who frequently draws on other-than-scholarly sources such as folk tales, African popular literature and divination as everyday social practice, to illustrate theoretical points, draws on a story about blind men attempting to describe elephants to critique the tendency in scholarly practice to over-emphasize some sources and subject positions over others as legitimate frames for study and analysis,

> What if elephants were unknowable? What if our conventional indicators of knowledge were inadequate for us to access and claim knowledge of elephants,

even when armed with optimally efficient senses? What if the reality of elephants were larger than could be fathomed by the senses? Would we, the scholarly community, then be reduced to keeping up appearances of claiming knowledge that we could never really access however hard we worked at it? How would we, given this hypothetical truth, relate to others equally involved in keeping up appearances, only in a different way and with claims different from ours? Would we confront and contest, or seek to understand and accommodate them? Would we invite them to a discussion of how to provide a level playing field for competing sources of ignorance? (Nyamnjoh, 2012, p. 81)

Inviting the diviner into scholarly reflection is an opening into another approach to understanding the dialectical nature of the relationship between past and present. Araguete (2017), examining the exhumation of mass graves from the Spanish Civil War, demonstrates how re-entering archeological remains acts to restore memories of loss and repression in a community of practice between archeologists, anthropologists and local residents, often the descendents of those whose remains and effects are found in excavation. Araguete considers these found objects and remains, along with archival documents, war remnants and stories recounted by descendents as instructive on the process of constructing new histories and socio-political claims about the past and present (2017, p. 15). The excavations in Spain are provisional and limited by a political climate which temporarily permits such a process to occur. Considering the production of truth under ideological conditions, demonstrates how "truth" is an emergent and ongoing endeavor. The voices which can "speak" in the archive, or are heard by those who access, are shaped by historical and cultural practices, which are then selected to speak further, extending their truth-claiming influence through works of scholarship and literature (Weidman, 2014, p. 37). To unsettle, extend or question who has voice, both in the archive, and in interpreting it. In this paper, divination becomes part of a wider assemblage of information to consider the violent past which produced the contemporary world, contributing to an "alternative archive in formation" (Aragüete-Toribio, 2017).

The Relational, Radical Participation, Dreams and Ships

Following Kovach (2009), this work is inherently relational, and so the rationale for this approach is introduced via the relational encounters with the seen and unseen worlds which preceded it and which are key to making sense of the story. Penny Bernard, whose work incorporates scholarly enquiry into the connections between beliefs in water divinities across Southern African contexts, with an auto-ethnographic account of her initiation as a diviner following a calling from these divinities, speaks to the deployment of a radical participation methodology driven by dreams as key guides to the research route. Bernard's dreams, a central feature of the training process for diviners, opened new avenues for understanding the phenomenon of water divinities (2010, p. ii). Samia Khatun (2018), whose dreams plotted her research trajectory across the Australian outback, traces an alternative history of Indian Ocean mobilities and the relational intersections of aboriginal and Bengali persons and religious structures, challenging the claims of superiority by European knowledge traditions over the epistemological traditions of colonized people. While ernard describes such an approach as a radical participation method, Khatun offers an approach which is more matter-of-fact in its deployment of a range of epistemological processes rooted in indigenous or colonized people's repertoire of "coming to know"—dreaming, poetry—as a means to track the material traces of marginalized histories.

I came to know Rev Thebe Shale following a move to the Eastern Cape for a job as lecturer in an anthropology department at a local university. The trajectory to this post had not only been informed by professional commitment to the craft of anthropology. A prolific lucid dreamer, my dreams had begun to take on a more insistent and directive quality while I was still living in London, mother to 3 young children, married to a South African whose lineages were even more diverse than mine, spanning Asian, African and European origins, living in social housing and with a precarious part time position on the margins of the academy. These dreams had explicitly directed me back to South Africa, the country I was born in, the first generation of my lineages—drawn from Northern European, Iberian and North African sources—to be so, though colonial routes had brought my mother as the child of an English doctor in the colonial service to what was then the British protectorate of Basotholand in the 1950s, and my father's people to the Portuguese colony of Mozambique in the early twentieth century, where he was born in 1940 in the slave trading station that became the city of Lourenco Marques, later Maputo.

As I surrendered further into the instructions coming from the dream world, a professional pathway opened alongside, with a post doc first taking me to Bloemfontein and then a job offer at an Eastern Cape University. Enquiries via divination into the deeper intentions of the dreams saw me connecting with a number of traditional healers. In one divination, ancestors came forward from the oceans, the ones who had sailed before, as the diviner described them "ruthless dark men, with hearts of gold." Through my

12 Speaking with the Sea: Divination and Identity in South Africa

Portuguese heritage, I knew that my ancestors may very well had had involvement with the voyages of the fifteenth, sixteenth and seventeenth centuries along the African coast, and across the Atlantic and Indian Oceans, opening sea trade routes for Europeans to exploit and trade in valued resources, including people. Further investigations revealed the intersections and overlaps of these sailors with Barbary pirates, whose crews encompassed many Muslim Iberians affected by the post-Christian re-conquest of the peninsula. Left in a state of precarity or exile, piracy offered a route to survival, or revenge (Murphy, 2013). The murky distinctions between trading and raiding on all these ancestral seafaring fronts, were part of an ushering in of an era of unchecked accumulation for its own sake and concurrent environmental and social crisis, "…a world organized by self-devouring growth" (Livingston, 2019, p. 1).

The dream and divination led trajectory which had brought my family to the Eastern Cape had dropped us in a town where settler-colonial heritage narratives celebrated the site of the first cross erected by Bartholomew Dias, a Portuguese trader whose attempts to open a sea route to the markets of Asia ended in Southern Africa waters, as emblematic of the "age of discovery," and shipwrecks as sites of European heroes and survivors. The divination which had revealed the presence of these sailor ancestors, angry at my shame of their problematic legacy, and demanding of a more fulsome version of their story to be reached, asked me to make offerings to them, and the spirits of the ocean with which they were allied and protected by, along this coast where the two shipwrecks this paper is concerned with occurred. I had through supporting a ceremonial cleansing for a close friend, met Ntate Thebe Shale, a former Catholic priest turned traditional healer of baSotho ancestry and initiated in that lineage, though born in the Eastern Cape and fluent isiXhosa speaker, attesting to the more porous nature of boundaries clearly delineated by provincial lines and ethnic descriptors. It was to him that I turned with questions about the stories I found as I explored accounts of these wrecks following the lead from the divination to go deeper into the ancestral connections which had brought me to this coastline, in order to access alternative frames of reference for the events of shipwrecks and their aftermaths to those found in the archive.

How might a speculative methodology such as the one outlined above, driven by divination, informed by a radical participation via dreaming trajectories and ancestral enquiry and held in check by the cautionary voices of subaltern studies, divine the stories of the ships that were wrecked? That carried enslaved people, and crews whose own trajectories were mired in the miseries of hierarchical societies whose elites enforced their privilege, and the expansion of their hold on the wealth that came out of trade route of expansion ("Discovery" was always about this, rather than romantic notions of human endeavor), with violence? How might such channeling engage those for whom rescue was not salvation, and whose presence sits in the dark spaces of where the archive is silent or sparse? How might we work with indigenous African modes of enquiry such as divination as an extension to the traditional tool kits of the humanities and social sciences, where indigenous frameworks of coming to know are a, "…methodological option for examining a problem" Kovach (2009). The problem posed here is two fold. Firstly, how to honor and speculate on the narratives that cannot be recovered from the wrecks of Portuguese slave ships along the South African Southern Indian Ocean Coast. Secondly, addressing the problem of colonial and racist approaches which have dominated accounts of shipwrecks and their survivors which are inherently violent in their silencing of other stories and ways of knowing, thus enforcing a narrow view of modernity as progressive, processual and centered on narratives which emerge out of triumphant European-elite accounts of history.

We approach these problems through engaging with the cosmological frame which informs indigenous knowledge about the ocean on this coast, along with accounts recovered from the wrecks of Portuguese slave shipwrecks, the Santo Alberto, Sao Joao Baptista and the Belem, as they relate to two characters in particular who appear on the sidelines of the archive. These are the stories of an enslaved person of South Asian descent, possibly the Malabar coast, known both as "Antonio" and "Mangabome", and a black Portuguese crew member "Diogo", whose new status as rainmakers among the indigenous groups they settled with is noted as an aside in survivor's accounts.

Many post-wreck trajectories, and in particular those ships which carried large cargos of enslaved people, did not follow the route of "survivor" finding "rescue." For many, the wreckage of ships provided a possibility for another kind of life, in an ontological reality, which at the time of the wrecks in the sixteenth and seventeenth centuries still arguably existed on its own terms in the period prior to the full realization of colonial domination and extractive violence which would have far reaching ripple effects on Southern African societies through the wars, colonial expansion, increased slaving activity and power struggles over the fruits of trade in the eighteenth and nineteenth centuries (Cobbing, 1988).

In this paper, which tentatively and experimentally considers how to bring historical accounts into conversation with traditional technologies of enquiry within African spiritual traditions, we consider two stories from the edges of

12 Speaking with the Sea: Divination and Identity in South Africa

the archives, whose post-wreck trajectories offer a glimpse of the alternatives to rescue considered by those on the margins of the societies which constituted the social world of the ships. In the archives, these possibilities appear as asides and fragments, but they offer a glimpse of alternatives to the post-wreck social groupings which formed seeking to re-constitute the racial and class hierarchies of onboard society. We are thus concerned with the alternative social trajectories wreckage provided to those marginalized groups onboard ships—enslaved people and the working crew described broadly as "Portuguese," but whose origins and identity affiliations would have encompassed the broader Iberian, Mediterranean, African and Asian worlds from which the identifiers "Portuguese" and "Spanish" arose in the post-Moorish era. In this time, forms of Christian power and ideology worked to reduce the scale and influence of a religiously and racially diverse population, in a process which can be linked to the formation of an idea of the European as religiously Christian and racially white (Barros, 2011). While colonial accounts of maritime history from the fifteenth to nineteenth centuries emphasize the European as the primary character, the ships which traversed the trade routes of the Indian, Atlantic and Pacific Oceans were populated by sailors whose origins lay in Africa, Asia and the indigenous of the Americas, as well as Europe, and frequently included enslaved people as cargo, and crew (Cassidy, 1959; Mellet, 2020; Smallwood, 2007). Accounts of these sailors, where they do appear, are rooted in European conceptual frameworks and ideologies which can only offer glimpses of the diverse trajectories, origins and cosmological viewpoints of those onboard European ships. The very category "European," as distinct, special, white, is also thus thrown into disarray, as the invisible threads are revealed (Mellet, 2020). The proposal to work with Portuguese slave shipwrecks, the waters they lie in, and the people who are part of the story, asks a number of distinct, but aligned, questions with implications for both African diaspora and Portuguese identity and heritage, as well as broader maritime history.

Who were the crews of these ships? How has maritime Portuguese heritage been constructed to invisibilize the multi-racial and cosmopolitan make up of crew as part of a more wide ranging post-Moorish ideological project to erase North African/Muslim/Jewish connections and entanglements, emphasizing European/Christian lineages only (Barros, 2011)? Where else has this white-washing occurred? Elizabeth Currie, drawing on pre-Columbian iconographic and ethno-historical information to explore the worldview of land and water held in the Andean region, "…a different kind of 'native space', the sea - above and below the waves - and how it might have shaped ethnic identities and consciousness." She draws on the case of Pedro de Cama, of the seafaring Manteño peoples of the south central Ecuador, "In his petition to the Crown of Spain, Pedro de Cama declared himself to be a sailor, diver and 'man of the sea' and this is how the Spanish understood him, as an Indian 'ladino' - conversant with Spanish ways and language - their 'friend', who provisioned the ships of their navy, who sailed in their galleons, and who, on occasion, even saved their ships from sinking. His account is firmly situated within a Eurocentric conceptual framework; his credentials are those of a seaman from within European colonial definitions of that role. However, from an ethnic Andean perspective, we are reminded that space should include more than mere earth-bound notions of territory, that the sea itself constituted its own framework of experience." In asking these questions, I do not wish to evoke a cosmopolitan paradise hidden from view by Eurocentric accounts of history, nor draw clearly defined distinctions between victims and perpetrators. The stories are far more complex. Like Sephardic Jewish Iberians, fleeing the inquisition, formed colonies in the Caribbean where they held enslaved persons as property, while also negotiating the colorism and prejudice of Northern European Jews (Vink, 2010). I do wish to highlight the far more diverse and complicated range of persons who participated in this society, one characterized by profound violence, injustice and greed, which in its erasure of the people of color who were part of its origins, constructed a mythology of the White European Explorer.

As well as the trajectories of those who survived and chose alternatives to "rescue" in Eastern Cape shipwrecks (Crampton, 2004; Hayward, 2020; Vernon, 2013), there is also an increasing turn in maritime archeology concerned with slave shipwrecks to acknowledge, honor and ritualize grief for those severed from their places of origin, exemplified in the work of the Slave Wrecks Project, as well as position analysis of the archeological within decolonial frames which decenter Eurocentric domination of how archival and archeological evidence is interpreted and retrieved (Boshoff et al., 2016; Franklin et al., 2020; Jean et al., 2020). The recent archeological dive of the São José slave shipwreck off the coast of Cape Town, a joint endeavor by the Smithsonian Museum of African American History and Culture in the USA, Iziko Museum in Cape Town and Divers With A Purpose, part of the broader Slave Wrecks Project, was notable for the inclusion of a practice of grieving and spiritual acknowledgment in the project design. Soil from present-day Mozambique, where the 400 slaves onboard the ship had been captured, was brought to the waters over the São José wreck site during a memorial ceremony honoring those who lost their lives or were sold into slavery and bringing their story back into public memory (Boshoff et al., 2016). To use such tangible elements as earth from a significant place and water as a means of fostering connection between this world and the other and also to provide passage for spiritual entities which may require relocation or acknowledgment, is rooted in traditional African healing modalities in which the connection with the Other World, mediated through ancestral and elemental entities, is foundational.

Racism and colonial violence is implicit in the archival record of shipwrecks along the Indian Ocean Coast of the Eastern Cape, and accounts of these wrecks and the journeys taken following them to obtain rescue. Thus the stories of Mangabome and Diogo appear on the margins of these accounts which privilege a European elite perspective on who the primary characters of the narrative are, and what their ultimate purpose is—always, rescue and a return to "civilization." How can we work with indigenous technologies of enquiry to preserve, protect and learn from underwater cultural heritage in Africa? Can we bring the worlds of modern methods, colonial archives and indigenous technologies, such as divination, into conversation and alignment?

Concerning the Ocean, Shipwrecks and Making Rain: A Divinatory Conversation

To others it might be a belief, to me it is what is real. There are people who live under the ocean." These are Thebe Shale's words as we begin our conversation, channeling the information that came via his experience as one who has been taken underwater, indigenous knowledge keeper and diviner through whom information was transmitted as we discussed the cosmology of the Ocean from the perspective of Eastern Cape indigenous knowledge. We began by talking about what local conceptions of the ocean encompass, and what this perspective can offer to peeling back layers of understanding as to the nature and significance of slave shipwrecks, and their aftermath, on this coast. "The patent of the world and the ocean. They said the ocean typifies numinous living. As apart from the world.

Thebe situated himself immediately in relation to the tensions of rationality, and what this means, for those who move between worldviews which are mutually exclusive, and for who the viewpoint of the "other" is a spectacle of belief or misunderstanding. Thebe, who prior to his calling and initiation into an indigenous medicine tradition, was a Catholic priest, who spent several years serving in Rome, moved between languages of Catholic theology and indigenous cosmology as he spoke about the ocean, and its spiritual significance. Over the course of our conversation, the issue of working with and connecting to the intangible remains of those who died in the shipwrecks, and those survivors who appear as asides in accounts of post-wreck journeys, came to the fore, along with concerns for their contemporary descendants and the unfolding local water crisis, and wider global ecological crisis.

We begin with a discussion of the world under the water, the cosmology of the ocean: "Toward the center of the earth there is so much heat, the formation of the oceans moves from the heat of the center. The water compresses the heat. Under the water mirrors what is on top." Thebe spent two days to a week under the water, when he was taken by the people who live under the water. He met the people there and was introduced to their world, an experience common in the experience of being called to work as a traditional healer, some of whose training and initiation takes place under the waters of rivers, pools and ocean, where the spirits who reside there will call them for initiations (Bernard, 2010). As he puts it, describing the mirror world that exists beneath the waves, "The Ocean is a life system that mirrors the life system on land. One must move to eat, to collect food, to live. The (the people) did not seem naked, I could not say what they were wearing but they were clothed."

We discussed the events of shipwrecks on a coastline where such entities are in residence, and where the spirits of and in the ocean maintain a strict moral code, and what the consequences of this would be for those who died, and those who survived, the wrecks,

> The people who have been in the shipwreck are transformed. They mutate. There are strict codes of conduct in the ocean. Some of the ships were wrecked because they were on the edge of two kinds of life, and were not welcoming or respecting the Other World. This makes the ocean angry, or rather to react, an upward move. The Ocean always has highs and lows. Those who died in the shipwrecks are not dead, they are assimilated into underwater society.

The process of assimilation is not one which is inevitable, and those invited to assimilate are offered a choice, "They must decide if they are to be assimilated into the underwater society, or stay on the fringes and become 'ocean wanderers', naughty, silly, cheeky, even evil ones who haunt the oceans." The ships which sink and are wrecked in the inter-tidal zones are especially troubled by this outcome of troubled spirits, "It is better to sink deep or be excavated." Thus, those wrecks which occur in inter-tidal zones are, from the perspective of this cosmological frame, more in need of intervention to honor, sooth and assist in moving on from the state of being in-between worlds. Thebe speaks further on the destinies of those involved in the shipwrecks at all levels of onboard society, and the various ways in which they

12 Speaking with the Sea: Divination and Identity in South Africa

may become settled or unsettled spirits of the sea, or inland "Those people who desire home are almost in the palm of our hands. We who are living here can extend our assistance to them to provide what they need because they are still human. We are able to assist them to come to freedom, to experience the fullness of their lives."

We speak of the particular experience of those who are enslaved onboard ships that are wrecked and whose bodies do not survive,

> Their spirits can move onto land, though it is rare to wander deep inland. It is a choice they make. Most of the enslaved ones remain as Ocean People, usually much more powerful, or weave their way into the rivers inland. They are very influential. With the aid of the rain, they can fly to distant places. They have an unassailable ability to move.

This "unassailable ability to move" is in stark contrast to enslavement, with its restrictions on the freedom to move and act. There is also a level of responsibility on the part of those living to provide assistance wherever possible to the enslaved victims of shipwrecks who may not have been able to find a home with the water people, and are in the in-between state, "We would not want those who were gone in the ocean, especially those who were forcibly onboard, to be perpetual victims. They need a robe to cling to."

The requirement to act, to tell the stories as they are divined and revealed on the edges of the archives, of the enslaved, the crew, the elites, those who went under with the ship and those who made it onto land, is foregrounded by Thebe, unsettling the distinction made between "survivor" and "victim." These insights corroborate efforts now made to connect in spiritual, as well as material ways, with slave shipwreck sites,

> Both those who went under, and those who came out alive in body. Both are survivors in a different perspective. Some fare well, some don't. Both sides see casualties. The casualties can cause havoc –lingering sadness, regret, anger – not anger per se, more precautionary, it is dangerous to play on that line. So it shakes us up a bit to wake us up a bit. The onus is on those who are aware to do their little bit, to provide peace, settlement. What we do to the ship, to the site of the wreck, will reverberate.

We turn to the issue of those who survived, those who made it onto land. The ones who survived well, and the ones who did not. Thebe emphasizes affinities with the water which assist in determining the good fortune, or not, of post-wreck trajectories on land, but also that this affinity with the water extends to those who remain in the water and are integrated into the underwater society. Once more, the distinction between living in the physical body, or being in the form of spirit, is porous.

> Those who survived well had an affinity with the water. Both those who went under, and those who came out alive in body. Both are survivors in a different perspective. Some fare well, some don't. Both sides see casualties. The casualties can cause havoc – lingering sadness, regret, anger – not anger per se, more precautionary. It is dangerous to play on that line. They shake us up a bit to wake us up a bit. The onus is on those who are aware to do their little bit, to provide peace, settlement. These people must not feel overlooked or outcast. They are not by accident. We are not by accident. We are regularizing and straightening up the crooked ways. What we do to the ship will reverberate.

The crooked ways are the skewed version of history held in the archives and also the long-term ecological consequences of the kinds of trade and accumulation ideology carried with those who orchestrated the voyages.

> We have wrong scales of measuring the scales of our actions. Ecological impact of colonialism linked to unchecked capitalism. This we feel as drought and lack of water in the Eastern Cape. The ships are the start of this. Diogo as the rain maker comes as a messenger, a warning against controlling nature rather than collaborating with nature. Be warned! Don't go further than that! There would still be hope if we just listen and ask for directions from Above and Below. The Ocean has more answers.

Becoming a Rainmaker

Diogo is a sailor onboard the Santo Alberto, a Portuguese slaver wrecked in 1593 near the Umtata River. We obtain a glimpse of Diogo through the lens of the account written by Francisco Vaz d'Almada, a Portuguese soldier and officer who had served in India, and was a survivor of another Portuguese slaver, the São Joao Baptista, wrecked in 1622 near present-day Canon Rocks. Gillian Vernon, an Eastern Cape historian who tracked the trail made by shipwreck survivors for her popular account "Even the Cows were Amazed" recounts the story, "They met a survivor from the Santo Alberto who assisted them, and heard another survivor was resident there. He was a Portuguese named Diogo, who was married with children and had the reputation of being a rainmaker. When they went to his house, Diogo refused to come out" (Vernon, 2013, p. 59).

Diogo briefly enters the archive but does not linger. Others who have stayed and made new lives following wreckage, eschewing rescue, also appear,

12 Speaking with the Sea: Divination and Identity in South Africa

and disappear, as d'Almada's account continues. Their presence is a testament to the multi-racial character of the social world onboard ships, refuting the narratives of white exceptionalism which characterize interpretations which draw on these historical accounts from paintings to popular histories exploring the terrain of shipwrecks and survivors, "Near the Mdumbi River they came across a fisherman, a 'black Portuguese' who lived in a hut on the coast. He showed them where to find mussels and they also gathered berries" (Vernon, 2013, p. 82). Another "black Portuguese" at the Kei river shows them where to ford safely to the opposite bank.

Survivors from a later shipwreck in 1635, the Belem, also met a Santo Alberto survivor described as an ex-slave who lived along the Mzimvubu River. Again, making a brief appearance on the edge of the archive, we are offered a glimpse of the ways in which post-wreck survivors who were not rescued fashioned new lives, entering local idioms of political power to become Rain Makers, which Thebe locates as gifts from the water, "Originally named Antonio, and described as a Christian Cubra (Indian Servant), he had been left as a very small child and was now called Mangabombe. He could understand and speak Portuguese, had a reputation as rainmaker and was said to be very wealthy" (Vernon, 2013, p. 119).

The accounts from the Belem indicate a recognition of the potency of rain making as a tool of power, as survivors seeking shelter and kudos in villages claimed it as their own while in Mangabombe's land. When locals asked if they could provide rain, and the fortuitous appearance of a thunderstorm at the same time occurred, these folk claimed the credit for the water, reaching for some of the power and authority invested in rainmakers. When these same "rain makers" asked for shelter from the storm, they were told to make it stop themselves as they could confer rain whenever they wished (Da Costa, 1999 in Vernon, 2013, p. 161). Perhaps, huddled and wet through the night, these people realized that they were not among the ignorant, to be manipulated and trifled with, and that claims of rainmaking came with the requirement to follow through.

The fragments that appear in archival narratives, and their popular interpretations, provide glimpses of the other stories that run alongside and with dominant narratives of shipwrecks and their aftermaths. The challenges of researching stories and knowledge that might exist in oral narrative and memory, appearing on the edges of archives, are an opportunity to utilize alternative methodological approaches which decenter the colonial archive.

I have asked Thebe about Diogo and Antonio/Mangobombe, the two who appear in shipwreck accounts and are noted specifically as rainmakers. My question concerns the possibility of the experience of surviving a shipwreck, which means to emerge from the ocean being one which might confer some authority or enhance gifts of working with water. The rainmaker was, and remains, a significant figure in African political and social life, one which encompasses both power and precarity, as rain is a critical factor in all aspects of sustaining agriculture, livestock and life itself (Colson, 1948; Landau, 1993; Livingstone, 2020; Murimbika, 2006; Sanders, 2008; Schapera, 1971; Wilson, 1959). As the unfortunate survivors of the Belem might have realized when they tried to claim the influence of a skill they did not have, failing to appreciate the socially significant realm within which rain making occurs. That they were asked by locals, according to the account that survives, if they held that power, suggests the possibility that the autochthonous inhabitants of the land saw some connection between the emergence from the sea as the survivors of wrecks, and the capacity to have the special affinity with water required by those with the power to command rain.

We explore cosmological concerns initially, but 40 mins into our session, Diogo appears, channeled through Thebe,

> There is a man who comes. Water is water. It has a capacity and an ability we have not even started to tap. Diogo was a good receptacle to the water element which connects to both the liquid state and the glacial state. The water in the ocean made him a prisoner – too much water there already. The water sent him out as an emissary to become a rain maker.

Thebe identifies an ancestral connection between Diogo and the Rain Queen Modjadji, traditional ruler of the BaLovedu people of the area of South Africa now known as Limpopo Province, considered a living embodiment of the rain goddess whose emotional embodiment of the qualities of rain guarantees that they return in the wet season (Murimbika, 2006; Semenya, 2013). A Rain Queen has occupied the BaLovedu throne since 1880, though the current Queen-in-waiting is embroiled in controversy over her ascendency, and the possibility of a BaLovedu King. "They share an ancestral connection. There is Give and Take in this transaction. He (Diogo) came to earth and was not welcomed." The Rain Queen's ascendency in 1800 post-dates Diogo's time by some 200 years, but, in our speculative way, Diogo's descendants, or journeys of Spirit into the interior, might have played some part in this story as the rain making skill commanded by Diogo that is suggested by his brief appearance in the archive found further expression through his descendent lineages.

Implicit in Thebe's commentary is a critique of the kind of society Diogo was born into, and the ways in which his personhood was shaped through his positioning within it. His gifts as a worker with water are not recognized or welcomed. But, in the drama of the shipwreck, the Ocean does see it, sending him out as an emissary. This turn of events, an aspect of destiny or *fado*, is in line with cosmic justice, and speaks also to the current climate crisis and drought in the Eastern Cape. As Thebe puts it,

> It is very great and very good. Almost salvific. Now is the time, not to complete, but a little start. There are dry seasons of gargantuan proportions in our world now. These water people are ready to help. The oceans will not run dry. The land will run dry, and we better do something about it. Even if Diogo is not looking at us, he is concerned for his offspring.

From the other world, Diogo speaks through Thebe to express concern about his descendants who now reside in a land of scarce water. There is also the issue of my descent, as the enquirer who has found the story, Thebe expresses that there is a familiarity between me and Diogo—a shared ancestry—that requires me to "Leave everything and go for it" with regard to exploring the possibilities of his story for my own.

Thebe is concerned that I am too stuck on a particular idea of what a person can be, looking for Diogo in the past, when he continues to communicate in the present, "You have an idea of a man, and what does not fit this idea is excluded. Diogo has a rare gift to make rain. We twist and turn to make the powerful agree with us, Diogo refuses to sell his soul in this way." Thebe is speaking to the moment in the archive when Diogo makes his brief non-appearance—refusing to come out of the house and new life he has made when the survivors of the Santo Alberto come looking for him.

Critique of the extractive capitalism and over-consumption put in motion by the early voyagers of ships like the Santo Antonio and Sao Joao Baptiste, is once more foregrounded in the narrative Thebe conveys, "The ship owners were just doing business, disconnected from the dirt. The Portuguese were the ocean masters in those days. Absolute power without responsibility. The aim and purpose now is justice for the untold stories. How do we understand justice? Normality. Fairness. Equity. We work with and for the equity of all living beings."

Can You See the Power of Water?

Thebe tells me that Diogo is also a great philosopher. Diogo speaks. "Are you afraid of the appearance or the reality of water?" He asks him, as hail falls in East London where Thebe is staying as we speak over the phone. Appearances can be over magnified, and we almost lose the context and reality of what it is because of what appears. Fact from Fiction. Appearance from reality. We cannot recover these stories with any certainty. And yet they speak. They speak for our benefit, so we can imagine a different way of inhabiting our lifetime, and generating other future possibilities.

Thebe continues to channel and interpret the messages that come through, "We have so few ears to listen, and so many other voices. Sometimes it can be frustrating! We are sure that the very things we are saying here have been said before and are forgotten. But we do our little bit." Building rapport with each other, and with them, the ancestors who call us. "Mangobombe cautions, be very careful when we talk about Ontology, things as they are, the thing as it is. You don't know a man, you have an idea of a man, and anyone who fits into that idea, I classify as man. You think you know, you just don't know," Thebe concludes, "the rest as we go, there is more to come, more to add. The essence is her. And so we move."

Conclusion

This chapter has briefly sketched out preliminary possibilities for a conversation between slave shipwreck accounts in the colonial archive and indigenous ontologies, in order to speculate on how the Sea might speak of these historical events, and their aftermaths. In doing so, the primary marker of "coming to know" is shifted from a European-elite defined epistemology, decentering the primacy of this way of knowing. As Kovach puts it, there is a riskiness inherent in this line of work, what she describes as feeling like we are "not allowed" to do this, an unease, "…born of a colonial history that shadows our being" (Kovach, 2009, p. 1). And yet, it is only through attempting to bring diverse knowledge systems into sites previously claimed for a euro-centric rationality that we might, "…move beyond the binaries found within indigenous-settler relations to construct new, mutual forms of dialogue, research, theory and action" (2009, p. 1). Here, divination and the expression of those who wish to be heard, alongside cosmological frames which introduce the Sea as a relational entity who speaks through processes such as divination out of which knowledge is generated, is both complementary to the knowledge in the archive, and critical of the silencing inherent in the colonial archive. To divine on fragments of stories and accounts of slave shipwrecks along the southern Indian Ocean coast, foregrounds the incompleteness of the archive, the violence implicit in its making and move toward an appreciation of the many ways we can come to know, and the complementarity of knowledge systems in coming to know.

References

Aragüete-Toribio, Z. (2017). *Producing history in Spanish civil war exhumations: From the archive to the grave*. Palgrave Macmillan.

Barros, M. (2011). Christians and mudejars: Perceptions and power in medieval Portuguese society. *Imago Temporis—Medium Aevum, 5*, 135–147.

Bernard, P. (2010). *Messages from the deep: Water divinities, dreams and diviners in Southern Africa* [Unpublished PhD thesis, Rhodes University].

Boshoff, J., Bunch III, L. G., Gardullo, P., & Lubkemann, S. C. (2016). *From no return: The 221 year journey of the slave ship São José*. Smithsonian Books.

Callaway, H. (1872). On divination and analogous phenomena among the natives of natal. *The Journal of the Anthropological Institute of Great Britain and Ireland, 1*, 163–185.

Cassidy, V. H. (1959). Columbus and "The Negro." *The Phylon Quarterly, 20*(3), 294–296. https://doi.org/10.2307/273057

Cobbing, J. (1988). The Mfecane as Alibi: Thoughts on Dithakong and Mbolompo. *The Journal of African History, 29*(3), 487–519.

Colson, E. (1948). Rain-shrines of the plateau Tonga of Northern Rhodesia. *Africa: Journal of the International African Institute, 18*(4), 272–283.

Crampton, H. (2004). *The sunburnt queen*. Jacana

Da Costa. (1999). *The Itinerário of Jerónimo Lobo*. Routledge.

De Boeck, F., & Devisch, R. (1994). Ndembu, Luunda and Yaka divination compared: From representation and social engineering to embodiment and worldmaking. *Journal of Religion in Africa XXIV, 24*(1–4), 98–133

Franklin, M., Dunnavant, J. P., Flewellen, A. O., & Odewale, A. (2020). The future is now: Archaeology and the eradication of anti-blackness. *International Journal of Historical Archaeology, 24*, 753–766.

Hartman, S. (2008). *Lose your mother: A journey along the Atlantic slave route*. Farrar, Straus & Giroux.

Hayward, J. (2020). Orality in the digital age. In R. Kaschula & H. Wolff (Eds.), *The transformative power of language: From postcolonial to knowledge societies in Africa* (pp. 277–303). Cambridge University Press.

Jean, J. S., Joseph, M., Louis, C., & Michel, J. (2020). Haitian archaeological heritage: Understanding its loss and paths to future preservation. *Heritage, 3*(3), 733–752.

Khatun, S. (2018). *Australianama: A South Asian Odyssey in Australia*. Hurst.

Kovach, M. (2009). *Indigenous methodologies: Characteristics, conversations and contexts*. University of Toronto Press.

Landau, P. S. (1993). When rain falls: Rainmaking and community in a Tswana village, c. 1870 to recent times. *The International Journal of African Historical Studies, 26*(1), 1–30. https://doi.org/10.2307/219185

Livingston, J. (2019). *Self-devouring growth: A planetary parable as told from southern Africa*. Duke University Press.

Livingstone, J. (2020). *Self-devouring growth: A planetary parable as told from southern Africa*. Duke University Press.

Mellet, P. T. (2020). *The lie of 1652: A decolonised history of the land*. Tafelberg.

Murimbika, M. (2006). Sacred powers and rituals of transformation: an ethnoarcheological study of rainmaking rituals and agricultural productivity during the evolution of Mapangubwe state, AD 1000 to AD 1300. Unpublished PhD Thesis, University of the Witwatersrand.

Murphy, M. N. (2013). The barbary pirates. *Mediterranean Quarterly, 24*(4), 19–42. https://www.muse.jhu.edu/article/532906

Nyamnjoh, F. B. (2012). Blinded by sight: Divining the future of anthropology in Africa. *Africa Spectrum, 47*(2–3), 63–92. https://doi.org/10.1177/000203971 204702-304

O'Hanlon, R. (1988). Recovering the subject: Subaltern studies and histories of resistance in colonial south Asia. *Modern Asian Studies, 22*(1), 189–222.

Peek, P. (2013). Religion in Africa and the diaspora: The silent voices of African Divination. Available at: https://bulletin.hds.harvard.edu/the-silent-voices-of-african-divination/#Notes. Accessed 03 October 2021.

Said, E. (1978). *Orientalism*. Pantheon.

Sanders, T. (2008). *Beyond bodies: Rain-making and sense-making in Tanzania*. University of Toronto Press.

Schapera, I. (1971). *Rainmaking rites of Tswana tribes*. Afrika-Studiecentrum.

Semenya, D. K. (2013). The making and prevention of rain amongst the Pedi tribe of South Africa: A pastoral response. *HTS Theological Studies, 69*(1), 1–5.

Sharfman, J., Boshoff, J., & Parthesius, R. (2012). Maritime and underwater cultural heritage in South Africa: The development of relevant management strategies in the historical maritime context of the southern tip of Africa. *Journal of Maritime Archaeology, 7*(1), 87–109. Retrieved February 24, 2021, from http://www.jstor.org/stable/43551371

Smallwood, S. (2007). African guardians, european slave ships, and the changing dynamics of power in the early modern Atlantic. *The William and Mary Quarterly, 64*(4), third series, 679–716. https://doi.org/10.2307/25096747

Spivak, G. (1988). Can the subaltern speak? In C. Nelson & L Grossberg (Ed.), *Marxism and the interpretation of culture* (pp. 271–313). University of Illinois Press.

Tedlock, B. (2001). Divination as a way of knowing: Embodiment, visualisation, narrative, and interpretation. *Folklore, 112*(2), 189–197.

Vernon, G. (2013). *Even the cows were amazed: Shipwreck survivors in South East Africa, 1552–1781*. Jacana

Vink, W. (2010). *Creole jews: Negotiating community in colonial suriname*. Brill

Weidman, A. (2014). Anthropology and voice. *Annual Review of Anthropology, 43*(1), 37–51.

Wilson, M. (1959). *Communal rituals of the Nyakyusa* (1st ed.). Routledge.

Part III

Environmental, Legal and Political Realities

13

Selo! Oral Accounts of Seychelles' Maritime Culture

Penda Choppy

Introduction

Selo! I remember this cry from my great grandfather's accounts of his life on the outer islands of Seychelles. Usually, when my great grandfather (or *Peper*, as we used to call him) shouted 'selo!', he was telling stories of how this cry that signalled the arrival of a boat, was nothing more than ghosts luring victims out into the dark. 'Selo!' is the creolized form of the English maritime cry that signals the sight of a sail, thus 'Sail ho!' The D'Offay and Lionnait dictionary defines it as the cry which signals the arrival of a boat in the outer islands (1982, p. 352). To me, this cry also defines the maritime culture of the Seychelles, from its first establishment, its history as a slave society and its economic activities, past and present, which are determined by its being an isolated archipelago of very small islands.

Seychelles has a total land area of only 445 km² which is distributed among 115 islands. However, the archipelago has an Exclusive Economic Zone (EEZ) of 1.37 million km², which makes it a very big territory indeed,

---

The research presented in this chapter is supported by the University of Seychelles, Mahe Island.

P. Choppy (✉)
University of Seychelles, Victoria, Seychelles
e-mail: penda.choppy@unisey.ac.sc

but made up largely of ocean (SMSP, 2015).[1] Accordingly, its economic activities are marine-centred, namely, tourism and fisheries. 21% of the Seychellois population are engaged in the tourism industry which accounts for about 25% of the country's GDP (Philpot et al., 2015, p. 32). Seychelles is known as a top destination for island getaways, and its main branding is the quality of its pristine white beaches and its blue, blue sea (Burridge, 2017). As for fisheries, about 4% of the population get their livelihood from artisanal, or traditional fishing, while the tuna processing industry is responsible for 15% of the country's GDP, with16% of the population working in the industry (Clifton et al., 2012; Philpot et al., 2015) (Fig. 13.1).

Naturally, the Seychellois's main source of protein is traditionally fish, and many of their traditional dishes are derived from the sea. With the outer islands too scattered to be permanently inhabited, none of the inhabited islands are large enough or flat enough to allow large-scale farming and animal rearing, which increases the population's dependency on the sea. Things have to come either directly from the sea, or across it, as most consumer goods have to be imported (Skerrett et al., 2019).

These are the conditions that the islands' first settlers had to contend with in 1770 (Campling et al., 2011, p. 17). However, it might have been these very conditions, combined with the economic and political activities of the time, that is, colonialism and slavery, which attracted these settlers. There are two defining moments in the history of Seychelles which determined

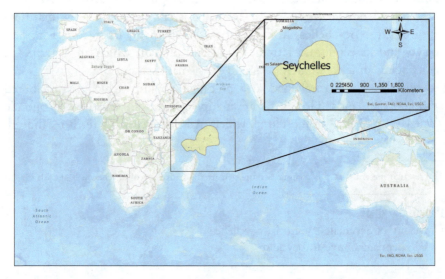

Fig. 13.1 Seychelles exclusive economic zone

the nature of its population and culture. First, its isolation and geographical position made it militarily and commercially strategic for the French, who wanted to use it as a provisioning station between the East (India, especially) and the coast of Africa (Nicholls, 2018, p. 58). To produce these provisions, slaves were imported from Madagascar and East Africa, but the islands also represented an opportunity for adventurous French colonials to enrich themselves through a plantation economy, again, using slave labour. Thus, a creolized culture emerged, of which a French-based creole is an example, which symbolizes both the merging of disparate cultures (African and European), and the insularity of the habitat where this language evolved, since eventually, even the descendants of the French colonials used it or where influenced by it (Choppy, 2020a, p. 60). Second, again for strategic reasons, the British wrested the islands from the French in 1814, and were responsible for enforcing the abolition of slavery in the region (Benedict & Benedict, 1982, p. 129). The archipelago's geographical location made it ideal for depositing the slaves which the British navy had liberated on the high seas while enforcing abolition (Benedict & Benedict, 1982, pp. 131–133; Campling et al., 2011, p. 17). This further increased the African genetic pool of the Seychellois population, and enriched its cultural heritage.

This heritage is reflected in the oral expression of the people, and the sea has a strong presence in this orature. For example, the proverb *Dibri lanmer pa anpes pti pwason dormi* (The noise of the sea does not stop the fish from sleeping), which in his book on creole proverbs, Raphaël Confiant has interpreted as referring to the ability of living beings to adapt to their natural environment (2004, p. 18). This is indeed creolization in process in that it speaks of adaptation. It also reveals the Seychellois' use of metaphorical language to philosophize about life. Another proverb associated with the sea is *Torti i mor pour son lakok* (The turtle dies for its shell), which might be interpreted as a reflection on the exploitation of slavery. The Seychellois' close relationship with the sea is also present in their riddles: *Brans lo delo?* → *Zourit* (Branches on water? → The octopus); and *Latab lo delo?* → *Lare* (A table on water? → The manta ray) (Diallo et al., 2001, p. 31). Both sea creatures have a special place in the Seychellois creole cuisine (Choppy, 2018, pp. 101–102). Good things are also deemed to come out of the sea in Seychellois creole folktales. In *Renn Mer* ('The Queen of the Sea', categorized as ATU 563 in the ATU classification system), a fish gifts three magic objects to a fisherman which frees him and his family from poverty (Carayol & Chaudenson, 1978, pp. 132–141). In the story of *Kader* (no ATU classification), the most beautiful wife fit for a king and won as a prize by the hero, comes from the sea (Diallo et al., 1982, pp. 72–81).

However, the genre of oral expression with the most abundant references to the sea is the *moutya*. This is a form of song and dance which was probably brought to Seychelles by African slaves and transferred into the creole language and performance tradition (D'Offay & Lionnait, 1982, p. 273; Choppy, 2018, p. 115). The *moutya* was revived by labourers working on the outer islands of Seychelles during the 1860s when copra became Seychelles' main commercial crop (IDC, Seychelles).[2] This was some two or three decades after the abolition of slavery, and by then, the planters who had once prospered through slavery now either had to employ a sharecropping venture with their ex-slaves, or try their luck in the outer islands in coconut produce or guano extraction. Naturally, going out to the outer islands to work on a contract was the more attractive option to the ex-slaves and their descendants, since they were paid very low wages on the mainland (Franda, 1982, p. 16; Campling et al., 2011, p. 18). The isolation of the islands, the hard life of labouring and nostalgic memories of loved ones back on the mainland, were the perfect conditions for the revival of the *moutya* and its evolution from the lamentation genre associated with the slavery period, to the more exuberant social commentary more akin to the Calypso genre of the Caribbean (Choppy, 2018). The *moutya* lyrics are in themselves, a testimony to the archipelago's maritime history and culture. It is thus fitting that we should examine these lyrics for a better understanding of this history and culture.

Some Methodological Considerations

A friend of mine once said that if you want to know about a people's culture, look at their folklore, because songs, stories, riddles and proverbs are the people's own account of their history. Having grown up in a very traditional community and having participated actively in the community's cultural practices and performances (in my teenage years, I was a member of my community's cultural troupe), it is easy for me to relate to my people's history and social practices through its folklore. This is thus the reason why I chose to discuss the maritime culture of Seychelles through certain aspects of its folklore. There are two specific corpuses that I find particularly relevant to a discussion of Seychelles maritime culture. These are (i) a corpus of *moutya* lyrics, comprising of a collection compiled in 1982 (Diallo et al., 1982) by the Oral Traditions Section of the Culture Department (now the National Heritage Research and Protection Section) and another collection compiled in 2007 (Estrale) by the Creole Institute of Seychelles, and (ii) a corpus of memoirs collected from elderly members of various communities on the three main inhabited islands of the archipelago, Mahé, Praslin and La Digue, between 2002 and 2004. The latter corpus was compiled by Odile de Commarmond and Colette Gillieaux, as part of the UNESCO Slave Route project (1994), and The International Year to Commemorate the Struggle Against Slavery, and its Abolition, in 2004. Neither corpus has been analysed extensively, and certainly not from the maritime angle.

The data has been analysed qualitatively through a thematic approach, focusing on two main periods of Seychellois history which illustrate the role played by the ocean in that history. These are (i) slavery and its aftermath, (ii) the period of commercial exploitation of the Seychelles outer islands from the mid-nineteenth century up to Independence in 1976. This data might be considered as the auto-analysis of the labouring class of Seychellois society (and as such, mainly slave descendants), as it represents the feelings and perspectives of a cross-section of the Seychellois population. I have used extracts from the memoirs of the elders (De Commarmond & Gillieaux, 2004) to support my interpretation of the *moutya* lyrics, which I feel enhances their authenticity. It is important to note that both sets of data are primary data. In the opening section of the discussion of the data, I have also used Peter Nicholls' Ph.D. thesis *The Door to the Coast of Africa: Seychelles in the Mascarene Slave Trade, 1770–1830* (2018) as grounded theory to support my arguments vis-à-vis the centrality of the ocean and islandness in the historical events that shaped Seychellois culture.

Section 1: Seychelles' Strategic Position in the Indian Ocean Slave Trade and Its Oral Legacy

There is an intricate relationship between Seychelles as a scattered archipelago, the slave trade, and the development of the *moutya* form. In the first place, the *moutya* form itself was transferred from Eastern Africa to Seychelles through the slave trade in the Indian Ocean region. Bollee's etymological dictionary traces this genre back to the Makua region of Mozambique (1993, p. 333). This tallies with Peter Nicholls' theory that the interest in this archipelago was heightened when the main supply of slaves for the Mascarenes (Ile Bourbon and Ile de France) shifted from Madagascar to East Africa around the mid-eighteenth century (2018, p. 52). Nicholls argues that in fact, French slave traffickers in the Indian Ocean were using the archipelago as a provisioning station for the slave ships going from East Africa to Mauritius and Reunion (p. 32). Citing Edward Alpers (2009, p. 176), he describes the slave trade (both legal and illegal) between Eastern Africa and the Southwestern Indian Ocean islands as 'a tightly interwoven commercial nexus' (Nicholls, 2018, p. 22). Seychelles played a very crucial role in this nexus because its numerous islands could be used as a 'sanatorium' and 'acclimatization station' for slaves (Nicholls, 2018, p. i). Nicholls cites several works, that though are not dedicated solely to this theory, give weight to his argument that Seychelles had a central role to play in the Western Indian Ocean slave trade, namely, Toussaint (1965), Nwulia (1981), Allen (2014).

Secondly, the accessibility of East Africa and Madagascar from the outer islands of Seychelles made it possible for Seychellois planters and traders themselves to participate in slave trafficking in this region, both before and after abolition. De Commarmond and Gillieaux's data, which are accounts of stories of slavery, passed down orally, seem to prove Nicholls' theories on several counts. For example, he states that the outer islands of the archipelago 'became the centre of a wide-ranging smuggling network that drew on the outer islands of the archipelago to move East African and Malagasy slaves predominantly to Réunion' (Abstract, p. i). A 71-year-old informant named Jeanbaptitia Lenclume from the district of Port-Glaud recounted the story of a man who had actually been captured by a passing schooner while working on the outer islands (the date of this incident is unclear). Apparently, he and other Seychellois labourers were working there, but it seems they were mixed with Africans who had freshly been brought there and could not yet speak Creole. To distinguish themselves from the 'savages', as they called them, they decided to greet the captain of the boat in creole. This resulted in them being returned to the island while those that could not 'speak' (rather like Conrad's 'savages' [1899], the inability to understand the Africans' languages was taken to mean that they could not 'speak' and were thus uncivilized) were considered fair game and kept on-board (De Commarmond and Gillieaux, pp. 57–58). This suggests that local traffickers from Seychelles were playing foul and stealing each other's slaves that were stationed on the outer islands. Several testimonies also suggest that Seychellois traffickers were catching slaves off the African coast or the Northern coast of Madagascar, for their own use or for sale (D. Malvina, p. 34; C. Adrienne, p. 27; A. Payet, p.2; F. Hoareau, p. 43). What is more interesting, is that some testimonies suggest that this illicit trafficking was still ongoing in Seychelles as late as the early decades of the twentieth century! Adele Payet's testimony is the best example of this:

> I remember the time of slavery. When I was five years old, in 1925, I lived at Côte d'Or on Praslin. The slaves belonged to a big proprietor, Mr. Loy... Apparently they came from Africa (Malindi, Lamu). It was Mr. François Morel who was in charge of everything to do with going to catch the slaves. Apparently, he put a flag on the beach and waited for the people to come. Then he gave them wine and made them drunk. Next, he put them on the boat and brought them to Seychelles. The black people who were drunk were tied on the boat.
>
> My husband's father once went with Mr. François Morel. Mr. Morel said that they were going to the outer islands. Then he changed his route to Africa. The trip took seventeen days in a schooner.
>
> When they got to Seychelles, the slaves were sold to the big proprietors (Choppy, on Marianne island and Jean-François Hodoul at Carana on Mahé).
>
> (De Commarmond & Gillieaux, 2004, p. 2).

These underground activities on the outer islands were complemented by the supposedly legal slave transactions on the inner islands, which Nicholls calls a 'large scale abuse of the so-called transfer system', whereby thousands of slaves were 'acclimatized' in Seychelles before being transferred to Mauritius (2018, Abstract, p. i). Based on evidence gathered from the works of Nwulia (1981), Scarr (2000) and Allen (2014), Nicholls suggests that these activities on the outer and inner islands of Seychelles played a huge role in the suspicious population increase in the Mascarenes and Seychelles in the post-abolition period (2018, p. 17).

Thirdly, and because of all this trafficking, the slaves' experience in itself provided the material for the composition of the *moutya* lyrics, as a form of lamentation, as will be illustrated forthwith. This experience ranges from the accounts of the slave trafficking itself to nostalgic memories of 'Grannter' (the African mainland), and the miseries of life in bondage.

An example of a *moutya* lyric which might be interpreted as the nostalgic memories of Africa is *Mon Lans Deger* (My War Spear), which was recorded and compiled in 1981 and was transcribed and published in a booklet in 1982 (Diallo et al., 1982):

Aswar mon napa doumi	I cannot sleep at night
Zot aswar mon napa doumi	Oh I cannot sleep at night
Kan mon mazin mon lans deger	When I think of my war spear
Nek delo i vin dan mon lizye	Tears come to my eyes
	(Diallo et al., 1982)

This song can only have originated from the memory of a first-generation slave, since there are no known records of the war spear having been a weapon or tool in Seychellois culture. It is a lamentation because the song evokes something that is no longer available to the singer: 'I cannot sleep at night I When I think of my war spear'. An obvious interpretation of this lamentation is a slave who had once been a warrior, and who longed for his life back on the African mainland where he had been respected. While he would once have fought against his oppressors with his spear, he is now a 'spearless' man. The other songs that follow will show the nature of that spearlessness.

The following song, *Ma Leoni ek Pa Zafa*, is a direct reference to slave trafficking as it depicts the arrival of new slaves at the main port. It is one of the most well-known *moutya* lyrics in Seychelles as it appears in both the 1982 compilation (Diallo et al., 1982) and Estrale's, 2007 compilation. The version below is from my own personal collection, from my days in the Anse Etoile Cultural Troupe in the late 1980s.

Navir Soubwannan i antre (3x)	The ship Soubwannan came in
Ki marsandiz ti anmennen	What merchandise did it bring?
Ki marsandiz ti anmennen (3x)	What merchandise did it bring?
Ma Leoni ek Pa Zafa	Ma Leonie and Pa Japha

It is important to establish the fact that *moutya* songs, though they depict events or practices in real life, are also compositions that are subject to their author's interpretation of things, as well as any mispronunciations or poetic license they decide to take! Since no ship of the name *Soubwannan* is known aside of this song, one might conclude that its author either mispronounced the name or invented it to avoid being too direct. Much of the messages passed during a *moutya* performance is in code. Part of the attraction is the skill of decoding the messages, and to do that, one must be an initiate of the gathering, in other words, a member of the community (Diallo, in press). That aside, the song recounts the story of a ship which arrives with human merchandize, named Ma Leonie and Pa Japha. This is thus a story about slave trafficking in Seychelles—something that is solid history in both the memory of the local people (De Commarmond & Gillieaux, 2004) and published documentation, for example, Franda (1982), Scarr (2000), McAteer (2008), to name just a few. Certainly, the story of Ma Leonie and Pa Japha is known in the community where I grew up, where, according to oral discourse, they have descendants.[3] They turn up in Gillieaux and de Commarmond's compilation of memoirs as well, being cited as slaves who were personally known to one Henri Descombe (De Commarmond & Gillieaux, 2004, p. 32). Why should these personages be remembered in oral discourse? What in fact is their story? There is a hint in the last two couplets of the song—the fact that they were brother and sister is linked to something that was whispered in the market during the jackfish season. Jackfish is cheap and would naturally attract the poorest people, who are also the practitioners of the *moutya* (Estrale, 2007; Diallo, 1983). Bearing in mind the commercial nature of slavery, and the role of the *moutya* as social comment (Choppy, 2006), one may well be excused for concluding that the song is about mating a brother to a sister. The conflicting values of the Catholic moral code under which the slaves were supposed to be instructed and the lack of respect by the ruling class for their stature as Christians and human beings are thus at the origins of the social conflict in this song.

In contemporary performances, *moutya* songs celebrate our culture and identity, but at the same time, they reveal the trauma of the slavery system. Yet this revelation is only possible if one knows the stories behind the lyrics. Thus, the *moutya* becomes a medium for encoding taboo subjects, and because it is performed and passed down through the generations, also acts as an oral archive.

The powerlessness (or spearlessness) of the slave and his descendants in the pre- and post-abolition periods have never been more clearly depicted than in the following song entitled 'Gran Blan' (*Great White Master*):

Gran blan, Msye mon bourzwa		Big white man, my master
Gran Msye, donn nou nou lavi	2	Great sir, give us our livelihood
Lot fwa mon ti mank noye,		The other day I almost drowned,
Gran Msye, ti ape rod ou bouyon	6	Sir, I was trying to catch fish for you
Mon pa'n fer sa par mon leker		I did not do this because I wanted to
Mon fer sa pour sov ou lavi	12	I did it to save your life
		(Estrale, 2007)

Unlike the previous song which had a definite unspoken content behind what was spoken, *Gran Blan* makes no attempt to couch anything in innuendo. The song typically reflects many of the stories told by the older generation of what life was like for landless slave descendants. Since the man, who should have been the head of the family, did not have his own land or boat, he had to work for the white owner, as both labourer and fisherman. The second couplet describes the average Seychellois' concept of slavery, which is to work from sunrise to sunset (De Commarmond & Gillieaux, 2004). Ironically, this man has to toil in the soil and risk his life out at sea to ensure that his master's table is well supplied, while his own children eat 'dry food' (rice without meat or fish). To top it all, his wife is prey to his master's lust. This is indeed, a spearless man—without the means of protecting the only real assets he has—his family.

The majority of the data collected by de Commarmond and Gillieaux confirm this kind of existence for slave descendants. I shall take only two samples, to illustrate two of the most crucial lack of this class of people, that contributed to their helplessness. One is the testimony of Amelina Mondon (De Commarmond & Gillieaux, 2004, pp. 8–16). She recalls different members of her family going to prison because of stealing one coconut or one breadfruit to appease their hunger. The worst one was her own mother who got caught taking one coconut from the plantation where she worked, and was fined a sum that she obviously could not pay, and then sent to prison, leaving her children to fend for themselves. Amelina was eight years old at the time, and she remembers that her mother was made to walk from the South of Mahé, where she lived, to the other side, almost in the North, where the prison was. The event was so momentous, she said, that people made a song about it. And all because she had no oil and needed one coconut to cook the fish she had prepared for her family. This is no less than Victor Hugo's *Les Misérables* (1862) set in early twentieth century Seychelles! As the head of her home, Amelina's mother's attempt to fulfil her family's basic need of food was undercut by her lack of property, and the colonial laws that favoured the landowning class. The second sample is the testimony of Antoinette Belle Colett (born in 1916). She speaks of women and girls who had to work in the forest where the plantations extended, and who were made to oblige the foreman, or the plantation owner himself. Some girls were taken to the '*Gran Msye ki annan larzan*' (rich men) by a 'matron', who claimed that they were being taken to work. And then, if they had a baby, she said, life just went on! (pp. 17–18). Whereas the previous testimony illustrated the lack of ownership of one's labour, this one illustrates the lack of ownership of one's own body. Such was the nature of slavery in this tiny plantation society, which has had a lasting impact on its present society. It has left a legacy of mistrust between men and women, creating a society of too many single-parent families and absent fathers (Chang-Him, 2010; Choppy, 2020b). It has also left a legacy of mistrust between employee and employer which even in present times affect labour relations and productivity (Benedict & Benedict, 1982; Campling et al., 2011, p. 8). This, however, is not any different to the legacy of slavery in the Caribbean on both counts (Borilot, 2014; Mulot, 2000; Reddock, 2004).

In discussing slavery, accounts of the 'Middle Passage' tend to refer to the Caribbean and the Americas only. The Indian Ocean crossings and slave routes are often forgotten, or are not known (Haring, 2005; Ottino, 1974; Vergès, 2007). And yet, slave trafficking by sea in the Indian Ocean region preceded the Atlantic Middle Crossings (Campbell, 2010).

As Nicholls' research shows, the Southwestern Indian Ocean islands, especially the Seychelles archipelago, were implicated in a tight network of slave smuggling and trafficking, which permitted the business to continue well after abolition. The testimonies from Gillieaux and de Commarmond's data support Nicholls' theories, and show that these sea crossings to capture slaves on the African coast were still going on well into the twentieth century. The *moutya* genre is a record of this history and a social commentary about it.

Section 2: The Evolution of the *Moutya* in the Outer Islands During the Post-Abolition Period

As already discussed in the introduction, the *moutya* genre really came into its own during the 1860s when it became the main form of entertainment for labourers working on the outer islands (Diallo et al., 1982). Producing copra, extracting guano and other such activities was backbreaking work and pretty dull, which are testified by the *moutya* lyrics themselves (Diallo et al., 1982; Estrale, 2007). Why should such conditions produce a cultural form as exhilarating and linguistically rich as the *moutya*? The answer lies in this quote from the History page of the Seychelles Island Development Company (IDC):

> "When the schooner disappeared over the horizon, there was no contact with the outside world for months, until the next ship was due"
>
> (IDC Seychelles, n.d.).

This meant that the workers had no means of knowing what was going on with their families back home. They had to become their own doctors, had to learn to provide for their needs from their environment to supplement the supplies brought by the boats. Most important, they had to create their own entertainment to relieve the drudgery of hard work and the worry about what was happening back home on the mainland. As David Pitt points out in his ground-breaking work, *Sociology, Islands and Boundaries* (1980), islands are centres of seminal ideas and cultural survivals (p. 1053). In this case, the workers of the Seychelles outer islands were not only contained by the physical boundaries of their islands, but they became islands in themselves. In such conditions, everyday occurrences are magnified, and often, social norms are forgotten, or hidden indulgences become impossible to hide. As Pitt puts it, 'boundaries do not only contain groups but permit all kinds of boundary crossing' (1980, p. 1055). Thus, all the goings-on between men and women, all the unfulfilled longings, all the nostalgia for home, and the daily struggles to survive on the islands, were expressed in the *moutya*, sometimes in direct form, most times in very intriguing metaphorical language. What is also very important to note is that the isolation of the islands also permitted a more exuberant indulgence in its performance as the restrictions which stifled its existence on the mainland did not apply. On Mahé and other inner islands, the *moutya* was banned in specific areas and could not be performed after nine p.m. (Seychelles Penal Code, 1935). This of course killed it, as a proper *moutya* only started at 6 p.m. and could last as late as the next morning (Estrale, 2007). In the outer islands, the *moutya* genre evolved from the lamentation style of the pre-abolition period and its immediate aftermath to the Calypso style of social commentary of the late nineteenth century and twentieth century.

This was the same period that the cry 'Selo!' also became magnified in significance in Seychelles Creole (Adam, 1987, p. 65). To the workers of the outer islands, it meant not only relief from the drudgery of hard physical labour, from the monotony and long periods of isolation, but also the possibility of news and supplies from the mainland, and even possibly, leaving the island for a brief respite. Indeed, a testimony from Bollée and Rosalie's collection of memoirs from Seychellois elders confirms that when that cry was heard, signalling the arrival of a boat, everybody had to stop what they were doing to help with the offloading and loading. Men, women, labourers, foremen, managers, everybody rushed to the jetty (2014, p. 28). I suppose this is why there are so many stories of ghosts and *malfezan* (*Mal faisants* [French], or evildoers) using this cry to lure out people in the night. The boats were also brought back to the mainland, the new custodians of the *moutya* and their transmission of it, as well as their stories of life in the islands. That is why, to me, the cry 'Selo!' represents so many things about Seychellois culture: stories, traditional economic activities, events and the names of boats and faraway places that turn up in songs and other forms of expression...

The following extracts from these types of *moutya* songs are testimonies to the lives of the labourers who worked on the outer islands during the specified period. Sometimes they sought a contract voluntarily as it represented a sure source of income, as testified by Cecilia Adrienne's memories of her father, whom she says went to the outer islands to escape poverty (De Commarmond & Gillieaux, 2004, p. 29). The first extract is from a song called, *Laprovidans* (Providence):

Providence is a platform reef in the Farquhar Group of the Seychelles outer islands. Its main economic activities are fisheries and copra production, the latter of which was only abandoned recently when Cyclone Bondo destroyed the plantation in 2006. Providence could only house a workforce of about forty people, which gives an idea of the closeness of the community who lived on it (IDC Seychelles).[4] The fact that the narrator of the song had to seek an advance from the company to fund his trip to the island is an indication of the extent of his poverty. It also reveals the nature of the outer islands as a means of survival for landless slave descendants and people of other ethnic origins who had no stable source of income.

Sometimes, the choice of going to work on the outer islands did not belong to the workers. Many of them were obliged to go wherever their employers required them to, as is testified by the following song:

Msye Langnen mon bourzwa | Ki kote ou anmennn mwan | Lilo Sen Pyer mon pa pou ale... | Beriberi i a touy mwan | Maladi anfle i a touy mwan.
(Diallo et al., 1982)
Mr. Lanier, my master | Where are you taking me? | I do not want to go to St. Pierre... | I will be killed by beriberi | The swelling disease will kill me.

Lanier is one of the family names associated with the history of the outer islands of Seychelles (Adam, 1987, pp. 53–54; Durup, 2008). Edouard Lanier was associated with Farquhar, of which he seems to have been the owner (Durup, 2008). Captain George Lanier is cited as having acquired the *Wanetta*, a schooner operating not only among the outer islands of Seychelles, but which also made trips to Nosi Be in Madagascar, Mombasa, and other ports in the region between 1925 and 1950. At the time that she was wrecked in 1950, it seems that Edouard Lanier had been her owner (Durup, 2008). It is not certain which specific Lanier is being referred to in the above song. At any rate, Edouard Lanier is cited as being concerned with social justice and equality (Adam, 1987, p. 54). However, the Lanier in the song seems to have inspired a certain degree of dread in the narrator, who was forced to go wherever his master decided. He is particularly weary of Lilot St. Pierre, where it seems people tended to suffer from Beriberi.

Other types of *moutya* lyrics give detailed descriptions of the work done by the labourers on the islands, and their suffering. The following song by the Anse Etoile Cultural Troupe (1986) seems to be a modernized version of an older song which appears in Diallo's 1982 collection:

Dan mon dou-z-an katorz an		When I was twelve, fourteen years
Pti mon manman dan lakour		Still my mother's child at home
Mon al plis koko lil Alfons		I dehusked coconuts on Alphonse
Mon al tir gwano Lasonmsyon	4	I extracted guano on Assumption
Kan Zipowa i arrive		When the Zipporah arrived
Mon debarke lo lasose		I stepped down on the pier
Mon lapo ledo in ize		The skin on my back was worn
Msye Marsel pe riye	8	Mr. Marcel was laughing

The song mentions specific economic activities that the narrator was involved in on the Alphonse Atoll and the island of Assumption, that is, copra production and guano extraction. As is often the case in *moutya* songs about the outer islands, he also mentions the boat that he was to travel back to the mainland on, the *Zipporah*. This was a schooner belonging to Abdool Rassool, a Persian from Mauritius, trading in Seychelles as Said and Co. (Scarr, 2000, p. 87). The narrator's age reveals the fact that child labour was a reality for Seychellois of the poorer classes (Ministry of Education and Information, 1983). The main topic of the song though is the grievances that the narrator feels against his employer. He complains that his back is worn (*Mon lapo ledo in ize*) from hard work, and Mr. Marcel, the employer, laughs at him and postpones the payment of his dues to the next day, which would be a bottle of wine. Though this might seem to be an exaggeration at first glance, Scarr's account of the *Zipporah*'s passengers who went to labour on the islands seems to confirm that they were at the mercy of the island managers and had no recourse to a magistrate once out of Port Victoria. They 'could only swim for it if they felt aggrieved' (2000, p. 87). Furthermore, Scarr gives a very telling example of one of the island managers' tongue in cheek attitude towards the labourers:

> "The men coming back from the Islands always have grievances when they land here… but these grievances vanish when they are sober again, and then they draw their money and no more difficulties are experienced."
>
> (Scarr, 2000, p. 87)

This suggests that alcohol featured somewhere in the equation, and that the men only got the courage to complain when they were inebriated. It is worth noting that inebriating their ex-slaves was one way the plantation owners and employers had of controlling their workforce and cheating them of their proper dues, after abolition (Choppy, 2018). In the older version of the above song, in which the narrator seems to be an old man reminiscing about his younger days, the sense of the employers treating the labourers (ex-slaves) as children comes out more clearly—'*Manrmay, manrmay, manrmay,* | *Mon pa pou peye ozordi*' (Children, children, children | I will not pay you today). The old man's grievance is etched in the last line—*Msye Marcel Lemarsan bizwen mazinen* | *Mon'n is later Lasonmsyon* (Mr. Marcel Lemarchand, you have to remember | That I pulled the soil on Assumption) (Diallo et al., 1982, p. 21). There is definitely a sense of being cheated here, and a plaintiveness about unrewarded labour ('Pulling the soil' is very likely a euphemism for extracting guano). One may well conclude that an analysis of this song could support the argument that this history of exploitation has contributed to the Seychellois labourer's mistrust of their employers.

One of the most characteristic aspects of *moutya* songs composed on the outer islands is their tendency towards misogyny. The song *Laprovidans*, for example, is a typical story of domestic trouble due to misogyny. A man seeks an advance payment to go work on Providence Island, taking his wife with him. However, as soon as he gets there, he suspects his wife of having an affair with the 'Komander' (administrator)—*Me si ou galan konmander dir mwan*| *Dir li pa bezwen touy mwan dan louvraz* (If you're the administrator's lover, tell me| Tell him not to kill me on the job) (Estrale, 2007, p. 64). There may have been two reasons for this kind of mind-set. One is that taking one's wife to the Islands was always considered risky by male workers because of the scarcity of women there, which automatically made all women the object of desire.[5] Two, it must be remembered that the transfer from plantation work on the mainland to the outer islands occurred shortly after abolition, and there was still a culture of exercising the '*Droigt du Seigneur*' among the white upper class who were likely to be the new entrepreneurs of the outer islands (Chang-Him, 2010, p. 139). Very likely, this attitude was imitated by the administrators and island managers, and anybody who had ascended to any form of power in these conditions. One may also consider the possibility that the misogyny was a legacy of the slavery period which made studs of men, and sexual objects of women (Chang-Him, 2010). As Diallo points out, the women were also likely to be castigated because the composers of the songs were male, and they attributed to women, their own standards of behaviour (Diallo, in press).

The following song, entitled *Manman Torti* (Mother Turtle) is an example of such misogyny:

Manman torti i koze	Mother Turtle speaks
Be mwan sa lannen mon pa ponn dan rido (bis)	This year I won't lay my eggs on the beach
Mon a kit rido ek madanm marye	I will leave the beach to the married woman
Sa lannen mon pou ponn an plennmer (3x)	This year I will lay my eggs out at sea
Pour kit rido pour madanm marye	I will leave the beach to the married woman
Mon oule kit rido pour madanm marye (3x)	I will leave the beach to the married woman
Baba i a pe kwi mon dizef ek ledo	Baby, she's cooking my eggs with her back

What is remarkable about this song is its excellent use of metaphor. The turtle, which is known to lay its eggs on the beach on most of the Seychelles islands, especially the outer islands, becomes the narrator. The composer of the song suggests that the turtle won't be able to lay its eggs because its place was taken by the married woman indulging in illicit sex. Illicit, because if it had been with her lawful husband, it would have been on their marital bed and not in the hidden recesses of the beach, where her activities overheat the turtle's eggs! Further on in the song, the male voice takes over the narration from the turtle and speaks of seeking an advance payment to go work on Denis Island, which implies, like the *Laprovidans song*, that the men indebted themselves to seek better opportunities for their families while their wives repaid them with infidelity. This is a common theme of the *moutya* genre, especially those songs composed on the outer islands in the pre-independence period.

Finally, the sea itself features in many *moutya* songs. In most cases, they record events that have marked the lives of people on the islands—for islanders know and understand the power of the sea. The sea can provide, but it can also be a dangerous foe. The following song, entitled *Vandredi Sen* (Good Friday) was recorded by Jean-Marc Volcy in 1993. It won an international award, sponsored by *Radio France Internationale* (RFI) and gave the Seychelles *moutya* its moment of fame on the international music scene. This song tells the story of some women who went out in a little boat which sank, causing them to drown. The song mentions Peros, probably referring to Peros Banhos in the Chagos archipelago. Both the Bollée/Rosalie (2014) and de Commarmond/Gillieaux (2004) compilations of memoirs show that Seychellois, Mauritians, and other island peoples in the Southwestern Indian Ocean rubbed shoulders on the copra islands of the region. The *moutya* compilations by Diallo et al. (1982) and Estrale (2007) also contain evidence of this. Some of the islands mentioned are Agalega, Juan de Nova, St. Brandon, Diego Garcia and other islands in the Chagos group such as Peros Banhos. As discussed earlier, the workers of these islands became the new custodians of this evolved form of *moutya*, and upon their return to the mainland, transmitted this heritage in their communities. This is what happened in this case. Volcy got the original form of this song from an old man who had probably worked on the islands himself, or it had been transmitted to him from another source who had been there (Ministry for Youth, Sports and Culture, 2017).

The song itself depicts many aspects of creole life and practices on the islands. For example, women are often called *pti zozo* (little bird) in creole songs, especially in the *moutya* form. It is often the case that boats are given the name of women, and in this case, it seems that the boat that sank was called *Pti Zozo Ble* (Little Bluebird). The exact location of the accident is mentioned—*Lapas Norde* (Northeastern Pass). Creole women are known for their resourcefulness, and it is possible that these women had gone out to fish on their own. The *Komander* (Administrator), a central figure in the running of a plantation, is the one who borrowed the plantation master's (Mr. Sauvage) boat to go to the rescue. The boat he used is called a *katyolo*—a word originating from Swahili (Bollée, 1993, p. 206). The vocabulary employed shows the multi-ethnic origins of the language and its creolization. Most of the words are recognizably of French origins, but some of them are specific to the region, for example, *Norde* is the creole pronunciation of *Nord-est* (North East) in French, and is listed as a maritime term in the D'Offay and Lionnet creole dictionary (1982, p. 278).

There is a very strong sense of pathos throughout the song, enhanced by the distress expressed by the narrator in the line 'Fanm Peros, travay labati lazournen' (Women of Peros, working hard all day), and 'lanmen lo latet' (heads on head), the latter which is a gesture in creole body language for great distress. The pathos rises to a crescendo in the last lines of the song— 'Pa koule, pa koule, pti zozo…' (Don't sink, don't sink little bird). The tragedy is made especially poignant by the fact that it happened on a very significant day for creole peoples of this region, *Vandredi Sen* (Good Friday). This reflects the Christian traditions of creole culture. The symbolical significance of the day has helped fixed the Peros tragedy in the minds of its witnesses, and thus fixed it in the Seychellois intangible maritime heritage through song.

Conclusion

As a living cultural practice and oral record of the Seychellois people's history, the *moutya* genre is an archive of Seychelles' history since its settlement. This includes its history as a country that is defined by its relationship to the ocean. The archipelago has played a more significant role in the Indian Ocean slave trade for example, than it has been given credit for. Peter Nicholls makes a very good case in this respect, in his research on the history of slavery in the archipelago. That the memoirs of the Seychellois citizens whose parents or grandparents have lived these events should coincide with both the theories put forward by Nicholls and the testimonies of the *moutya* songs are more than validation for his work. The strategic position of the islands on the spice route to India and other parts of Asia during the European exploration and exploitation of the area between the sixteenth and nineteenth centuries is also a very important aspect of the archipelago's maritime history.

In the twenty-first century, it is this strategic position and the economic implication of the archipelago's large expanse of territorial waters that are given particular importance in the country's political and economic vision. In spite of its land area of only 455 km^2 and its population of less than 100,000 defining it as a small island state, Seychelles has been referred to as 'a big maritime nation', by virtue of its Exclusive Economic Zone (EEZ) stretching across 1.4 million Km^2 (Safe Seas, 2017). In putting forward a bid for a non-permanent seat on the United Nations Security Council in 2012 (Amla, 2014), and in adopting global leadership strategies for SIDS in the domain of climate change, insecurity and poverty (Athanase & Bonnelame, 2016), Seychelles has in fact attempted to act as a big nation. In the pre-pandemic period before Covid-19 hit the global economy, the country's economic success did make it possible to aspire to being a global leader (Amla, 2015). However, the Covid-19 epidemic and its negative impact on tourism and related industries has made it clear that Seychelles is too dependent on its tourism industry (World Bank, 2021). In its drive for economic success in the past, this creole nation has tended to overlook the richness of its culture, derived from its colourful past as a slave society and European colony. Consequently, in the search for economic diversification, investment in the creole culture which can be exploited through cultural industries has not been seriously considered (Lepathy, 2020). This means that all the rich sources that can be tapped in the Seychellois oral traditions have also been neglected. And yet they can contribute so much to the discussions on identity and nation-building.

The reason behind the reluctance to pay more attention to oral traditions and the central role it plays in manifestations of the creole culture is the same in most creole societies—plantation creole societies still bear the trauma of slavery (Choppy, 2018). In Seychelles, the *moutya* form is a particular example of a resilient African heritage that in spite of its stigmatization and oppression in the past has still managed to become the nation's most popular form of expression. Jean-Marc Volcy's success story with his song, *Vandredi Sen* (1993) is testimony to that. This song celebrates the lives and struggles of the Seychellois creole nation as a maritime nation that has been shaped and formed by plantation slavery. The *moutya* generally, can tell us many stories. The song *Vandredi Sen*, tells a tragic story of loss, of the ocean's ability to provide as to take away life. And yet, the people who suffered this loss manifested their resilience by turning their grief into a buoyant song. *Navir Soubwannan* tells the story of people who survived the Indian Ocean Middle Passage; *Mon Lansdeger* laments over lost symbols of manhood and power—and yet all these songs are now vibrant symbols of the Seychellois' creativity and their ability to turn adversity on its head. Such is the power of creolization, especially in such insular spaces as is represented by water-bound communities like the labourers who worked on the plantations of the outer islands, under very harsh conditions. These stories can be used to advantage, whether to make new creations and develop the country's cultural industries, or to learn more about this creole nation's incredible journey to its present sovereign and proud state.

Notes

1. Seychelles Marine Spatial Planning, 2015. http://www.seychellesmarinespatialplanning.org/.
2. Island Development Company Ltd. (n.d.). *IDC Seychelles - History*. Retrieved from http://www.idcseychelles.com/history.html.
3. Personal communication with Jean-Claude Mahoune, anthropologist and local researcher. October 2015.
4. Retrieved from IDC Seychelles, Providence. http://www.idcseychelles.com/providence.html.
5. Personal communication—Marcel Rosalie (Conducted research with elders who had experience of working on the outer islands).

References

Adam, J. L. (1987). *Histoires varies des Seychelles*. Roma Flash.
Allen, R. B. (2014). *European Slave Trading in the Indian Ocean, 1500–1850*. Ohio University Press.
Alpers, E. (2009). *East Africa and the Indian Ocean*.
Amla, H. (2014). 'Seychelles calls for United Nations reform at G77 summit'. *Seychelles News Agency*, Friday 20 June [online]. http://www.seychellesnewsagency.com/articles/781/Seychelles+calls+for+United+Nations+reform+at+G+summit (Accessed 25 March 2019).
Amla, H. (2015). 'Seychelles moves to World Bank's rich list as income per capita increases'. *Seychelles News Agency*, Friday 3 July [online]. http://www.seychellesnewsagency.com/articles/3263/Seychelles+moves+to+World+Banks+rich+list+as+income+per+capita+increases (Accessed 2 September 2021).
Anse Etoile Cultural Troupe, Folkloric songs (transcripts), (1986).
Athanase, P., & Bonnelame, B. (2016). In Seychelles: UN Leader commends island nation's leadership on global issues. *Seychelles News Agency* (Weekend edition) Sunday 8 May [online]. http://www.seychellesnewsagency.com/articles/5120/IN+SEYCHELLES+UN+leader+commends+island+nation%27s+leadership+on+global+issues (Accessed 25 March 2019).
Benedict, M., & Benedict, B. (1982). *Men*. California University Press.
Bollée, A., & Rosalie, M. (2014). *Parol ek memwar, Récits de vie des Seychelles* (2nd ed.), Bamberg: Kreolische Bibliothek.
Bollée, A. (1993). *Dictionnaire étymologique des créoles français de l'Océan Indien*. Buske Verlag.
Borilot, V. C. (2014). En mal de mots: représentations de la figure paternelle dans les littératures de la Caraïbe et des Mascareignes. PhD thesis, The University of Iowa [online]. http://ir.uiowa.edu/cgi/viewcontent.cgi?article=5473&context=etd (Accessed 02 September 2014).
Burridge, G. (2017). A Local's Guide to Seychelles. *National Geographic – Travel*. https://www.nationalgeographic.com/travel/article/travel-guide-seychelles (Accessed 02 September 2021).
Campbell, G. (2010). Slavery in the Indian Ocean World. In G. Heuman, and T. Burnard (Eds.), *The Routledge History of Slavery*, Routledge Handbooks Online. https://www.routledgehandbooks.com; https://doi.org/10.4324/9780203840573.ch3 (Accessed 01 September 2021).
Campling, L., Confiance, H., & Purvis, M. T. (2011). *Social Policy in the Seychelles*. Commonwealth Secretariat.
Carayol, M., & Chaudenson, R. (1978). *Lièvre, Grand Diable et Autres: Contes Créoles de l'Océan Indien*. EDICEF.
Chang-Him, F. (2010). *Cohabitation: A Christian Response to Ménage in Seychelles*. M.Phil. thesis, University of Wales.

Choppy, P. T. (2018). *Attitudes to slavery and race in Seychellois Creole Oral Literature.* Masters thesis, University of Birmingham. https://pdfs.semanticscholar.org/a576/1bd65390e4e26a90843d7b653d5c295cb62d.pdf (Accessed 21 February 2019).

Choppy, P. T. (2020a). From local Creoles to global Creoles: Insights from the Seychelles. *Small States & Territories, 3*(1), 57–70.

Choppy, P. T. (2020b). 'Women in Seychelles' in *Oxford Research Encyclopedia: African History.* Oxford University Press.

Choppy, P. T. (2006). *The Seychelles Moutya as a theatre prototype and historical record*, Creole Institute: unpublished document.

Clifton, J., Etienne, M., Barnes, D. K. A., Barnes, R. S. K., Suggett, D. J., & Smith, D. J. (2012). Marine conservation policy in Seychelles: Current constraints and prospects for improvement. *Marine Policy, 36*(3), 823–831.

Confiant, R. (2004). *Le Grand Livre des Proverbes Créoles.* Presses du Châtelet.

D'Offay, D., & Lionnet, G. (1982). *Diksyonner kreol-franse.* Kreolisch Bibliotek.

De Commarmond, O., & Gillieaux, C. (2004). *Memories of Slavery*, Ministry of Education, Seychelles: unpublished document.

Diallo, A. (in press). *Chansons Moutya et Berceuses de Seychelles*, Seychelles: Departement de la Culture.

Diallo, A. (Ed.) (1983). *Contes, Devinettes et Jeux de Mots des Seychelles.* Editions Akpagnon/ACCT.

Diallo, A., Rosalie, M., Essack, G., & Labiche, E. (2001). *Zedmo Sesel* (Riddles of Seychelles). National Heritage.

Diallo, A., Rosalie, M., Essack, G., & Labiche, E. (1982). *Sanson Moutya* (*Moutya* songs). Culture Division.

Durup, J. (2008). The Lovely Schooner, Wanetta. *Seychelles Weekly.* http://www.seychellesweekly.com/October%203,%202011/top5_wenetta.html (Accessed 30 August 2021).

Estrale, M. (2007). *Tradisyon Oral Dan Kiltir Seselwa: Moutya* (Oral traditions in Seychellois culture: The *Moutya*). Creole Institute: unpublished document.

Franda, M. (1982). *The Seychelles: Unquiet Islands.* Westview Press.

Haring, L. (2005). 'Eastward to the Islands: The Other Diaspora'. *The Journal of American Folklore*, 118(469), (Summer), pp. 290–307. http://www.jstor.org/stable/4137915 (Accessed 03 November 2019).

Hugo, V. (1862). *Les Misérables.* A. Lacroix, Verboeckhoven & Cie.

Island Development Company Ltd. (nd). IDC Seychelles - History. http://www.idcseychelles.com/history.html (Accessed 02 September 2021).

Lepathy, M. A. (2020). The Creative Industries in Seychelles have been neglected for too long. *Seychelles Nation*, 5 June 2020. Retrieved from https://www.nation.sc/articles/4897/the-creative-industries-in-seychelles-have-been-neglected-for-too-long (Accessed 02 September 2021).

McAteer, W. (2008). *Rivals in Eden: The History of the Seychelles, 1742–1827.* Pristine Books.

Ministry of Education and Information. (1983). *Portrait of a struggle.* Seychelles National Printing Company Ltd.

Ministry of Youth, Sports and Culture. (2017). Torch Bearer of Seychellois Music. Retrieved from https://www.pfsr.org/uncategorized/jm-volcy-4/ (Accessed 01 September 2021).

Mulot, S. (2000). « *Je suis la mère, Je suis le père !* » : *L'Enigme matrifocale. Relations familiales et rapports de sexe en Guadeloupe*. Thèse. Ecole des Hautes Etudes de Sciences Sociales de Paris.

Nicholls, P. (2018). *The Door to the Coast of Africa: Seychelles in the Mascarene Slave Trade, 1770–1830*. PhD Thesis, University of Kent. https://kar.kent.ac.uk/67029 (Accessed 25 May 2019).

Nwulia, M. D. E. (1981). *The History of Slavery in Mauritius and the Seychelles, 1810–1875*. Rutherford.

Ottino, P. (1974). *Madagascar, les Comores, et le sud-ouest de l'océan Indien*. University of Madagascar.

Philpot, D., Gray, T. S., & Stead, S. M. (2015). Seychelles, a vulnerable or resilient SIDS? A local perspective. *Island Studies Journal, 10*(1), 31–48.

Pitt, D. (1980). Sociology, Islands and Boundaries. *World Development, 8*, 1051–1059.

Reddock, R. E. (Ed.) (2004). *Interrogating Caribbean Masculinities: Theoretical and Empirical Analyses*, University of the West Indies Press, 4.

Safe Seas. (2017). 'Maritime Security in Seychelles' [online]. http://www.safeseas.net/wp-content/uploads/2017/01/Concept-Note-4-Maritime-Security-in-Seychelles.pdf (Accessed 02 September 2021).

Scarr, D. (2000). *Seychelles since 1770: History of a Slave and Post-Slavery Society*. Hurst and Company.

Seychelles Marine Spatial Planning, (2015). On the leading edge of Marine Conservation and Climate Adaptation [online]. http://www.seychellesmarinespatialplanning.org/ (Accessed 23 August 2021).

Seychelles Penal Code. (1935). *'Drums Regulation'*. Chapter 93, Section 183. Waterlow and Sons Ltd.

Skerrett, A., & Skerret, J. (Eds.). (2019). *Seychelles, Ships and the Sea: The story of the lifeline of Seychelles*. Mahe Shipping Co., Ltd.

The World Bank. (2021). Seychelles: Country overview [online] https://www.worldbank.org/en/country/seychelles/overview (Accessed 01 September 2021).

Toussaint, A. (1965). Le trafic commercial des Seychelles de 1773 à 1810. *Journal of the Seychelles Society, 4*, 20–61.

Vergès, F. (2007). Indian-Oceanic Creolizations: Processes and Practices of Creolization on Réunion Island. In C. Stewart (Ed.), *Creolization: History, ethnography, theory*. Left Coast Press.

14

Waking up to Wakashio: Marine and Human Disaster in Mauritius

Rosabelle Boswell

'Blue' heritage consists of that assemblage of tangible artefacts, intangible cultural practices, and landscapes that are important to the sustainability of identities and livelihoods connected to the oceans and coasts. In a Post-2015 UN Task Team document on sustainable development, it is stated that culture, an enabler and driver of development, does not prominently feature in the UN SDGs. Moreover, while culture and the relevance of a culture-centric Africa is emphasised as a key aspiration of the Africa Agenda 2063, national governments in the Southern African Development Community (SADC) region still emphasise the economic benefits of the Blue Economy. Mauritius, a SADC country with a sizeable African diaspora population, is one case where this contradiction is especially important. This chapter, therefore, draws on past ethnographic research in Mauritius (2016–2019), commemorations of the MV Wakashio disaster (August 2021), and a key interview (August 2021) to reflect on the complexities of blue heritage in this Indian Ocean country with a sizeable African diaspora population.

The research in this chapter is supported by an NRF UID Grant number 129662.

R. Boswell (✉)
Nelson Mandela University, Gqeberha, South Africa
e-mail: rose.boswell@mandela.ac.za

The discussion proposes that the impacts of the MV Wakashio oil spill in July 2020 may be more complex and damaging than presently realised by the local population. These are the immediate impacts of the disaster (noted in this chapter), the long-term economic and health effects of the oil spill discussed by health researchers and environmentalists, as well as apparent and increasing marginalisation of local coastal communities, in particular small-scale fishers. Specifically, the discussion shows that not only have the rights of small-scale fishers in Mauritius in general, and in the oil spill affected areas in particular, been steadily eroded over the years, but that the Wakashio disaster has now created conditions for the further marginalisation of fishing dependent families. The chapter also proposes that the story of the Wakashio disaster goes beyond the details of the event itself. It is an event symptomatic of larger, more pernicious processes of exclusion and is indicative of the growing disregard for vulnerable populations and precarious coastal livelihoods.

The chapter is organised as follows. The first part introduces and discusses the concept of 'blue' heritage as it relates to the life of coastal dwellers and those who rely on the sea for their livelihoods. The next section offers a brief introduction to Mauritius and the place of small-scale fishers in the country's socioeconomic landscape. The third and fourth parts of the chapter outline the Wakashio disaster and national government response. The final section of the chapter discusses how, a year later, inhabitants of the affected regions are responding, and provides information on their situation. The research presented in the paper uses a mixed-method approach, drawing on secondary data from academic papers on pollution and environmental conservation in Mauritius, to media articles and video recordings after the disaster, as well a key qualitative interview with a sea-dependent entrepreneur and activist, directly involved in managing the social and economic impacts of the Wakashio disaster. The research forms part of a larger, multi-country project in Africa and its nearby diaspora which seeks to investigate the richness of oceanic and coastal heritages and cultures, and the complexities and challenges facing them.

Mauritius History and Blue Heritage

Mauritius is a multicultural country of some 1.3 million people. According to Eriksen (1993), Mauritians speak some 22 languages, are adherents to at least four major religions, and can trace their ancestry to India, Africa, Europe, and China. The country is described in historical (colonial) accounts,

14 Waking up to Wakashio: Marine and Human Disaster in Mauritius

as being *terra nullius* (empty land) prior to its settlement by African slaves, Indian indentured labourers, and European slave owners and colonials. While colonial accounts of slavery 'silenced' the contributions of Africans to the making of Mauritian society, more recent revisionist accounts have revealed a vibrant, interethnic, and dynamic world in the slave period and shortly thereafter (Allen, 2017; Nwulia, 1981; Teelock, 1998). These accounts reveal that Africans who had been forcibly relocated to the island settled, at first, on the east coast of the island, in a port town called Mahebourg. The town is named after the first French governor of Mauritius, Mahe de Labourdonnais, it is also the place where the English seized control of Mauritius from the French in 1810. It was at Mahebourg that the first slaves were bathed and prepared for sale to plantation owners. The tangible heritage of this past is still visible in the town, in the form of stone slave baths, and a local cemetery containing graves from the slave period and the names of settlements, described in Mauritian sugar estate parlance as *camps* (labour camps). In addition, there is, a little further along the coast, a monument commemorating the abolition of slavery (Fig. 14.1).

The demographics and landscape of Mauritius changed dramatically however, after the arrival of Indian indentured labourers in 1834. The Indian labourers not only sought to express their beliefs through the construction of tangible religious monuments and temples they also increased the population, changing the ethnic landscape. In time, many Indian descendants also

Fig. 14.1 Monument to Commemorate the 170th Anniversary of Abolition in Mauritius (*Source* the author, 2016)

pursued alternative livelihoods beyond the plantations, with some of them choosing to blend subsistence fishing with land cultivation. Most of the sugar plantations were located close to the coastline, and this, also, facilitated such strategies. In an early account, Robertson indicates that,

> …sugar cane was introduced by the Dutch but not cultivated until 1735. It now occupies the greater part of the coastal districts of Savanne and Grand Port on the Southeast, the plains of Flacq on the east, and much of the northern district of Rivière du Rempart and Pamplemousses. (1930, p. 338)

In the early 1900s, the island also received traders and merchants of Muslim and Chinese origin. The latter were prominent in coastal villages, where they brought imported consumer goods to the island society. In brief, today, Mauritius is a thoroughly creolised society (Vaughan, 2005), in the sense that its inhabitants share a lingua franca, *Kreol*, participate in the exchange of culinary practices, and engage in interethnic marriage and social groups, developing hybridised beliefs and values.

Despite such intermingling and social exchange, ethnic identification has remained a factor in Mauritius and, in some cases, historical patterns of ethnic group clustering and geographical location have persisted. In an earlier account (Boswell, 2006), I posited that the island's early colonial agricultural *camps*, the Best Loser electoral system, racism, and government efforts to maintain cultural roots deepened ethnic affiliations. However, and as I argue here, identity is also shaped in other ways. When slavery was abolished in 1835, African slaves left the plantations and migrated to urban centres and the coast (Carter & D'unienville, 2001). Many slave-descended Africans began new and independent lives as fisherfolk along the south and west coast of the island. In part, some settled in these areas because they had been living in the forested areas of the southwest part of the island, having formed slave maroon communities there prior to abolition. By living along the coastline, some African descendants acquired an alternatively located identity, one closely associated with the sea and coast. Those who remained close to urban centres such as the capital, Port Louis, and joined the workforce there, as dockworkers, builders, artisans, and entrepreneurs, also came to appreciate this coastal aspect of identity. As my earlier ethnography (Boswell, 2006) showed, Creoles or African descendants came from towns and cities to the coast to relax, be entertained and to socialise (Fig. 14.2).

There were also those African and Indian descendants along the southeast coast of the island, in Mahebourg and Grand Rivière Sud-Est, who lived close to estuaries and rivers. These waterways influence sociality and identity, in ways still remembered by those interviewed in 2016. It is for this reason that I argued that,

14 Waking up to Wakashio: Marine and Human Disaster in Mauritius

Fig. 14.2 Map of the villages and towns of Southeast of Mauritius (*Source* https://www.mapsland.com/africa/mauritius/detailed-road-map-of-mauritius-with-cities-and-villages accessed 23/08/2021)

> By turning to the sea to deepen understanding of identity in Mauritius, one might be able to consider issues of identity in terms of scale, depth, border, area, and temperature, in other words, other means of 'measuring' identity. One might also be able to perceive the salience of the sea itself in the making of islander identity. The waters are oceanic (vast), and there are tidal pools, reefs, eddies, and open sea. The environment of the water is also diverse. It is sonorous, viscous, and tidal. Perceiving identity from a marine perspective and including maritime ethnography in assessments of identity, one might propose that racialized and mixed identities are incredibly diverse and diversely (where feasible) measurable. Identities ebb and flow like tides, with personal, communal, and social shifts. (Boswell, 2019, p. 467)

When one considers, however, the perception and conservation of heritage in Mauritius, it becomes evident that the country's government perceives heritage as a territorialised asset and a part of national, tangible artefacts. In other words, there is still a grounded concept of heritage. Moreover, heritage is also treated as a legacy that one passes from one generation to the next

(Garcìa-Canclini, 1995). In the case of Mauritius, heritage is also at risk of becoming an ethnic legacy, useful to narratives of fixed identity.

Peckham states that '"heritage", globally, carries two related sets of meanings (2003). On the one hand, it is associated with tourism and with sites of historical interest that have been preserved for the nation…on the other hand, it is used to describe a set of shared values and collective memories; it betokens inherited customs and as sense of accumulated communal experiences that are construed as a "birthright" and are expressed in distinct languages and through cultural performances' (Peckham, 2003, p. 1). Thus, heritage has two purposes. However, in both instances, 'cultural heritage does not exist, it is made' (Bendix, 2009, p. 255).

In the context of Mauritius, I propose that heritage is primarily useful in government construction of national (ethnic) narratives of the nation. The cultural nature of the society however also necessitates, as Peckham argues, the sharing of values, collective memories, and sense of belonging. Heritage in Mauritius is therefore complex, because not only is it deliberately constructed by a state which faithfully documents and preserves it, but it is also something made by local people. This means that cultural practices and values are subject to multiple processes of 'heritagisation', and that heritage becomes a multiply valued asset of enduring historical and national importance. Unfortunately, though, the authorised heritage discourse prevails, and is presented for public and tourist consumption as the distilled, publicly approved identity of the Mauritian nation. As I show further on, the making of heritage has become even more complex and multi-layered. Today, authorised heritage discourse in Mauritius now appears to be tied to projects of gentrification and re-racialisation and these processes, along with the environmental disaster of Wakashio, are further alienating locals. To clarify (and as explained in more detail at a later point in this chapter), coastal areas are systematically being 'whitened' and gentrified, while local black communities are being shifted to livelihoods beyond the coast and constrained to perform identities that cohere with, and conform to, the narrative of re-racialised landscape. I return to this point in greater detail below.

The inscription of the Apraavasi Ghat (the landing place of Indian immigrants) and the west coast mountain of Le Morne (the site of a mass suicide by formerly enslaved Africans who feared that British soldiers were coming to take them back into slavery, shortly after the declaration of slavery's abolition), on UNESCO's World Heritage List (WHL), are two examples of public heritage in which ethnic identity is emphasised. To be fair, these heritages have only recently been accepted as narratives of the Mauritian nation. I use the word 'recently', here, because for a long time, the histories

14 Waking up to Wakashio: Marine and Human Disaster in Mauritius 231

and cultures of slavery and indentured labour were not considered acceptable as heritages of Mauritius. Colonial and European histories were foregrounded and celebrated. Discourses of the latter were still found in both the Blue Penny Museum in Port Louis, and in the Naval History Museum in Mahebourg between 2016 and 2019. One might ask, then, what is the connection between Aapravasi Ghat, Le Morne, the Blue Penny Museum, and the Naval Museum? The simple answer is that all of these sites offer a state-accepted, sanctioned, authorised national heritage discourse.

Herzfeld has another plausible explanation. He argues that nation states are unlikely to foreground that which compromises their public identity (Byrne, 2011). His argument may apply to Mauritius. It took decades of lobbying and global discussion for slave heritage to be perceived as important to the public heritage in Mauritius. Today, Mauritians' common heritage of the sea is overlooked. Several reasons may be offered for this, the most compelling being that existence of a common heritage compromises the state discourse of ethnic distinctiveness. A more complex explanation would be that recognition of the common heritage of the sea would shift attention to very valuable resources, the ocean, and coasts. Let us follow the more complex explanation.

Since the 1970s, when the Mauritius government sought to diversify the economy so as to leverage different sources of income and advance economic development, it has actively sought to rebrand the country as a hub for foreign investment. In the last 20 years, that process has involved building luxury property development schemes for a foreign, and elite, tourist clientele. To emphasise the cultural dimension of locals' connection with the sea and their fundamental, that is, human and heritage rights to it, could severely compromise both the property development goals and tourism revenue.

It would appear, as Herzfeld (in Byrne, 2011) and Xia (2020) have argued for other societies, that ordinary citizens in Mauritius have tried to shift to culturally intimate engagements with heritage that do not form part of the authorised heritage discourse. With direct reference to blue heritage, Mauritians have continued, or tried to continue, a private, social and recreational engagement with the sea, despite the increasing pressure on them to relinquish the island's beaches to tourism. But, as I have shown over the years, the hold that locals have over the sea and coasts (as well as their intimate relationships with the marine), remains a tenuous one, and continues to be eroded. That process of erosion has accelerated since the Wakashio disaster.

Among other interventions, the national government is now setting up various inland livelihood options for those no longer able to live with or from the sea. As I show in the next section, however, the marine cultural (and intangible) heritage of Mauritius is significant and indicates cross-cultural elements including beliefs, fishing practices, fishing technologies and cultural-linguistic contributions (Boswell, 2021). The inland livelihood options offered by the government cannot be a substitute for the coastal, marine livelihoods that people pursue, because what the state is offering cannot restore rapidly eroding significant cultural knowledge embedded in fishing practices, technologies, and the substance of life at the coast.

The coastal areas of the island are key to the island's tourism. Tourism in Mauritius remains mostly recreational, and the east coast of the island, from Rivière du Rempart (in the north-east) to Mahebourg in the south-east is already dotted with luxury hotels. Writing about anticipated tourist arrivals from 1 October 2021, a news media source indicated that Mauritius was anticipating about 100,000 tourists and that this amounted to about 3000 tourists per day (Khodabux, 2021). Bhuckory (2021) states that, in Mauritius, tourism 'revenue… dropped to 578 million rupees ($13.6 million) from 16.08 billion rupees a year ago, according to the Port Louis-based Bank of Mauritius. Revenue in June fell to 20 million rupees from 383 million rupees'. This translates into a *96 percent* drop in revenue in the first half of 2021, a catastrophe due to the Covid-19 pandemic.

Recreational tourism is, therefore, not only a most important source of foreign direct investment for Mauritius: it is one where no further losses can be sustained (Bhuckory, 2021). Furthermore, tourism remains important because a historical elite that still holds much power in Mauritius (Salverda, 2015) continues to invest in it. The first stage of luxury property development for instance, benefited those European-descended Mauritians who owned large tracts of land along the coastlines. They benefited from a property development programme entitled the Integrated Resorts Schemes (IRS). This programme had many stringent requirements for those seeking to participate in it, including the stipulation that potential investors would need to have access to vast tracts of land. Subsequently, the government put forward similar schemes with more flexible requirements for investment that then benefited a wider pool of landowners: however not only do, the historical (white) elite families still hold considerable economic power (Salverda, 2015), they are still, in present times, using historically acquired collateral for big investments along the island's coastline. A quick glance at the board of directors of the Beau Vallon Company, of which Riche-en-Eau, a sugar estate that spans from Grand Port to the Midlands (3500 hectares), close to Mahebourg provides one example of this phenomenon. In days gone by, those associated

14 Waking up to Wakashio: Marine and Human Disaster in Mauritius

with the company would have been regular visitors at the chateau Riche-en-Eau, a massive colonial home built on an expansive site overlooking both cane-fields and the sea.

In recent years, the same company has pursued a Property Development Scheme (PDS), along the same lines as the IRS described earlier. This PDS is meant to recreate an idyllic coastal life for foreign investors. The most expensive property on offer to them is priced at approximately Euro 1.7 million and offers investors the opportunity to live an 'authentic' village life on grounds previously part of the Riche-en-Eau sugar estate. This is an estate that previously had slaves, and slave descendants. Thus, coastal areas of the island and the southeast coast discussed here, have become zones of prime real estate, in which historical sugar estates where slavery flourished have now become property development schemes.

Since the late 1990s, the national government's Board of Investors (BOI), more recently known as the Economic Development Board (EBD), has approved the construction of many luxury development residences and complexes along Mauritius' coastlines to provide high-end accommodation for foreign investors. It is noted that many of the luxury development residences and high-end hotels are located close to the country's marine parks and fishing reserves. This suggests one of two things. Either these biodiverse sites (and the ecological ethos which they represent) are of high value to these high-end tourists, or, alternatively, that the national government is, in collaboration with private investors (some of whom are descendants of the historical elite), specifically 'controlling' local community use of marine resources in those areas deemed to be of value to the high-end tourist industry. What some may dismiss as cynical and unfounded speculation is supported by the fact that luxury development residences and property development schemes aimed (Boswell, 2008) at creating a 'supportive' and visually appealing socioeconomic setting for a high-end tourism environment. In other words, there has been a strange and 'perfect' confluence of gentrification, whitening, and the creation of prime real estate along Mauritius' coastlines. This phenomenon is not unique to Mauritius. Across the Caribbean, similar processes are being played out, which led one author, Polly Pattullo, to lament in 1996, that some Caribbean peoples now feel as aliens in their own land.

In the following part of the discussion, I offer some insight into the lives of people who were resident in Mahebourg and its environs in 2016, after the processes of gentrification described above and five years before the MV Wakashio disaster. I follow this with an analysis of the immediate effects of the bulk carrier's grounding off Pointe d'Esny, before then moving on to discuss contemporary challenges and experiences. My purpose, here, is to demonstrate that the inhabitants have a rich tangible and intangible 'blue' heritage in Blue Bay and in the southeast coastal villages of Mauritius, and to show how, in the wake of gentrification as it has been emerging in Mauritius and, now, the Wakashio disaster, this blue heritage is being severely compromised.

Marine Parks, Fishing Reserves, and Fieldwork

Mauritius has two marine parks (Balaclava on the northwest coast, and Blue Bay on the southeast coast), an estuary reserve (Rivulet Terre Rouge), and six fishing reserves (Poudre d'Or, Port Louis, Black River, Post Lafayette, Trou d'Eau Douce, and Grand Port). The last three of these are located along the east coast of the island. Blue Bay, an ecologically pristine lagoon off the southeast coast was declared a Marine reserve in 1997. A document describes the unique features of the bay as follows:

> The Blue Bay Marine Park, located in the South East of Mauritius was proclaimed a National Park under the Wildlife and National Parks Act 1993 in October 1997. It was declared a Marine Protected Area and designated a Marine Park in June 2000 under the Fisheries and Marine Resources Act 1998. In January 2008, it was officially nominated as the second Wetland of International Importance (RAMSAR Site) for Mauritius. The total area of the Marine Park is 353 hectares; it includes the lagoon starting from Pointe Corps de Garde as its northernmost point up to Pointe Vacoas, its southernmost point and extends about one kilometre seaward from the reef crest. The depth of the park varies from 1 to 150m metres. The Blue Bay Marine Park is known for its diverse and rich fauna and flora especially the corals, mainly for a brain coral of diameter 6-7metres. 108 species (33 genus) of coral, 233 fish species, 201 species of molluscs were inventoried in 2012.
> Blue Bay being a popular tourist spot and the most favourite beach in the southern part of Mauritius is extensively used for recreational purposes. It is estimated that more than one hundred thousand visitors including Mauritian and foreign nationals visit the park every year. The various recreational activities that are carried out in the park are: (i) scuba diving and snorkelling, (ii) non-motorised surface water sports such as wind surfing, sailing, water skiing, paddle boats, kayak, (iii) swimming, (iv) boating activities such as glass bottom boats, boats transporting divers and snorkellers, boat transporting visitors into and outside the boundaries of the park, (v) recreational fishing with pole and line along part of the coast, and (vi) fishing using pole and line and basket trap beyond the fringing reef. (Hurbungs & Mohit, n.d., pp. 2–4)

14 Waking up to Wakashio: Marine and Human Disaster in Mauritius

The purpose of this declaration, therefore, was to recognise and protect the unique marine biodiversity of the bay. In particular, the bay has a wealth of marine life, as well as well-developed and diverse coral, much of which was at risk of being degraded by ocean pollution and acidification in the years prior to the declaration. According to the Mauritius Fisheries and Marine Resources (Marine Protected Areas) Regulations 2001, no commercial, polluting, extractive, or otherwise ecologically compromising activity is permitted in the Bay (Government of Mauritius, 2001). Only limited, recreational and artisanal fishing is allowed. In Sect. 6(2) of the Regulations, it is noted that 'A person may fish with a bait gear in a Fishing Reserve if he holds a licence issued under Sect. 31 of the Act authorizing him to do so'. Fisherfolk are not permitted to use fish traps or to beat the water to attract the attention of fish for a catch. Similarly, the use of boats in the marine reserve is governed by particular rules, and the entire area is zoned to ensure that human movement in the area is limited to less fragile sections.

In 2016, I conducted anthropological field research in Pointe d'Esny and Mahebourg, on the south-eastern coast of Mauritius. The research conducted at that time provided a preliminary framework for the larger project on 'ocean cultures and heritage' funded by the South African National Research Foundation, the first phase of which coincides with the first five years of the UN Ocean Decade. While in Mauritius in 2013, I had heard about a village called Ville Noire and that it was possibly founded in the 1700s, at the time when slavery began in Mauritius. I was curious about Ville Noire, and felt that anthropological research there would enrich my understanding of slave history in Mauritius, adding to the doctoral and postdoctoral work I had conducted in both Mauritius and elsewhere in the southwest Indian Ocean region (Fig. 14.3).

After a day of interviews in and around Mahebourg during the fieldwork session in 2016, I would 'finish' the day by walking along the beach where I was staying, to Blue Bay beach. There, I would engage with locals and visitors who had come to the beach to relax, or, as one interviewee put it, '*pou kasse stresse*' (to break stress). As usual, I found young children playing in the shallow, clear waters of the bay, their parents (for the most part) sitting on woven mats under trees lining the beach, while some men on the beach would sit and play cards at a makeshift table. Unlike in the case of Le Morne on the west coast of the island, I did not witness people wading in to collect shellfish from the bay. I later learned that this was because the bay was a marine protected area, where people could not collect any seafood from the lagoon without a permit.

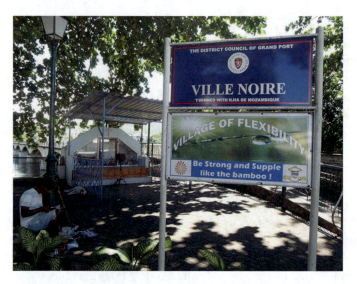

Fig. 14.3 Ville Noire village. The village's slave history is indicated via its twinning with Ilha de Mozambique, and the allusions to efforts to overcome enduring inequality (*Source* https://www.lemauricien.com/actualites/ville-noire-ancre-dans-son-histoire-et-tourne-vers-lavenir/275203/ accessed 23/08/2021)

However, on going further 'inland' to Ville Noire and interviewing people there, I found that some women still collected shellfish from the river (*Rivière la Chaux*) that flowed through the village. In particular, the women collected *mangwak/mongwate* which, the women say, is key to survival in difficult times. The shellfish was, and remains, an important source of protein in those seasons when the fishing catch is poor, or when the fishing season is concluded. Moreover, it appeared to be gathered by women, while men went out to fish. One of the women interviewed said,

> We used to wake up before dawn to go and collect it [*mongwate/mangwak*]. It was important to get there before the tide came in and there was a way to do it. Our mother taught us and my sisters and I waded out into the cool waters at first light, our feet sinking in the mud and sand of the [tidal] zone to collect the shellfish. We knew those waters as if they were fields. You couldn't just go in there to collect it. You had to know where to go and you had to be careful when removing it from the rocks. They are so fragile you know, that a clumsy hand can easily lead to the shell breaking and spoiling the flesh. Not many people know this of course and today people just go to the shop to buy stuff. In my day though, we went out at first light. I remember that time so well. The sky was light but the sun had not yet risen and we were always the first ones there. (cited in Boswell, 2019, p. 471)

14 Waking up to Wakashio: Marine and Human Disaster in Mauritius

Reflecting on recent (pre-Wakashio) access to the sea and to seafood, a fisherman, interviewed in 2016, indicated that,

> My life has changed. There was a time when I was young and you know in those days I could go out in my little boat and catch up to 200 pounds of *carangue* [trevally] in a day. Now, well now you're lucky if you get 50 pounds on a good day. You see, the problem is that people don't understand the seasons. It took me four full years to learn how to be a fisherman. I didn't just go out there on a boat. But these days, people just buy a boat, get a license, and think that they will know how and where to fish. The sea is like a map. Underneath it has coral, stone, valleys, and hills. The water moves over it differently at different times of the day. You may be out there and suddenly the wind will change. If you can't read the wind you will get lost at sea. That happened to me once and I was lucky to have someone rescue me. (cited in Boswell, 2019, p. 471)

One day I took a glass bottom boat ride across the bay. As noted earlier, only those with the relevant tour operator or fishing permits can access the bay's deeper waters. The tour operators on the boat were a father and son. While the father steered the boat, I spoke with the son, who had been given the task of explaining what marine life and other marine features I was seeing through the glass bottom of the boat. He was about to launch into a rehearsed English script of the visit, when I stopped him and explained that he could continue in *Kreol*, because I understood the language. This seemed to make him more relaxed and he proceeded to describe the different kinds of coral we were seeing below. It soon emerged that there are specific *Kreol* words used to describe the different corals in the bay and that young men, like the son of this tour operator, enjoyed a multiply situated existence with the sea. The son explained how most of his time after school is spent at sea, scuba diving, snorkelling, swimming, fishing, night hunting for crab, and joining leisure catamaran tours as a helper on weekends. This rich, embodied existence with the sea came to a halt in 2020, when the MV Wakashio ran ashore off the reef, just beyond Pointe d'Esny.

The Wakashio Disaster

'On 25 July 2020, the Japanese ship, MV Wakashio, ran aground the coral reef off the eastern coast of Mauritius. The vessel discharged more than 1000 tonnes of oil into the island's pristine lagoon including its Blue Bay Marine reserve' (Boswell, 2020, online). The spill of Very Low Sulphur Fuel Oil

(VLFSO) near Pointe d'Esny in Mahebourg on the east coast of the island spread, to varying degrees, from Rivière du Rempart in the northeast part of the island, to Mahebourg in the south. According to Asariotis and Premti (2020, online), the grounding of the vessel occurred close to the Blue Bay marine park and two internationally protected wetland sites, as well as a small atoll (Ile aux Aigrettes) which has endemic bird species and other biodiverse fauna and flora.

The Pointe d'Esny site is also known for its association with the history of the first Dutch landing in Mauritius in the early 1700s, as well as the 1810 battle between the French and the English, which resulted in Mauritius becoming a British colony. The seas nearby contain historic shipwrecks that are about 200 years old. Asariotis and Premti (2020) argue that the catastrophe has consequences not only for the natural environment but also for human and other animal health. For marine species, oil spills can physically smother the animals, cause chemical toxicity, produce ecological changes in the marine environment, and can lead to the loss of shelter or habitat.

Regarding events precipitating the disaster, it was initially proposed that MV Wakashio was compromised by poor weather off the coast of Mauritius. Subsequently, when interviewed, the captain of the vessel claimed that the crew had been on board for more than a year due to the global Covid-19 pandemic and that a decision had been made to steer the vessel closer to Mauritius to obtain WIFI signal, so that crew members could contact family. Apparently, a crew member was celebrating his birthday. However, this story was soon disputed, as several media sources revealed that access to the internet is possible beyond the reef, and that the vessel had not needed to come closer to Mauritius' shores to obtain internet access. It was also revealed that most ships travelling such routes and distances have access to satellite technology and internet connectivity. More importantly, in a preliminary report, the Panama Maritime Authority (AMP) stated that it was poor seamanship which led to the grounding of MV Wakashio (The Maritime Executive, Online). The key officials on duty on the bridge of the vessel failed to supervise and monitor the navigation equipment, appeared not to hear/notice the warning calls from shore authorities in Mauritius, and did not apply good seamanship practices that would have involved analysing the situation of the vessel followed by the taking of action appropriate to avoid the accident. As of this writing, a year after the accident, however, the final report of AMP however, has not yet been released. Key documentation and recordings remain inaccessible to the investigators, and some of the vessel's crew, including its captain, are still being held in custody in Mauritius. Worse, and according to the statement of the interviewee offered next, the captain of the vessel was apparently

14 Waking up to Wakashio: Marine and Human Disaster in Mauritius

interviewed further. This was when it was revealed that he had completed similar journeys past Mauritius on at least seven occasions in the past, which suggests that he would have known the risks of coming too close to shore with the bulk carrier. BP, the supplier of fuel to the MV Wakashio, denies any responsibility for the disaster, stating that the fuel sold to the charterer, Mitsui OSK Ltd (MOL), was of a standard acceptable to the International Maritime Organisation (IMO).[1] Discussion of these matters remains ongoing (Fig. 14.4).

Around the fifth of August, it was noted that the ship was sinking, despite the local efforts that had been made to steer it off the reef. Providing an overview of the government's response, the United Nations Office for the Coordination of Humanitarian Affairs (OCHA) stated that,

> The [Mauritius] Government is leading the response and has established the following coordination mechanisms: the National Crisis Management Committee (chaired by the Prime Minister) meets daily in the afternoon to review operations and provide strategic guidance; the National Oil Spill Coordination Committee (chaired by the Director of Environment) meets daily in the morning to review progress, assess the situation and needs, and plan work for the next 24 hours; there is also a National Emergency Operations Command chaired by the Commissioner of Police and a dedicated Coordination Committee for the Salvaging of the Vessel (chaired by the Director of Shipping).

Fig. 14.4 Extent of the MV Wakashio Oil Spill, August 2020 (*Source* https://giscan.com/monitoring-the-impact-of-a-shipwreck-the-case-of-the-wakashio-bulk-carrier/ accessed 10/08/2021)

> A team of environmental experts from the International Tanker Owner's Pollution Federation Limited (ITOPF) and Le Floch Depollution—the international contractor appointed by the Protection and Indemnity Club (P&I), insurers—are onsite in Mauritius and are preparing an action plan for clean-up and restoration of affected sites. (OCHA, 2020, online)

Then, on 15 August, the vessel split in two, releasing oil onto the coral reef and into the lagoon. Reflecting on the government's decision to seek to float the vessel, a Mauritian ecologist, Sunil Dokwarkasing, said that the.

> delay in trying to address the problem of this wreckage on our reef is, I would say, grossly negligent by our government. Their primary concern was to refloat the ship – no one considered the danger represented by the 3800 metric tonnes of oil on the ship. No one considered or seemed to care about the substantial risk this would present to the island and the lagoon; the government sat on the file for more than 12 days without making the decisions that should have been made. (Kingdom, 2020, online)

Internationally, certain states rallied to assist in alleviating the immediate impact of the disaster. Specialists were sent form Japan and France to deal with the clean-up process. There are four international conventions which can provide the basis for medium to long-term compensation to countries affected by oil spills. These are the Civil Liability Convention (CLC), the FUND Convention (an additional layer of funding support), the Bunkers Convention and the Hazardous and Noxious Substances (HNS) Convention, which is not yet in force. According to Asariotis and Premti (2020, online),

> Mauritius is a State Party to the IOPC FUND regime (International Convention on Civil Liability for Oil Pollution Damage (CLC) and the International Convention on the Establishment of an International Fund for Compensation for Oil Pollution Damage (FUND), as amended in 1996) –which would have provided liability and compensation of up to 203 million SDR (approx. 286 million USD) for this incident (also covers reinstatement of the environment). The 2003 Supplementary Fund Protocol provides even higher liability and compensation, up to a maximum of 750 million SDR (around 1.05 billion USD) per incident, but has not been ratified by Mauritius.

In a more detailed analysis of the international legal framework dealing with oil pollution damage from oil tankers, Asariotis et al. (2012) note that claims for damages can only be lodged with the insurer and shipowner, that there is a time limit for the claim (preferably within three to a maximum of six years after the accident), and that from the IOPC Fund, a maximum of about 300 million US dollars can be claimed. Claims can be dismissed if it is found that the claimant (i.e. Mauritius) did not follow protocol (or was negligent) in respect of guiding and communicating with the stricken vessel, or if there was naturally bad weather causing the accident, or if the accident resulted from acts of sabotage or war (Asariotis et al., 2012, pp.50–76).

Far from the international legal landscape, and horrified at the event and the impending disaster for their country's shores, volunteers gathered from across Mauritius to construct booms and mop up the oil. From the University of Mauritius, Karishma Daworaz and Arun Ramluckhun spoke about fishermen who risked their lives and went out to sea to set up the booms. Ramluckhun noted that the task was difficult, and it would have been preferable to avoid this catastrophe. He added that after a day's worth of cleaning up, 'we would go to bed at night with our eyes and throats burning and it was really uncomfortable'. A few days after the oil spill, Daworaz said, 'we had dolphins and whales all dying. Crabs were running to the shoreline and there was a dreadful stench of rotting marine life not long after' (*l'express* YouTube 25 July 2021, online).

Bagasse, as the fibrous remains of sugarcane stalks are called, were collected from nearby operating sugar estates (Riche-en-Eau and Bel Ombre) to fill the makeshift booms. Shade cloth was also provided by textile firms to house the *bagasse*. Mauritians were also encouraged to supply hair for the filling of the booms. Many had their hair cut, and then supplied the cuttings to the boom-makers. The process was captured by a number of photographers, both amateur and professional, showcasing the humanitarian aspect of the disaster and the willingness of Mauritians to collaboratively respond to the disaster. Looking back on the crisis a year later, Mauritians remarked that they had collectively responded to 'save the lagoons', beaches and fishing spots (Figs. 14.5 and 14.6).

Local Voices, Long-Term Consequences

Reflecting on the consequences of the oil spill a year later, those interviewed by the Mauritius media (l'Express Youtube, 25 July 2021) had the following to say,

Fig. 14.5 Volunteers carry booms for the Wakashio oil spill (Photo by Daphney Dupré, 2020)

Fig. 14.6 A volunteer helps to clean up after the Wakashio oil spill. Photo by Daphney Dupré, 2020

'*Bé après panne kapav fer narnier. Ki pu fer?*' [afterwards we could not do anything, what could we do?]?

'*47 ans monne vivre ar la mer – pu nu, ti ène premier fwa sa. Kans maré nwar la vini, nu business affectée* [for 47 years I have lived with the sea – for us it is the first time we are experiencing this. When the black tide came, our businesses were affected]'

'*Amène en mofinne are nu. Dimoune pa le aster, banians pas le prends. Ziska ler enkor ena de l'huile, manguier enkor ena de l'huile. Zwitre pe mor, mongwate pe mor*. [It brought bad luck to us. People don't want to buy (fish), the banyans

14 Waking up to Wakashio: Marine and Human Disaster in Mauritius

[middle men] don't want to take our fish. There is still oil. The mangroves have oil. Oysters are dying, shellfish are dying].'

'*pwason gagne li moyen, pwason rester. Government ti nu ène ti l'argent, après sa fini, bizin rouler*' [fish are there but much less, and fish stay behind, that is they are not sold. Government gave us a stipend but after that finished, we had to fend for ourselves].

'*péna pwason. Pwason ine quitte paye ine aller. Dimoune pas pu desann Mahebourg pu aster pwason. Dimoune per*' [there is no fish. The fish has left the country. People are not coming to Mahebourg to buy fish. People are scared].

An inhabitant of the region, a Creole businessman and a citizen committed to coastal justice, had the following to say when I interviewed him in August 2021 (I interviewed him in *Kreol* to better understand the nature of blue heritage in Mauritius, and the potential long-term challenges posed by the grounding of MV Wakashio. The following extracts from the interview are translated from *Kreol*:

I have lived in Mahebourg for more than 40 years. My father was a fisherman but as I was growing up, I knew that I didn't want to become a fisherman, I wanted to try something different but to still be with the sea. You must excuse me I have a problem with my one ear. When I was younger, the doctors told me that I would have problems with my ears one day, because I love to scuba dive. I enjoy all things to do with the sea, swimming, sailing, fishing, and diving. I know that in the area where I live there are about 150–250 fishermen.

Our communities were already being affected by Covid-19. We had not been working for four months when the Wakashio accident happened. The government closed the lagoon for six months and provided a solidarity grant of Rs. 10,200 to the affected families. This money was not enough, people had a lot of commitments, therefore there was a lot of stress. So, for almost an entire year, tour operators, taxi owners, street food sellers and fishers were badly affected by the double problem of Covid-19 and Wakashio. The people sat at home and worried first about Covid-19 and then the long-term financial effects and finally the pollution caused by Wakashio. Some people are especially worried about the long-term consequences of dealing with the oil, especially the high risk of cancer. People are also really worried about the seafood because of oil contamination and the risk that potentially contaminated seafood, can cause cancer.

For many people the beach is a place to release stress or, as we say in Mauritius, *casse stresse*. People swim, scuba dive, snorkel – but now not many people are doing this. They are scared. So, there is the immediate effect of the grounding of the MV Wakashio but there is also fear, a psychological damage which the government must consider in its plans for compensation and support.

Up to now (August 2021) there is still no formal report from government regarding the water quality in the Mahebourg lagoon. No-one really knows if it is safe to go into the water. Lots of people have started to live like our forefathers did a century ago. They are scared of eating fish even if it is the thing that is the easiest to get. So they are cultivating chickens in their backyard, growing vegetables and just staying away from the shore. I even stopped my son from swimming and going windsurfing, it was difficult, imagine making a teenage boy not go out to swim.

Before Wakashio, fishermen working 6 days a week, could earn up to Rs. 24,000 because they earn up to Rs. 1000 per day. So now, the fishermen are getting only half of their income. The other thing you need to consider is that the fishermen has to put fuel his boat, he has to buy mesh to mend his fish traps etcetera, all these input costs add up. Before, the fisherman's day was divided in two in the afternoon he might well take up another job that will allow him to add to his income. With Covid-19 and Wakashio, this became very difficult because jobs were already scarce.

It is also well known that if you come from a coastal area it is difficult to get work. There appears to be discrimination against coastal inhabitants. Our people look for jobs at the airport but we are told that there is no transport for locals even though you see that transport is being provided for inhabitants from the north. There appears to big promises made during the election times, but our people are forgotten when it comes to getting jobs. The other thing is that coastal peoples are mostly fishermen and labourers, so many are uneducated and it is difficult for these families to ensure that the next generation moves into a more lucrative position. When elections take place, they are likely to accept support in kind from the candidates, instead of really asking what that candidate can do for them.

As for me, I wanted to become a coastguard, a sea policeman but I did not make it even though I had the education. I think I did not have the 'backing', so I had to do my own small business to support myself and family. There is still a lot of corruption in Mauritius. One must have that beautiful little envelope that you can pass under the table. If you don't have this, the chances are you will not progress. But there are also other worrying phenomena, where you see people that have no money actually spending money in a way that does not help the family to move forward, to *sorti dans zotte petrin*.

There is also a reluctance to engage in activism for coastal justice because there is fear. People want to ensure that they are not associated with actions for claims, or to pose as victims because they do not want to be seen as "opposing" the government who is telling them that everything is alright. Government has said that they will give a compensation of Rs100,000 to the fishermen. But this is a pittance if we consider the health issues that may still come. We still don't know what government has received or what it will get for the claims it is making against the owners of the Wakashio. We know that the Japanese

14 Waking up to Wakashio: Marine and Human Disaster in Mauritius

government has given a lot of money just to "say sorry" and that money does not even include the claims that government must still make.

What I find strange is that government waited 12 days to attend to the vessel. I am not sure what was happening but to me it sounds like if you put a tank of petrol in your house and I tell you that is dangerous, you come back to me and say, 'no, don't worry, its fine, what do you know? I am an expert'. We would have been much better off if we had people in Mauritius who know how to handle environmental disasters, but we don't. No-one screamed or shouted about this. Everyone let government and the experts handle the matter. In the end, it was volunteers who came to remove most of the oil from the shores.

I want to tell you something else. On the day of the grounding of MV Wakashio, a few of us were standing on the shore looking out to sea. We could see that the vessel was moving. We came back the next day and it had changed direction. When we saw this we enquired if we, as people who own boats could somehow help to perhaps advise on how to steer it off the reef. We were told that the "experts" are busy with it. But I want to tell you something. I am a child of the sea. When a boat changes its direction on the reef, it means it is not stuck at all or, that only a piece of it is stuck. This means that it is not entirely difficult to remove it. This was a "floating boat".

My grandfather always said, it is easy to say come home and I'll give you a fish dinner tonight. It is much harder to say come with me and I will teach you how to fish. What I mean by this is that in Mauritius and certainly in our poor communities, people are willing to take monetary compensation provided by the government. They are less likely to challenge government and to ask why it is that they are being given money or to ask government to justify the amount they are being given.

The entire tourism value chain was already being affected by Covid-19 but when Wakashio happened, the closing of the beaches of the southeast affected the taxis, the tour operators, the small beach restaurants, the suppliers of food to the restaurants and hotels on the southeast coast. For me, I used to pay about Rs200,000 worth of boat fuel per month but with the closure of the beaches, I didn't buy fuel. There was also the issue of not having money for a whole year, so we didn't buy from shops and grew our own garden and didn't buy from the local market, so many people suffered, not just those directly involved in the tourism industry.

Lastly what I don't understand is that the government has opened the lagoon after several months to allow the fishermen to fish but, there is a sign at the beach that says "do not swim in the water". Reflecting on this we are wondering. How can it be safe to fish in these waters but not safe to swim? What happens if tomorrow I bring tourists to swim here and something happens to them, they have a skin reaction or they fall sick?

Regarding that YouTube interview that you watched, I wanted to say that when the comment was made about the persistence of oil in the lagoon, the ecologist in the interview answered that he had recently taken a sample just below the sand and that he had still found oil. You ask me why I think the environmentalist from the government offered a different view, why he claimed that when his team took samples there was no oil to be found. Well, all I can say is that sometimes, a person is employed and paid to say things that they do not want to say. Unfortunately, if the government does not open the tourism season in October 2021, it will be chaos in the country. And, there is a high risk that even if we open, there may be consequences that may lead to other forms of chaos, as I have already described.

You have also asked me what I think of the rumours that the MV Wakashio was perhaps carrying drugs and that is why so much effort was being made to float the boat and send it on its way. I will ask you, what has government done in the last year to dispel this rumour? From what I see, nothing has been said and so that remains a great mystery. We do not know whether there were drugs on the Wakashio. Nothing else seems to explain to us why it needed to come so close to our shores. From what we hear, the captain of the vessel had done this route via Mauritius seven times. I can't understand then, how it is that the vessel got stuck on the reef, if he had sailed such a massive vessel, seven times on this very route.

Right now though, people are not thinking about the long-term health and pollution consequences of the Wakashio. People are thinking about the money that they can get in compensation, and many fishermen families don't even understand the quantity of money which is due to them. Worse, they don't understand the true value of the sea they think that money can buy them that value. I would say that it is because people have been so poor for so long.

For me and the people of Mauritius who live with the sea today, it is a part of our body. I grew up with the sea and I would be sad tomorrow if I could not touch the sea and enjoy it. The most relaxing thing you can do to relieve stress, is to go to the sea. What I know is that the salted water of the sea is also healing. When you have a wound and you go into salt water, the wound heals faster, it is also good for people with arthritis. For me, salt water is like the water that you drink. Without this water you die.

This multi-layered interview, like the conversations gleaned from the media interviews cited above, reveals the palpable fear of locals as they are now compelled to return to the sea to fish. The interviews also reveal the seminal role that the sea plays in the lives of coastal inhabitants. For the key interviewee, not only does the sea provide him with a livelihood, it provides him and his family with multiple forms of leisure and an embodied sense of identity, which have now all but disappeared because of the Wakashio disaster. In this regard, when we consider blue heritage, we must remain aware of the fact that small-scale fishers, too, live complex lives. There are important generational distinctions, in that while a grandfather and father might choose to pursue small-scale fishing, a subsequent generation might engage in tour operation or, other tourism ventures but still 'live' with the sea. The key interview also reveals that blue heritage stretches beyond the coast, as people from inland are drawn to the coast and sea for stress relief and healing. The seaside also provides important opportunities for socialising and intergenerational bonding. These social ties have been eroded after Wakashio.

The issue of ongoing pollution and contamination is also critical. Although the Wakashio oil spill is less than the oil spills in France (1978) and the Gulf of Mexico (2010), it affected three ecologically sensitive areas. Of concern is that the oils released contain Polycyclic Aromatic Hydrocarbons (PAHs) which are known carcinogens for both marine animals (Seveso et al., 2021) and humans (Abdel-Shafy & Mansour, 2016).

Conclusion

In the last few years, Mauritius' coastline has become prime real estate. As one luxury property site puts it, 'blue is the new black'. This means that there is a financial incentive for government to shift small-scale fishers from their original source of livelihood, fishing, to alternative, inland, livelihoods: it also means that fish are declining in the lagoon (Government of Mauritius, 2018). To continue attracting FDI, the government also appears to be pursuing a 'development' plan of continuing to shift coastal peoples inland.

The Wakashio disaster has produced an additional set of problems. Pollution may mean that coastal areas are no longer prime real estate and knowing this, foreign investors seeking new 'green' horizons may not want to invest in a place that may no longer be pristine.

Third, the Mauritius government is a signatory to key conventions that will permit it to claim for damages caused. However, it will be interesting to see how the government frames its claims, especially if it only considers damage to the natural environment and neglects the quantum of blue heritage lost. Those living close to the sea engage with it by eating fish and other seafood, swimming, diving, sailing, fishing, and snorkelling or, alternatively, simply by sitting on the beach and looking at the horizon. People describe themselves as children of the sea (*zenfants la mer*).

A year later, the waters appear to be cleaner but, as some ecologists and locals argue, there is still oil in the mangroves and under shallow layers of sand in the lagoon. In September 2021, the key interviewee cited in this article, let me know that the Mauritius government is offering fisher families Rs. 52,000 each in compensation (approximately 900 Pounds Sterling) for damages incurred by Wakashio. But, as discussed here, the damages appear to be far greater and government is eligible to claim for a vastly greater amount.

The conservation and inclusive sustainable use of the oceans involves close attention to historic livelihoods and sensitivity regarding the meaning and value of such livelihoods to local communities. Blue heritage, like other forms of terrestrial (tangible or intangible) heritage, cannot be summarily alienated and replaced with this or that putative alternative. This is so because living and working with the sea is much more than a matter of income: it is a way of life, and an integral part of selfhood's constitution and identity. The Wakashio disaster will be felt by Mauritians for many years to come, and its tragedy will be experienced through the unfolding of its health and pollution effects. The Mauritian government has a duty to come clean with its citizens, and to reveal the true impact of the disaster. As it is, coastal citizens have just woken up to Wakashio, and many have not yet realised that they are in a nightmare that will not end soon.

Note

1. BP specifically states that the oil sold was graded ISO-8217–2020.

References

Abdel-Shafy, H. I., & Mansour, M. S. M. (2016). A review on polycyclic aromatic hydrocarbons: Source, environmental impact, effect on human health and remediation. *Egyptian Journal of Petroleum, 25*(1), 107–123.

Allen, R. (2017). Ending the history of silence: Reconstructing European Slave trading in the Indian Ocean. *Revista Tempo, 2*(6), 295–313.

Asariotis, R., Lavelle, J., Benmara, H., & Premti, A. (2012). 'Liability and Compensation for Ship-Source Oil Pollution: An Overview of the International Legal Framework for Oil Pollution Damage from Tankers.' *Studies in Transport Law and Policy no. 1*. https://unctad.org/system/files/official-document/dtltlb20114_en.pdf (Accessed 25 August 2021).

Asariotis, R. and Premti, A. (2020). 'Mauritius oil spill highlights importance of adopting latest international legal instruments in the field', UNCTAD, Article No. 58 of UNCTAD Transport and Trade Facilitation Newsletter N°87. https://unctad.org/news/mauritius-oil-spill-highlights-importance-adopting-latest-international-legal-instruments (Accessed 12 August 2021).

Bendix, R. (2009). Heritage between economy and politics: An assessment from the perspective of cultural anthropology. In L. Smith, and N. Akagawa (Eds.), *Intangible Heritage*. Routledge.

Bhuckory. (2021). Mauritius First-Half Tourism Revenue Plunges, Expects Rebound. https://www.bloomberg.com/news/articles/2021-08-21/mauritius-first-half-tourism-revenue-plunges-expects-rebound (Accessed 23 August 2021).

Boswell, R. (2006). *Le Malaise Creole: Ethnic Identity in Mauritius*. Berghahn Press.

Boswell, R. (2008). Integrating Leisure: Integrated Resorts Schemes in Mauritius. *Africa Insight, 38*(1), 11–23.

Boswell, R. (2019). The Immeasurability of racial and mixed identities in Mauritius. In P. Aspinall, and Z. Rocha (Eds.), *The Palgrave International Handbook of Mixed Racial and Ethnic Classification*. Palgrave Macmillan.

Boswell, R. (2020, September 6). Mauritius must protect vulnerable communities from the effects of the oil spill. The Conversation. https://theconversation.com/mauritius-must-protect-vulnerable-coastal-communities-from-the-effects-of-the-oil-spill-145411 (Accessed 26 August 2021).

Boswell, R. (2021). How Covid-19 is transforming the heritage of speech in Mauritius. *Journal of African Diaspora Archaeology and Heritage, 9*(3), 255–276.

Byrne, D. (2011). Archaeological Heritage and Cultural Intimacy: An Interview with Michael Herzfeld. *Journal of Social Archaeology, 11*(2), 144–157.

Carter, M., & d'Unienville, R. (2001). *Unshackling slaves: Liberation and adaptation of ex-apprentices*. Pink Pigeon Books.

Eriksen, T. H. (1993). *Common denominators: Ethnicity, nation-building and compromise in Mauritius*. Berg.

Garcìa-Canclini, N. (1995). *Hybrid cultures: Strategies for entering and leaving modernity*. University of Minnesota Press.

Government of Mauritius. (2001). *Fisheries and Marine Resources (Marine Protected Areas) Regulations 2001 GN 172/2001/The Fisheries and Marine Resources Act 1998*. Government of Mauritius.

Government of Mauritius (2018). *'Performance Audit Report: Moving Towards Sustainable Artisanal Fishery in Mauritius', Ministry of Ocean Economy, Marine Resources, Fisheries and Shipping*. Government of Mauritius.

Hurbungs, M.D., & Mohit, R.D. (n.d.). Template for Submission of Scientific Information to Describe Ecologically or Biologically Significant Marine Areas. https://www.cbd.int/doc/meetings/mar/ebsa-sio-01/other/ebsa-sio-01-mauritius-en.pdf (Accessed 20 August 2021).

Khodabux, R. (2021, September 14). Réouverture des frontières sans la quarantaine - Dr Sharmila Seetulsingh-Goorah: "Nous allons tout droit au suicide"', *Defimedia.info* https://defimedia.info/reouverture-des-frontieres-sans-la-quarantaine-dr-sharmila-seetulsingh-goorah-nous-allons-tout-droit-au-suicide (Accessed 18/08/2021).

Kingdom, S. (2020). Mauritius Oil Spill – Pictures, maps, details. https://africageographic.com/stories/mauritius-oil-spill-pictures-map-and-details/ (Accessed 15 August 2021).

L'Express. (2021, July 25). Naufrage du MV Wakashio: Un ans après un souvenir encore vif. https://www.youtube.com/watch?v=ng21YLrlqQ0 (Accessed 23 August 2021).

Nwulia, M. (1981). *The History of Slavery in Mauritius and the Seychelles 1810–1875*. Associated University Press.

Pattullo, P. (1996). *Last Resorts: The Cos of Tourism in the Caribbean*. Ian Randle Publishers.

Peckham, R. S. (2003). Introduction: The politics of heritage and public culture. In R.S Peckham (Ed.), *Rethinking Heritage, Cultures and Politics in Europe*. I.B. Tauris.

Robertson, C. J. (1930). The Sugar Industry in Mauritius. *Economic Geography*, 6(4), 338–351.

Salverda, T. (2015). *The Franco-Mauritian Elite: Power and anxiety in the face of change*. Berghahn Press.

Seveso, D., Louis, Y. D., Montano, S., Galli, P., & Saliu, F. (2021). The Mauritius oil spill: What's next? *Pollutants*, 1(1), 18–28.

Teelock, V. (1998) *Bitter sugar: Sugar and Slavery in 19th Century Mauritius*. Mahatma Gandhi Institute.

The Maritime Executive. (2021). Panama cites improper charts and poor seamanship in Wakashio accident. https://www.maritime-executive.com/article/panama-cites-improper-charts-and-poor-seamanship-in-wakashio-accident (Accessed 20 August 2021).

United Nations Office for the Coordination of Humanitarian Affairs (OCHA). (2020, August 17). MV Wakashio Spill Update no. 4.

Vaughan, M. (2005). *Creating Creole island: Slavery in eighteenth century Mauritius*. Duke University Press.

Xia, C. (2020). Fluctuation between AHD and cultural intimacy: Heritagisation of a historic private house in Qingtian, China. *International Journal of Heritage Studies*, 26(7), 1047–1060.

15

Blue Heritage Among Fishermen of Mafia Island, Tanzania

Mariam de Haan

Off the South-East coast of Tanzania, at the mouth of the Rufiji River, lies the Mafia archipelago. This archipelago is made up of eight islands, namely Mafia, Chole, Jibondo, Juani, Bwejuu, Nyororo, Mbarakuni and Shungi Mbili (Moshy, 2016). In 2019, I conducted six months of ethnographic fieldwork on the largest island, Mafia. While living there, I was able to understand and analyse the importance of the sea and its animals to those who live there. The people admire the wonders of their marine environment, while also being wary of its dangers. The sea is seen as a place where one can relax. This was particularly evident to me when my friend, who, after a long stressful day, would often suggest that we "go dump our worries into the sea". Along the coast, we would find other families and children who had gone to cleanse their worries in the water. Not only does the sea cleanse the worries of people, it is believed to clean itself too, in between the monsoons. It is during these times when one can find what the sea hides within its currents, currents that are feared by mothers who warn their young girls never to enter the sea during high tide. These dangers are confirmed by the dead bodies that wash up on the shores of Mafia island. People will debate as to whether those bodies have come from far off places like Somalia or Comoros, or whether they are those of fishermen who strayed too far out to sea. The islanders of

M. de Haan (✉)
The British Library, London, UK

Mafia have multiple stories about fishermen who have gone crazy at sea after getting mysterious illnesses. Others are said to have let their greed take over as they went further and further into the sea, only to have the current carry them away from the archipelago. However, it is also these fishermen who, after a long day of work, like to "back flip" from their boats into the sea. Taking turns to push and splash each other with the waves, they frolic on the shore in an almost childlike manner, as if to show the youth which they have lost while working at sea. This chapter will explore the concept of blue heritage by showing how the fishermen of Mafia have altered their language, perception of time, and sense of community, in ways that include the sea and its animals.

The Time and Language of the Wind and Moon

The wind and moon play a large role in the movement of the ocean. The wind dictates the direction of currents and strength of the waves, while the moon pushes and pulls the tide. The moon's phases also create neap and spring tides. The waters of Mafia are affected by the monsoon trade wind system, in which the northeast monsoon, known as *kaskazi* in Tanzania, blows from December to April, and the southeast monsoon, known nationally as *kusi*, blows from June to October. These monsoons affect the climatic and ocean conditions with water temperatures that range from 25 degrees Celsius to 31 degrees Celsius, and with air temperatures ranging from 20 degrees Celsius to 32 degrees Celsius (Moshy, 2016). However, the influence of kaskazi and kusi monsoons, as of the neap and spring tides, are not just ecological: they are visible to those whose daily lives they alter. For example, many purse seine fishermen prefer the kaskazi period because it is then that "the sea is calm" and thus more bountiful, while in the kusi period it becomes exponentially harder to catch fish (de Haan, 2020, p. 52).

During the *kaskazi* period, the day of the fisherman begins at around 8 a.m. and ends at about 6 p.m. During kusi, however, fishing begins much earlier at 3 a.m. and ends at about 6 p.m. At the same time, the difference between neap and spring tide heavily influences ones access to fish. This, in turn, influences daily life on Mafia. I still vividly remember the times when Mafia endured neap tide. Fish is the staple protein of the island, especially for those who are less well off (as many of my research contacts were). The end of January 2019 was a particularly difficult time during my fieldwork because the island was already reeling from the beginning of kaskazi, and then found itself in the grip of the neap tide. Many of the fishermen I knew were grounded under such weather conditions. Since fishermen were unable to go out to sea, fish stocks in the town dwindled. At one point, the only fish one could find was fish that had been cooked and stored the week before. Due to the sheer influence that the wind and the moon have on the ocean, fishermen do not separate the three elements: rather, they view them as one interconnected entity. This is especially evident in the way that they talk about the ocean. In this section, I explore the way that fishermen relate the ocean to the moon and wind in order to explore the interactions that fishermen have with this part of the environment.

The picture was taken during neap tide at the beginning of kaskazi. Prior to taking the picture, I had spent the whole day in the government offices and so I was taken aback upon seeing how far the sea was. The government offices are right next to the ocean, and, usually, the waves are right at the edge of the green tables during low tide. During high tide, they are right under the white and blue boat that can be seen to the right of this picture. During neap tide, fishing is limited in Mafia: this is because the gross fluctuations in tide make it difficult to reliably conduct fishing activities (Fig. 15.1).

The fishermen who are at liberty to do so will usually wait until the tidal variations are less intense. Others will migrate to areas in mainland Tanzania, like Kilwa, especially if, as in the picture above, it is also the beginning of kaskazi. Those who have little financial leeway, and who are unable to migrate chose to fish regardless. Beyond the tidal periods of spring and neap tide, the phases of the moon influence night fishing in particular. When conducting

Fig. 15.1 Image of disparities that can occur in the water level during neap tide (*Source* Author's own image)

night fishing the main catch is these tiny fish known as *dagaa*.[1] The majority of night fishermen use a *mashua*, a plank boat with a motor, and two dinghy boats. The main boat carries approximately 20 up to 40 men, along with the purse seine nets and scoopnets. The dinghies carry one man in them who is equipped with a wick lamp, a metal or ceramic plate,[2] and a piece of coloured, semi-transparent plastic. When out at sea, the dinghies are dropped off at approximately 50 metres away from the main boat. The man inside the dinghy uses the plate to direct the light from the lamp downwards and into the sea. Every now and then, he will change the direction of the light to attract the dagaa towards the light. After he sees the school of dagaa, and if there is no cellphone coverage, he covers the light with the piece of coloured plastic. At the sight of the signal, the main boat raises its anchor and makes its way to him. When the main boat is close, the man in the dinghy raises his anchor. Meanwhile, the men on the main boat lower the net and encircle the school of dagaa. Once this is done the men on the boat pull at the net so that it tightens at the bottom, creating a pouch. Then, the fishermen can scoop up the fish with their scoopnets. This entire process is impossible during the full moon: the moonlight means that fishermen cannot use their lights to attract fish into the nets, nor can they use the signal lights. During the full moon there is so much light in the night sky that the fish are not attracted to the lights that the fishermen use, so night fishermen choose not to go out to sea during this period. As the moon plays such a huge role in everything they do, fishermen do not use the Gregorian calendar, opting instead for the lunar calendar. Therefore, when creating appointments with fishermen one must know the lunar calendar date as well. This has caused an interesting phenomenon in Mafia as a whole, because everyone on the island uses a fusion of the Gregorian and lunar calendar in their day to day lives, with a preference for the lunar calendar. By using the lunar calendar fishermen are also able to determine the phases of the moon in advance. The fusion of calendars also reflects the presence of blue heritage in Mafia. It is highly likely that this practice was born out of necessity by fishermen, and was then adopted by the people of Mafia because it was the easiest option. At the same time, the existence of this practice acknowledges the ocean and its importance to the planning of life in the archipelago. Therefore, although the moon affects fishing activities, this is not a major hindrance, because citizens and fishermen can prepare themselves for the full moon's return. The wind, be it during the monsoon and not, is a much less predictable phenomenon.

The wind affects every single fishermen (albeit to different degrees) in Mafia because of the way it changes the waves and visibility. The fishermen who use boats with sails are affected by the wind because they need it to enable their movement. Those who have small boats, particularly dugout canoes, are incapable of battling the large waves. The fishermen with larger boats also struggle with the waves, but are able to overcome this problem. In addition, it is very rare, if not impossible, to find a fisherman whose boat is equipped with a GPS device. Instead, some fishermen use their phones to determine where to go if lost. However, this too can lead to problems because the cellular telephone network is hard to find and connect to at sea. Due to the complications of the cellular telephone network, and the lack of finances with which to buy other forms of technology, fishermen must learn how to read the waves from their experienced colleagues. They have particular terms to describe the state of the waves: the term *kucharaza*, for example, is used to describe the "cha cha" sound that a wave makes when it hits the reef. When a boat's captain hears that sound, it indicates to him that there is a reef nearby, and that helps him to steer his boat accordingly. Another term that fishermen use to describe this phenomenon is *kuchapa* which is the Swahili[3] word for "to beat", or "to hit". This is used to refer to how the waves hit or beat the reef where the area is shallow. In other words, it is through sounds and visual information that they are able to tell what is occurring underneath the water. Visual information is especially handy, because with it one can determine from afar the location of a reef plateau or a sandbank, by observing waves breaking on these features. Fishermen also use visual evidence to determine the wind speed. As one fishermen states:

> "*Unaona kama vile vipovu povu, wenyewe vile tunaita wanafunzi. Ule ndo Upepo ule*". ("You see that white on top of the waves that looks like foam, those of us who understand, we call it *wanafunzi*. That is the wind"). (de Haan, 2020, p. 100)

The word *wanafunzi* means "students", but there is a deeper context to the way that fishermen use it. It refers to the primary school uniform that students wear throughout Tanzania. The shirt (and headscarf, if the student choses to wear one) is white, while the pants and/or skirts are dark blue. The waves that fishermen describe using *wanafunzi* all have white crests on top, with the blue of the ocean underneath. The term is a handy way for fishermen to memorise this weather condition for future reference, when it becomes necessary to know how safe it is to go out to sea in relation to the intensity of the wind. During the kaskazi period, the wind poses another challenge: that of decreased visibility. When the wind blows during this period, the ocean turns a strange brown colour, and carries with it a particularly strong scent. One fisherman explained this event as the "water having dust" (de Haan, 2020, p. 101). When seen from the sky this brown substance looks like giant oil spills on the ocean. Tour guides attribute this brown sea to algae, while the rest of the people in Mafia simply describe it as "the ocean cleaning itself" (ibid.). They describe it as a ritual which occurs every kaskazi, when the sea cleanses itself of all unwanted and unhealthy objects which it casts towards the shore. This is also why, according to them, plastic and seaweed are even denser along the shoreline during this time (Fig. 15.2).

The kind of fishing that occurs around Mafia island relies less on advanced technology and more on the knowledge that fishermen pass on to each other when they learn about the ocean. This has two results. Firstly, fishermen are heavily influenced by the fluctuations in the weather. During neap tide, the tidal variations make fishing more complicated, while night fishing is futile during full moon. As noted above, the sheer influence that the moon has on fishing is such that fishermen use the lunar calendar instead of the Gregorian. Ultimately, this is a practical measure which helps fishermen to be aware of the phase that the moon is in, and to plan their activities accordingly. When it comes to the wind it is much harder to plan, however, and as such fishermen

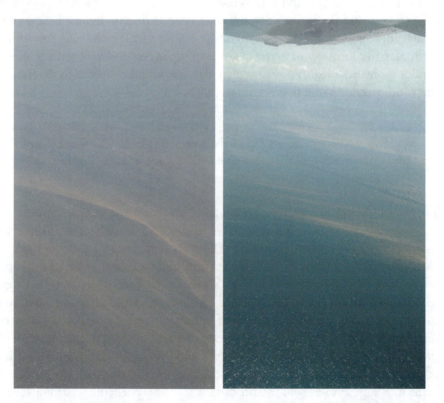

Fig. 15.2 Pictures taken off the coast of Kilindoni during a flight from Mafia to Dar es Salaam on 15 February 2019

must use aural and visual indicators to determine the strength and speed of the winds. The second result of Mafia's fishermen's minimal use of advanced technology is that they are forced to interact more deeply with the environment, to the extent that they need to adapt their lives in accordance to the fluctuations of the environment. This manifests itself in a multitude of ways. By structuring their days and appointments with a calendar that is based on the moon, the fishermen of Mafia have broadened their language to include terms that describe the waves and the movements that they encounter while out at sea. Fishermen with less experience learn about this interaction from those who are more proficient in the language of the sea. The language used to describe the sea can at times attribute human traits to the ocean. The fact that the people of Mafia believe that every *kaskazi* the ocean cleanses itself, and casts all evils like plastic and unwanted seaweed out towards the shore. This makes it seem like the ocean adheres to rituals as humans do. This belief is part of the blue heritage of Mafia as a whole, because it is a sign of Mafia's culture being linked to the oceans. Moreover, the fishermen's alteration of language all reflects the blue heritage present within Mafia. The alteration of Swahili words like *kuchapa* and *wanafunzi* gives the words new meanings which relate to the sea. These meanings are then passed on and inherited by other fishermen, creating new knowledge that is formed in relation to the ocean. These connections become even more intricate when looking at the way that fishermen refer to fish.

The Intelligence of Fish

When looking at the way that the fishermen of Mafia describe marine animals, the Swahili proverb "*hasira za mkizi, tijara ya mvuvi*" (the anger of the fish is the gain of the fisherman) comes to mind. The fishermen view marine animals as much more than a means to gain food and profit. Rather, they share with them a relationship that is similar to that of predators and their prey. Fishermen refer to fish as cunning and capable beings. Just as an orca uses its intellect to catch the penguin, so too do the fishermen, when they gather knowledge about the behaviour of fish in order to capture them. During our interviews, fishermen would often use imagery and metaphor to explain their interactions with the environment. Analysis of this imagery and metaphor demonstrates the importance of the marine environment to their fishing activities and, ultimately, their identity. Through understanding this, one can also see the complexities of blue heritage in Mafia.

The fishermen of Mafia describe fish as cunning and clever animals. This can be seen in the terms that they use. Two terms that are frequently used to describe the behaviour of fish is *wanakimbia* (they run away) or *hawapatikani* (they cannot be found). It may seem that *hawapatikani* implies that the fish are purposely hiding from the fishermen, but this is usually used to describe the way that the wind influences the behaviour of fish. As one fisherman explained:

"Kwa maana hiyo sasa upepo ukiwa kaskazi hata kwenda baharini utafautaji wa samaki unakuwa mgumu kwasababu maji yakiwa na upepo unakuta na samaki yeye chini hatuli. Anakwenda shagala bagala". ("This means that when the *kaskazi* wind is blowing even the act of looking for fish is difficult because when the water is being affected by the wind you will see that the fish within the water also do not stay still. They start to go this way and that and completely haphazardly"). (de Haan, 2020, p. 102)

When conducting scuba diving observation trips, it became clear to me that the fish behave differently during windy days, for the simple reason that the current is stronger on these days. Therefore, one can see fish who usually swim in an idyllic calm manner, swim in a manner where they frantically switch directions to avoid the current. The term *wanakimbia* is used more frequently to describe both the movements of the fish and an awareness that the fishermen perceive in the fish. Fishermen use it to explain how fish migrate to Mafia, and why they prefer it to the Zanzibar archipelago. It is also used to argue that the lack of fish is not due to overfishing; instead, it is argued, the fish are leaving Mafia. Finally, the same term is used to describe the behaviour of fish towards hook and line fishermen. Due to the fact that hook and line fishermen use a few hooks that have bait on them, the fishermen will describe how you can only catch fish that are hungry and gullible enough to latch on to the hook. In addition, when targeting a school of fish, the fisherman can only catch a few before the rest of the fish figure out what is going on. Therefore, fishermen believe that fish have more than enough intelligence to be aware of their activities, and determine ways to evade them. As one can imagine, this evasion affects, in turn, the way that hook and line fishing occurs. One hook and line fisherman explained this during his interview:

"Ni sawa mvunaji matunda, yakiisha hapa si itabidi uende sehemu nyengine ukavune upate. Ndo ilivyo. Yaani wewe muda wako nikuhama tu". ("It's just like someone who harvests fruit, when there is no more fruit in one area you

need to move to harvest in another area. That's exactly what it is like. The entire time you are just moving from place to place"). (ibid., p. 103)

The comparison that this fisherman makes to the harvesting of fruit is quite interesting and is among one of the ways that fishermen emphasise the importance of the marine environment and display their knowledge about it.

The way that fishermen use metaphors and imagery to tie their activities to the environment says a lot about how they see their position within this entire ecosystem. Within his interview one fisherman stated that "when looking at Mafia, the ocean is our farm" ("*ukizingatia Mafia, shamba ni pwani*") (de Haan, 2020, p. 103). However, there is more to this statement than meets the eye. First of all, *shamba* is a term that refers to a plot of land that can be used for both agriculture and animal husbandry. Secondly, when looking at the terms that signify the marine environment in Swahili, *pwani* is the broadest of them. *Pwani* encompasses the shore, the entire ocean and all marine animals[4]: it is a term whose closest English equivalent would be the marine ecosystem. With this statement, he reiterates the importance that the ocean and all its elements hold for all fishermen. Another fisherman takes this a step further as indicated by our exchange below.

> "*Kwasababu huku kulikuwa kuna mavuvi kuliko Zanzibar. Yaani hapa kaskazi kama hivi hakuna mavuvi kwasababu kuna upepo sana.*"
> Mariam: *Mavuvi yaani samaki wako wengi?*
> "*Ndio. Samaki wako wengi, dagaa wako wengi.*"
> ("Because here there is more fishing than in Zanzibar. I mean, right now, during kaskazi there is no fishing because there is a lot of wind."
> Mariam: By fishing, do you mean there is a lot of fish?
> "Yes. There is a lot of fish. There is a lot of dagaa"). (ibid.)

This fisherman's language is particularly interesting because he uses the term fish and fishing interchangeably. There is even ambiguity with his use of *mavuvi* because it carries two meanings. The first meaning is simply the plural of fishing, of which the English equivalent would be "fishing activities". Secondly, *mavuvi* means the knowledge that fishermen have acquired from their work experience. The Swahili word for fish is *samaki*. The fact that he uses *mavuvi* instead of *samaki* implies that he believes that his fishing activities and fishery knowledge are tied to, if not thanks to, fish. He continues to do this throughout the interview, but at one point he clarifies between fishing, the wind and fish.

> "*Watu wanatafuta uvuvi unapovulika. Wengi wameondoka. Wameenda Tanga, Somanga, Kilwa, Nyuni kwasababu mwezi 18 kutakuwa hakuna mavuvi tena.*

Sio kwamba samaki na dagaa hakuna lakini upepo ni mkali sana." ("People look for fishing where they can fish it.[5] Many have left. They have gone to Tanga, Somanga, Kilwa, Nyuni because during the eighteenth month[6] there will be no more fishing. That does not mean that there are no more fish and dagaa but the winds become significantly more intense"). (ibid., p. 104)

Here, again, he uses fishing and fish interchangeably. The first time that he says fishing he refers to fish. This is clear because the word used directly after this is *unapovulika*, which translates to "where it can be fished". However, when he uses it the second time he refers to the act of fishing itself, which he then decides to clarify by stating that it is not the fish that prevent fishing during the eighteenth month, but rather the wind. Through this language, he is not just linking the act of fishing to fish, but he is also attributing the knowledge and expertise that he has acquired as a fisherman to fish. When this is coupled with the examples of the fishermen who compared the ocean to a farm, it becomes clear that the fishermen of Mafia place great importance on their position within the marine environment.

The fishermen do not view fish as animals that lack intelligence: this is abundantly clear when looking at the terms, metaphors and imagery they use to describe them. Fish are seen as incredibly aware of the changes in wind, which then influences their behaviour. Fish, the fishermen of Mafia believe, are also aware of fishing activities, and base their migratory choices on this awareness. Fish are seen, furthermore, as animals who are capable of having preferences on where to live. Hence, fishermen will state that fish prefer Mafia to the Zanzibar archipelago. This piscine intelligence influences, in a multitude of ways, the interactions that fishermen have with the fish they hunt. Firstly, it is perceived that fish are capable of evading the traps that hook and line fishermen place. Secondly, fishermen are forced to learn the behaviour of fish in order to determine how much profit they will obtain on any given day. Thirdly, some fishermen even go as far as to accrediting the knowledge that they and other fishermen have to fish themselves. This means that fish are not just seen as a means of gaining profit, or as mere prey, but rather that they become an extension of the fisherman's intelligence. This extension is a direct cause of the daily interactions that fishermen have with these animals. It also shows how they play a part in the fishermen's blue heritage. The interaction between fishermen and fish has created new meanings and deeper value of the animals and their home. The fishermen see the ocean as something of great importance to their lives, hence why they use such broad terms like shamba and pwani to describe it. Common perception assumes that fish are simple animals, but fishermen describe them as highly intelligent. For some fishermen that intelligence exceeds their own, such that they choose to learn from fish, and credit their knowledge to fish. Having said that, to understand the depth that marine animals play within the blue heritage of fishermen in Mafia one should look at one fish in particular: the whale shark (Figs. 15.3 and 15.4).

Fig. 15.3 Sign located next to the bus stop at Kilindoni. The Swahili phrase underneath the picture of the whale shark reads "let us protect the whale sharks. Their existence is our benefit" (*Source* Author's own photo)

Fig. 15.4 Whale shark statue located next to the airport. https://www.booking.com/hotel/tz/mafia-island-tours-amp-safaris.en-gb.html?activeTab=photosGallery (Public Domain) accessed 21/08/2021

Whale Sharks: The Befriended Foe

Whale sharks, or *potwe* as they are commonly called in Swahili, are the largest fish in the world, and although they can be found in multiple places throughout the world, Kilindoni is one of the only places where they stay almost permanently. The reason for this is primarily due to their biology. Their habitat is tropical waters that lie near the Equator, and whose surface level temperatures range between 18–30 degrees Celsius. Whale sharks are filter feeders, and eat a diverse array of nekton and plankton, the production of which is ideal in waters like those found off the coast of Kilindoni. They feed by opening their mouths and sucking in water that is rich with their prey. Whale sharks spend a majority of their time feeding, and, therefore, their movement can be influenced by the prevalence of their prey in a given area where they feed, as well as by changes in wind, water temperatures and other environmental factors in that area. In the waters around Mafia, they can often be found next to the harbour, and they are often spotted at 0–5 metres deep during *kaskazi*. During this time, they are found feeding on a kind of pelagic macro-plankton known as sergestid shrimp. During the *kusi* months of February to mid-March, they change their prey, and dive into deeper waters, making it harder to observe them (Lind, 2018). One fisherman who fishes at night summed up the relationship between fishermen and whale sharks:

> *"Kwa mfano yupo samaki anaitwa potwe. Huyo ni kivutio cha utalii kwa Mafia. Potwe anatuathiri. Hii nyavu amechana mara tatu. Wao wanasema ni sisi tunamchokoza potwe ili afanye yale. Kwetu sisi potwe ni rafiki. Aliyopo potwe ndo wapo samaki. Hatuwezi kumuathiri potwe kwasababu potwe ni rafiki wa wavuvi."*
> ("For example, there is a fish called the whale shark. They attract a lot of tourism to Mafia. The whale shark harms us. This net has been torn by them three times. They say that it is us who tease the whale shark to the point that they do this. To us the whale shark is a friend. Wherever the whale shark is, there is fish. We cannot harm them because the whale shark is the friend of the fishermen"). (de Haan, 2020, p. 105)

Through this quote two things become evident about the status of whale sharks in relation to fishermen. Firstly, whale sharks are a tourist attraction in Mafia. Secondly, although the whale shark causes considerable damage to fishing gear, they are considered a friend rather than a foe. This section explores these two factors further to display the value that whale sharks represent to the fishermen of Kilindoni, Mafia.

Although research data shows that whale sharks feed off the coast of Kilindoni because their prey is easily accessible in this area, fishermen hold to their own perceptions of why this breed of fish stays close to the town. During one of my participant observations trips to the fishermen's camps, I decided to ask the fishermen why they thought that whale sharks did not swim towards the marine park boundaries. One of them told me that "*potwe anatupenda sisi na anapenda vurugu. Ndo maana yupo Kilindoni, kwasababu huku ni kama uswazi*" ("The whale sharks like us and they like commotion. That's why they are at Kilindoni, because the ocean here is like *uswazi*") (de Haan, 2020, p. 105). The use of the term *uswazi* is particularly interesting: there is no one word that can translate its meaning into English. It denotes a neighbourhood that resembles the favelas of Brazil, where houses are crowded closely together, and are connected by winding streets. However, living in the *uswazi* does not hold, on Mafia, the same stigma as living in a favela might do in Brazil, because the majority of the island's people live there. It is seen as a cultural melting pot where the traditions of the past can easily be continued, as the older and younger generations are in close contact. It is hectic and chaotic because a lot of people live in close proximity with each other. The antonym of *uswazi* in Mafia is *uzunguni*, or "where the white people live". Therefore, just with the use of the term *uswazi*, the fishermen are including whale sharks within their community. They share with them an ocean where a hubbub persists, one coming from intense boat traffic, competition for fish, whale shark tours and the whale sharks themselves. However, the sheltering of whale sharks by fishermen goes beyond pure altruism. As a marine park employee explains:

> "*Wavuvi huwa wana tamaa. Wanakimbia kundi la samaki linalokaa karibu na potwe. Potwe huwa anatembea na dagaa kwahiyo wao wanaenda kumkamata huyo samaki*". ("The fishermen are greedy. They run after the fish that swims close to the whale sharks. Usually, whale sharks swim with dagaa so the fishermen go to catch those fish"). (ibid., p. 106)

Previous research by Sophia Lind (2018) confirms these claims have been in Mafia for a while. She also clarifies that fish, specifically dagaa, exploit the same plankton that whale sharks do. However, there is also a perception among the fishermen and whale shark tour guides that this relationship has a more symbiotic nature. They explained to me that not only do the fishermen look for whale sharks to point out the presence of fish, but, also, that the whale sharks also look towards fishing boats to know where to access plankton. Although this is more a narrative than a factual claim, its existence implies that both fishermen and the people who work with whale sharks see a deeper relationship between the whale sharks and fishermen. This cements the cultural importance of the animals, by placing them within the community of fishermen, something that factors into the perceptions that fishermen have of them and of marine animals at large. This perception becomes particularly clear when looking at the physical interactions that whale sharks have with fishermen.

The perceptions that fishermen hold towards whale sharks are quite revealing, and are based on several conditions. However, before delving into these conditions it is important to note that the only fishermen who are influenced by the existence of whale sharks are those who use purse seine nets. This is because whale sharks get caught in such nets, and then tear large gashes in them when trying to escape. The majority of my participant observation was spent learning how to repair the damage that the fishermen testified was caused by whale sharks. After setting their net, when they notice that a whale shark is caught in the net, they jump into the water so as to physically steer the animal out of the net. Due to how often this occurs fishermen are able to tell whether or not the whale shark is "stubborn" through its body language. As expressed by one fishermen when he says:

> "Kwa mfano akieka mkia chini na kichwa juu huyo ni mkaidi wakutoka kwahiyo hapo kazi huna tena. Either nyavu anaichana au anatoka lakini anatoka kwa papara. Kwasababu silaha yake ni mkia, anatibua tibua. Anaweza kupiga boti mpaka ikasema basi kwa mkia tu. Kwahiyo nyie mkifiriji dagaa munaikosa. Hio ndo changamoto ya kazi". ("For example, if the whale shark puts their tail down and their head up, they will be stubborn to leave the net so there you no longer have work to do. Either they will tear the net or they will get out with a lot of disarray. Because the whale shark's weapon is their tail, they destroy, destroy. They can hit a boat until the boat says stop, with just their tail. So if you are all celebrating the dagaa, you will lose it. That is the challenge of our work"). (de Haan, 2020, p. 108)

The fishermen I spoke to all considered the presence of whale sharks as one of the greatest challenges they face in their careers. Despite viewing the whale shark as a nuisance, the fishermen also repeated that they would never harm these sharks because they regard them as their friends, because the sharks are helpful to them, and additionally (or alternatively) because sharks are created by God. Therefore, ultimately, fishermen view whale sharks as befriended foes. They are foes because of the damage that they cause to their nets, which they then have to spend hours repairing. At the same time, the fishermen see these sharks as friends, because they see a lot of benefits in sharing the same waters with them. My use of the term friend throughout this section is purposeful because it is the term that fishermen use to describe whale sharks. Not that this should be taken at face value: fishermen call whale sharks their friends because they view them as such, and try to treat them as such. The connection that fishermen have with whale sharks goes to the extent that they include them within their community. However, this relationship is an unstable one. Competition between the fishing industry and the whale shark tourism industry places it at more risk.

Currently, the relationship between whale shark tours and fishermen is one of toleration and co-operation, as fishermen will help tour guides by indicating where or whether they saw whale sharks on that given day. However, the increase in tourism activity puts a strain on this relationship because fishermen feel threatened. As one fishermen states;

"Asikudanganye mtu, Mafia kukosa bahari sio kweli. Bahari inaingiza. Sio chengine. Potwe wanasema wanaingiza lakini bahari ndo inaingiza sana kuliko potwe". ("Let no one deceive you, Mafia would not survive if it were to lose fishing.[7] The sea brings income. Nothing else. They say the whale shark brings profit but fishing brings more than the whale shark tourism"). (de Haan, 2020, p. 106)

Although, the claims that fishing brings more economic revenue than whale shark tourism is debatable, the two industries hold different weights. Fishing in Mafia affects a multitude of people. On separate occasions I was told of events that exemplified the importance of fishing in Mafia. The first event was recounted to me by a fishermen who talked about a time when fishing boats migrated from Mafia due to heavy regulation, with the effect that hunger swept over the entire community. Secondly, an employee from the WWF told me of an event that confirmed his belief that 99% of the people of Mafia are fishermen. As he describes it:

"Baada ya muda fulani, kama kipindi cha mwezi huu (mmoja) hakuna umeme Mafia. So kama hakuna umeme Mafia, maana yake hakuna mahala pakuzalisha barafu. Kwamba mvuvi akija ataweka samaki wake kwenye barafu kwa ajili ya wanunuzi wanataka kupeleka kwenda Dar es Salaam. Ataweka kwenye barafu ili asafirishe. Wanunuzi wa samaki wakaisha. Wanunuzi wengine wakakimbia kwasababu hawana cha kutunzia. Baada ya muda wavuvi wakaacha kwenda kuvua. Biashara zote zilicollapse. Zote! Iwe ya nguo, iwe ya nini zote zilicollapse, mpaka hata pengine za saloon…Nafikiri uliona wa mama wananunua samaki na wanakuja kuuza pale si ndio? Wanalea family…Kwahiyo ninapokuambia mimi wakisema serekali asilimia 80 wao wako sahihi kwasababu kuna watumishi wanaopata mshahara. Lakini mimi kwa jamii ya Mafia kama jamii ya

Mafia asilimia pengine naweza kusema mpaka asilimia 99 ni wavuvi ila hawajijui kama ni wavuvi. Wao wanajijua kama ni wafanyabiashara". ("After some time, there was no electricity in Mafia for about one month. So if there is no electricity in Mafia it means that there is no place to create ice. As in, there is no ice to create a cooling system for fishermen to use when they transport their fish to Dar es Salaam. They need ice to transport their fish. So the buyers of fish no longer bought the fish. After a while the fishermen stopped fishing. Soon every single business collapsed. All of them! Whether it was a clothing store, whether it was whatever it collapsed, even the saloon…I think you saw the women buying fish and selling it alongside the road, right? Those women have families to take care of…So when I tell you that government employees are right when they say that approximately 80% of the people of Mafia are fishers, I say so because they do not include people who obtain a salary. However, when the community of Mafia is concerned I can say that even 99% are fishers but they just don't know it. They think they are business people"). (ibid., p. 91)

Thus, the profits and influence of the fishing industry penetrate into all industries on Mafia. Furthermore, an issue in the fishing industry cannot just harm but debilitate all the other industries on Mafia, because it has a ripple effect on them. Thus, fishing may not grant the same economic revenue as whale shark tourism but it does hold vital economic importance to the society.

The value that the fishermen of Kilindoni place upon whale sharks is influenced by the increases in tourism, and in the interactions they have with the animal. The damage that fishermen attribute to whale sharks is undeniable. When asking them about whale sharks, fishermen will complain about the large gashes within their nets that the whale sharks have caused, as well as alluding to the sheer strength the animals possess. However, amidst these complaints fishermen will point out that they are friends with the animal, and they retain knowledge concerning the body language of whale sharks. Furthermore, through the terms that they use they include the whale sharks within their community, and use narratives to explain a relationship of synergy between the two species. However, the future of this relationship is uncertain. Currently, whale shark tourism is tolerated by fishermen because they share the seas with the tourists and their guides, and do not object to telling the guides where they have spotted whale sharks. In spite of this, fishermen express concern about the fact that (as they see it) whale shark tourism is given priority above fishing. At the same time, each industry holds an important place within the economy of Mafia. Tourism, be it from whale sharks or elsewhere, is indeed very lucrative, but fishing is a much more far reaching industry of both societal and cultural importance in Mafia. Tourism, or rather the competition that tourism presents, influences the relationship that fishermen have with whale sharks, which in turn affects the value they place upon whale sharks. Due to the close and daily contact fishermen have with whale sharks, the value they place upon them is of particular significance to Mafia's blue heritage.

Discussion and Conclusion

If one is to analyse the blue heritage of Mafia, one has to also reflect on the status of the sea and its animals in Swahili culture in general. I use Swahili not in the sense of the ethnic group which scholars have stated spans from the borders of Somalia to those of Mozambique, but rather in reference to the language Kiswahili. It is not my intention to group the people of Mafia into the Swahili ethnic group because, ethnically, the people of Mafia do not categorise themselves as Swahili. Instead they speak of "the people of the coast", and of the similarities and differences that the people of the coast share, including the cultures within the islands of the Mafia archipelago. Among those similarities are these peoples' shared reverence of the ocean through myths, legends and poetry. Within Jan Knappert's "Myths and Legends of the Swahili" (1975) there is the story of how the dolphin was created solely for the purpose of fetching Prophet Suleiman's sacred ring. Another legend tells of a learned fisherman who trapped a spirit using a knowledge of the stars. In the collection "Wasakatonge" (2013) the Zanzibari poet, Muhammad Seif Khatib, touches on a diversity of topics from the rich biodiversity of the reef to the feeling of reaching land after hours travelling at sea as well as the futility of fishing in neap tide. In a more recent poetry anthology from an array of Zanzibari poets the poet Said S. Mussa writes about how the fish feels after being caught by the fisherman, reflecting on its fear, and, finally, reproaching the fisherman for "taking away their rights and respect" ("*haki yao mepotea, na heshima kuondoka*") (2017, pp. 19–20). In a poem before this, Mussa shares his own reverence for the ocean, describing the beauty and dangers that lie within it. However, veneration of the ocean is not unique to these poets, nor to Zanzibar. Mahmoud Ahmed Abdulkadir, a poet from Lamu, speaks of the ocean in an almost intimate nature when he states that:

"Bahari kwangu mimi mbali nakuwa naipenda kwa ajili ya mazowezi, lakini piya bahari inanipa utulivu wa moyo na mapumziko ya roho baharini nahisi raha ambayo haielezeki labda ni kwasababu kuwemo baharini inaikumbusha nafsi yangu wakati nilipokuwa tumboni kwa mama nilipokuwa nikiogeleya katika maji yaliyonizunguka hali ambayo akili haikumbuki wakati huo lakini bila shaka roho na moyo unakumbuka hiyo hali labda ndio sababu yakuhisi utulivu ninapo

kuwa baharini". ("Beyond the fact that I love exercising in it, the sea grants my heart serenity and allows my soul to rest. In the sea I feel a happiness that cannot be explained, and maybe it is because when I am in the sea it reminds my spirit of the time when I was in my mother's womb. When I was swimming in water that surrounded me. It is a state that my mind does not remember but without a doubt my soul and heart remember it, and maybe that is why they feel serenity and peace when I am in the sea"). (Raia, 2019, p. 237)

Furthermore, it is important to note that the Swahili word for sea or ocean, *bahari*, is especially important within poetry because it carries two meanings. To fully understand the meanings of *bahari* one must understand the etymology of the word. *Bahari* comes from the Arabic word بحر (bahr). Within the Arabic language, the root of a word stays the same yet the meaning of it changes depending on the vowels placed upon the root. There are various meanings of bahr in Arabic, namely sea, river, noble man and poetic metre, depending on the vowels placed upon the b - h - r (ب - ح - ر) root. Therefore, it is no surprise that *bahari* means sea or ocean and poetic metre in Swahili. However, *bahari* can also mean categories of poetry, depending on the metre, rhyme and length of the poem (ibid.). As a result, not only is the subject of the sea important within Swahili poetry but it is tied, through its double meaning, to the genre itself. This connection between Swahili literature and the ocean shows just how prevalent blue heritage is within Swahili culture. Consequently, it should not come as a surprise that the ocean carries such importance to the people of Mafia, who see the ocean around them as a place where they may relax and throw away their worries. Those worries, they believe, will be cleansed by the ocean itself when it performs its ritual cleaning during *kaskazi*. The people of Mafia have also altered their way of viewing time, employing the lunar calendar in order to adapt to the ocean's movement. This is mostly due to the fishermen, who use the lunar calendar to plan their fishing activities. Indeed, it is through analysing the identity and interaction of fishermen with the sea and society that one can see how complex the blue heritage in Mafia truly is.

The culture and cultural heritage of the fishermen of Mafia revolves around the marine environment and the animals that call it home, and this is evident when analysing the practices, language, values, beliefs and community of fishermen. The practice of merging the lunar and Gregorian calendar acknowledges the prevalence of the moon within the activities of fishermen. Through understanding the phases of the moon, the fishermen are able to determine, and predict, the tidal variations of the ocean. At the same time, the use of the lunar calendar is also an alteration of the fishermen's perception of time because it is common to use the Gregorian calendar in Tanzania. Thus, each new fishermen must adjust the way they view time to include the moon and ocean. These alterations can also be seen in the language that fishermen use to determine the wind speed, and with which they study the topography of the ocean. The fishermen have changed the definitions of Swahili terms like *kuchapa* and *wanafunzi* to denote the presence of a reef and the speed of the wind. Moreover, their fishermen's language has been expanded to understand the body language of fish and whale sharks. Fishermen are able to determine how fish will behave on a given day depending on climatic conditions. They hold to the belief that fish are clever and cunning, while other fishermen go as far as to attach their knowledge of fishing to fish. In terms of whale sharks, fishermen have studied them so much that they think they can detect, from their body language, when whale sharks will be stubborn. They believe that they share a symbiotic relationship with this species. They follow the whale shark in order to acquire the dagaa that swim near them. At the same time, they believe in the narrative that whale sharks pinpoint their boats in order to acquire the plankton and nekton that dagaa eat. This narrative might be false but its accuracy is less important than the fact that it exists. The fishermen of Mafia believe that they hold a friendship with whale sharks, one in which both species (sharks and humans) work with each other to hunt their elusive prey. They believe that the whale sharks do not go to the marine park because they like sharing an ocean that is as crowded as the *uswazi*, where most of Mafia's fishermen live. By comparing the parts of the oceans where whale sharks call home to their own homes, the fishermen are including whale sharks in their community. As with most communities and friendships, this interspecies relationship is not without its struggles, but every fishermen that complained to me about whale sharks also stated that they would not harm their friends. Most also expressed the importance of protecting both these marine animals and the marine environment in general. Therefore, the reason why the existence of the narrative that whale sharks and fishermen possess a symbiotic relationship is significant, because it shows the immense value that fishermen place upon the marine environment. It signifies just how deep rooted their blue heritage is.

I would like to conclude this article with a sentiment that fishermen would consistently express throughout my interviews with them. All the fishermen I interviewed talk about the environment but they also all explained environmental phenomena using God. The fishermen I interviewed were Muslim. Therefore, in part, this may come from the Islamic principle of seeing God within His creation. The Qur'an is full of references to the natural world and imagery which attaches God's power to the natural world. For example,

Chapter 55 Ar-Rahman, talks about several natural occurrences after which it repeats the question "so which of the favours of your Lord would you deny" thus referring to those occurrences as Allah's favours. Part of this chapter delves into the ocean, as the excerpt below details:

مَرَجَ الْبَحْرَيْنِ يَلْتَقِيَانِ . بَيْنَهُمَا بَرْزَخٌ لَا يَبْغِيَانِ . فَبِأَيِّ آلَاءِ رَبِّكُمَا تُكَذِّبَانِ . يَخْرُجُ مِنْهُمَا اللُّؤْلُؤُ وَالْمَرْجَانُ فَبِأَيِّ آلَاءِ رَبِّكُمَا تُكَذِّبَانِ . وَلَهُ الْجَوَارِ الْمُنشَآتُ فِي الْبَحْرِ كَالْأَعْلَامِ . فَبِأَيِّ آلَاءِ رَبِّكُمَا تُكَذِّبَانِ

(He released the two seas, meeting [side by side]; between them is a barrier [so] neither of them transgresses. So which of the favours of your Lord would you deny? From both of them emerge pearl and coral. So, which of the favours of your Lord would you deny? And to Him belong the ships [with sails] elevated in the sea like mountains. So which of the favours of your Lord would you deny?). (The Qur'an, n.d., pp. 19–25)

If one reads the Qur'an, it becomes clear that the ocean plays a huge part in Islam: but the way that the fishermen of Mafia talk about the ocean goes beyond their Islamic beliefs. Even though it is clear to fishermen that the wind and the moon influences the ocean, "the movement and speed of its waters are God's own secret" (de Haan, 2020, p. 109). The animals that depend on the ocean are the creations of God. When one goes out to sea "you just pray to God that you get something". "Therefore, each day you depend on God to grant you something at the time when God decides" (ibid.). When talking to the fishermen of Mafia, it seems as if the environment is not just a reflection of God's power, miracles, and secrets, but something that deserves the same reverence and respect as God themself. Whereas with other Muslims, the environment is just a reminder of Allah, with fishermen it becomes an extension of their belief in Allah. Understanding this allows one to understand, also, just how vital the marine environment and marine animals are to the cultural heritage of Mafia's fishermen.

Notes

1. Fish which fall under the groups of anchovy, herring, sardinella, thryssa, pellona and silverside are all referred to as dagaa. Those of which are popular and often found in Mafia are from the anchovy, herring, sardinella and silverside fish groups.
2. Usually, metal plates are used for this work, because they are cheaper than ceramic plates.
3. In most usage, Kiswahili refers to the language and Swahili refers to the language and people. Personally, I do not agree with this practice, because I believe it

comes from the (colonial) tradition of oversimplifying the identity of the people who speak Kiswahili, seeing them as one single ethnic group, categorised under the umbrella term Swahili. Hence, I use Kiswahili and Swahili interchangeably in this chapter.
4. *Pwani* can also be used to describe a region that is close to the sea. For example, the administrative region of Tanzania that Mafia falls under is called Pwani. However, this is not being referred to in his original statement.
5. The translation of this statement seems clumsy because it is a literal translation: it really is "where they can fish it". *Unapovulika* means 'where they can fish it': only if the fisherman had said *unapopatikana* would 'where they can find it' have been more appropriate. By translating this phrase as "where they can fish it", I am making it easier for the reader to see how he uses fish and fishing interchangeably While this translation of that quote may be confusing at first sight, this is due to the language that the fisherman uses. I felt it important to convey that within the translation, because other fishermen in Mafia have altered their Kiswahili to reflect their connection to the sea.
6. This is according to the lunar rather than the Gregorian calendar.
7. *Bahari* means the sea in Kiswahili. However, he uses the term fishing and sea interchangeably.

References

de Haan, M. (2020). *Employees of the sea? Fisher identity formation in Mafia Island*. African Studies Centre.
Khatib, M. S. (2013). *Wasakatonge* (9th ed.). Oxford University Press Tanzania.
Knappert, J. (1975). *Myths & legends of the Swahili*. Heinemann Educational Books Ltd.
Lind, S. (2018). *Papa Potwe: An assessment of the socio-political relationship between people and whale shark around Mafia Island*. Masters. Norwegian University of Life Sciences.
Moshy, V. H. (2016). *The effects of socio-ecological changes on the livelihoods of fishing communities in Mafia Island, Tanzania*. Ph.D. thesis. Norwegian University of Life Sciences.
Mussa, S. S. (2017). Malenga wapya. In M. M. Mjawiri (Ed.), *Malenga wapya* (Vol. 12). Oxford University Press.
Raia, A. (2019). Angaliya baharini, mai yaliyoko pwani: The presence of the ocean in Mahmoud Ahmed Abdulakadir's poetry. In *Lugha na Fasihi*. University of Naples.
The Qur'an, n.d., pp. 19–25.

16

Blue Heritage in the Blue Pacific

Jeremy Hills, Kevin Chand, Mimi George, Elise Huffer, Jens Kruger, Jale Samuwai, Katy Soapi, and Anita Smith

The Blue Pacific

The Pacific Ocean is a lived, imagined, studied, interpreted, and travelled cultural space. Pacific Islanders' relationship with the ocean is 'routed' in genealogies, ancestral relations, and spiritual and totemic connections with marine creatures. Voyages conjured, anticipated, completed, and multiplied, have established and followed sea lanes creating networks across vast areas and through millennia. As such, the ocean is a place where knowledge that is essential for sustaining life has been learned and transmitted across generations, archipelagos, and islands. The ocean is tied to the land, the sky, and people, in a cosmological wholeness; it is not a separate entity, it hosts both visible and invisible life (see, for instance, Wilson, 2010). It is an ethical space, where harsh lessons are learned if and when respect and rules are flouted; it

---

The research presented in this chapter was contributed without direct funding, except for the lead author who was funded by UK Research and Innovation (UKRI) through the Global Challenges Research Fund (GCRF) (Grant Ref: NE/S008950/1 One Ocean Hub).

J. Hills (✉)
University of South Pacific, Suva, Fiji
e-mail: jeremy.hills@usp.ac.fj

K. Chand · J. Kruger · K. Soapi
The Pacific Community, Suva, Fiji

is a place of dreams, of inspiration where composers and orators go to find chants and songs; it is where the dead travel to, and where their spirits live on (e.g., Geraghty, 1993).

There are many examples of intimate and spiritual relations between men, women, gods, and sea creatures, including migratory animals such as whales, dolphins, sharks, and turtles, some of which traverse huge swathes of the Pacific (e.g., Leblic, 1989; Luomala, 1984; Oliver, 1974). Knowledge of these relationships and other cultural connections with the ocean is held within Pacific communities. Some of this heritage has been documented by early anthropologists and scholars, (mostly non-Indigenous), but it continues to be largely unrecognised in ocean policy and in the protection and management of the ocean and its resources (e.g., Ruddle, 2008). Reasons for this are multiple, and include the authority given to Western scientific knowledge and its methods, and the associated marginalisation of different kinds of knowledge and approaches (Vierros et al., 2020). Knowledge in Pacific communities is, essentially, passed on through stories, myths, legends, and oral traditions that are not considered credible according to Western scientific approaches, and which seem difficult to operationalise under the rubric of that paradigm. Such narratives are part of many other forms of cultural expressions including chants, dances, body ornamentation, sand drawings, and complex language which are commonly 'encoded'. 'Decoding' these expressions is something that requires careful and patient listening to those knowledge holders, and respect for their active participation in the generation of culturally informed and appropriate approaches to policy, including oceans policy.

Additionally, much Indigenous or traditional knowledge has been endangered or lost due to the impacts of colonisation and contemporary globalisation, and is in need of revival and renewal. Another challenge is that much of the documented and written material about ocean knowledge is getting

---

M. George
Pacific Traditions Society, Anahola, HI, USA

E. Huffer
IUCN Oceania Regional Office, Suva, Fiji

J. Samuwai
Pacific Islands Forum Secretariat, Suva, Fiji

A. Smith
La Trobe University, Melbourne, VIC, Australia
e-mail: anita.smith@latrobe.edu.au

'old' in both science and policy terms and needs to be revised using contemporary and especially Indigenous perspectives. This requires the engagement and authority of contemporary knowledge holders.

Revival of Heritage

Pathways to making cultural knowledge integral to ocean management are, nonetheless, opening. Two areas that have been essential to this movement are: (1) the revived and renewed focus on traditional navigation, which provides insight into the deep understanding of the ocean held by Pacific Islands communities including, in particular, the navigational expertise of people of communities along diverse and remote routes and networks of Oceania, and (2) the recognition of the cultural values associated with the ocean, which proceeds through formal global processes such as the inscription of World Heritage sites in the region. Both of these areas illustrate how traditional, ancestral, and cultural knowledges remain an inherent part of Pacific peoples' relationship with their ocean world, and how they remain relevant to ocean policy and management today.

Ancestral Voyaging Knowledge

Research conducted by Indigenous communities and scholars demonstrate the intimate and ongoing understanding of the ocean held by experts in ancestral voyaging knowledge, whose knowledge has been passed on through generations. Ancestral navigation experts have learnt to read the natural signs of local, regional, and trans-Oceanic seaways, and have used their knowledge of interrelated patterns of ocean and sky phenomena along inter-island seaways to create mental models of the systematic interrelationships of natural phenomena. Navigators learned to observe the movements of wind, sea, and stars that signal the change of seasons, the locations of land, the alteration of the weather in coming days, weeks, and years. They know, for example, that when the wind blows from a particular position on the horizon, then particular swell patterns and currents will form, certain animals will migrate or move in and out to sea from islands, plants will flower or bear fruit, and celestial bodies will rise, set, or reach their zeniths or nadirs. All these knowledges empowered Pacific peoples' ancestors to traverse that ocean's vast spaces, monitoring and sharing resources that make life on each island sustainable.

Ancestral voyaging knowledge is also expressed in many cultural forms known to expert practitioners, such as the place names cited in oral traditions and dance performances, and the characters and themes presented in them, e.g., songs, dances, carvings, rituals, and poems that honour or demonstrate the mistakes made and the lessons learned by the pan-Pacific Polynesian Story of Lata, the first person to accomplish construction of a voyaging vessel and a knowledge system for navigating to distant islands where particular plants and animals can be found (George, 2020a). The names of places identify the particular sea routes, seasons, and stars that are appropriate for certain voyages, or (for example) the locations and methods for gardening, fishing, or harvesting sea turtles. The names of reefs tell those who know them what is to be found at those locations, when to go to them, what alignments between those locations and other islands should be followed, and which rising star one should steer by. Some land and sea-mark names signal bearings that navigators should align with the rear and front of voyaging vessels. For example, the flowering of the *Crihimahi* plant on Taumako in the Solomon Islands signals the season of the *Te Ngatae* trade winds and of ritual events and performances, and voyages occur because winds blow from positions that fit particular seasons. Names also convey the sensation of what a navigational sign *feels* like, such as what the temperature is in a particular seaway, or how it is to experience the swell called *Te Poroporo*, when it causes a navigator to feel like tripping forward when the vessel meets the short steep swells that are reflected back after long swells have met an island or reef. Oral traditions are performed at appropriate times, such as before sleep, during a particular season, as part of a ritual, when certain stars should be observed, or when one is planning a garden. Chants, songs, poems, dances, and shapes, carvings and styles of rope lashing on vessels contain information and instructions about how to build, sail, and navigate. Images, such as the bird carving that supports the mast, and the crisscross pattern of a lashing that is called 'earthoven' are also characters in stories and art forms. Such stories often contain plot twists and have 'trickster' characters whose behaviours present problems that a navigator may encounter, and possible solutions to those problems.

For the seagoing navigator, the methods preserved in an ancestral knowledge system are mutually reinforcing and redundant. When one method is not sufficient, another can be used, such as when stars cannot be seen, the patterns of ocean swells can be felt. Competence in one or two navigational methods does not provide the safety or redundancy of competence in a complete system.

The patterns, complexities, and subtleties of ancestral navigation systems were 'the first to be lost' when colonisation by Europeans, world wars, and globalisation in general shut down inter-island networks and voyaging decreased (Lewis, 1972, p. 78). It is relatively easy for people to learn discrete navigation methods, such as following a 'path' of stars that rise and set in the same position around the horizon on a known route. But when clouds render the stars invisible, it is almost always possible to feel swell patterns that signal the direction and distance to islands, or when the sea is calm, birds can show the way to land.

Four systems of navigation that have been described in Pacific literature demonstrate how the knowledge used in each of these systems is shared between all the systems and mutually reinforces and demonstrates intimate knowledge of the ocean environment.

Te NohoAnga Te Matangi (TNTM)

This is the 'wind positioning system', as taught by Polynesian Master Navigator Koloso Kaveia of Taumako, in the Solomon Islands. It is one of many circular images reported across the Pacific that have been called 'wind compasses'. Kaveia's model of *TNTM* names 32 wind positions equally spaced around the horizon, with each position having a partner position located directly across from it, bisecting the circle. Each wind position is linked with sidereal, seasonal, meteorological, and oceanic, including swell patterns and currents, sea routes, oceanic lights (including *te lapa*) and behaviours of animals, plants, clouds, etc. *TNTM* experts know how these patterned phenomena correlate and calibrate with each other (George, 2012, 2020b).

Sidereal 'Compass'

Sidereal navigation systems include a circular array of celestial movements, positions, and relationships. Navigators in the Caroline Islands link these various elements with weather, calendrics/timekeeping, seasons, wind positions, currents, *et cetera*.

The Carolinian 'Sidereal Compass' is modeled in a stick and stone diagram of 28–36 positions in which stars rise and set around the horizon (Gladwin, 1970; Holton et al., 2015; Low, 2013). Many 'star compasses' have been reported across Oceania, but there has been little description of the detailed movements and relationships of celestial bodies with other phenomenal patterns.

Etak

Etak is a Carolinian navigation system that divides each inter-island route into 6 sections, each of which is also called an *etak*. In this deictic system, the navigator is a fixed reference point about whom the islands and ocean move. This model is inherently incompatible with cartography, which describes unmoving landmarks. The *etak* navigator monitors the movements of celestial bodies as reference points for intermediate islands or 'ghost' islands along the route. If the intended route cannot be followed, other related routes can be used. Stormy conditions, or, alternatively, good sailing weather occur, can be inferred when certain stars are observed to rise or set (Akimichi, 1980), or when the wind blows from particular positions, when there are particular swell patterns and seasons, and when ancestral signs are seen by a lost navigator (Gunn, 1980).

Swell Patterns

Swell patterns can be described with visual models such as the diagrams drawn in the sand by Kaveia of Taumako, or with stone configurations of the swell patterns that will be met at sea, such as is used on Tarawa Island to teach young students of navigation, who are made to sit down inside as if they were in a voyaging vessel, or the 'stick charts' created by Marshallese navigators to teach navigation students what the swell patterns are and how they change when the seasonal winds change. These mnemonic images represent refracted and reflected swell patterns, and demonstrate the ways in which they interact with each other in complex patterns around and between islands. These models are not maps and are not used at sea. Learning swell pattern navigation is only accomplished through prolonged sea-borne training, training that is visual, haptic, and spiritual. Swell patterns correlate with position and distance away from island(s), wind strengths and bearings, weather, calendrics, currents, rising, and setting positions of celestial bodies (Genz, 2008; George, 2020b).

What these different knowledge systems demonstrate is an understanding of the complex dynamics of the ocean, land, and sky, and their associated flora and fauna. In the past, only those who could be trusted with respecting and honoring this knowledge could be trusted with it. Today, if ocean natural resource management is to be successful and meaningful in the Pacific, then acknowledgement of the knowledge holders, support for them to reopen seaways and networks that unlock the ancestral knowledge contained in them, and investment in communications with knowledge holders in remote communities and co-creation between them and scientists to realise projects that document the contribution that navigational knowledge and all aspects of ancestral voyaging knowledge can make to ocean resource management.

Formalising Ocean Cultural Values

The Ocean Declaration of Maupiti, endorsed in November 2009 by 15 country representatives present at the UNESCO Pacific World Heritage meeting in French Polynesia, states that for many Pacific communities 'there are sacred and intrinsic links with land, sky and ocean [which] constitute . . . a fundamental and spiritual basis of existence . . . the Ocean is their identity, way of living, values, knowledge and practices that have sustained them for millennia' (UNESCO, 2009).

The cultural dimension of the Pacific Ocean is recognised in several international Conventions, including UNESCO's World Heritage List, in programmes of the United Nations and UNESCO, and by international non-government organisations active in the protection of cultural and natural heritage (Smith, 2017). These are illustrative of the depth and breadth of use of the ocean, which is seen as a pathway connecting communities, a resource for sustaining life (one whose use is shaped by customary tenure and traditional knowledge and practices), and as a shared, spiritual, storied, and ancestral place.

The transnational cultural dimensions of the Pacific Ocean can be recognised in two World Pacific sites, those of Papahānaumokuākea in the northwestern Hawai'ian islands, and Taputapuātea, in French Polynesia. Those dimensions can also be observed in the case of a third site, in Micronesia, that is currently nominated for inclusion on the World Heritage list. The significance of these sites underlines the global importance of the cultural values that Pacific peoples share across national borders and vast distances of open sea, and urges the recognition of their traditional knowledge and use of the ocean, both historically and today.

Papahānaumokuākea, a vast marine protected area in the remote Northwestern Hawai'ian Islands, was inscribed on the World Heritage List in 2010 as the first World Heritage cultural seascape. The site expresses the cultural associations, ancestral and continuing, of Native Hawai'ians with the ocean and with the Indigenous communities of Central Polynesia. The shared beliefs and ancestral connections of Hawai'ian peoples and communities in Central Polynesia, as evidenced in cosmologies, genealogies, and oral histories, bear an exceptional testimony to the strong cultural affiliation between Hawai'i, Tahiti, and the Marquesas. This relationship has resulted from long periods of voyaging over the vast distances between these island groups (Kikiloi et al., 2017). These living traditions are aspects of a wider East Polynesian cultural continuum that is also expressed in tangible form, in the unique architecture of and cultural practices associated with monumental stone structures of heiau or marae in East Polynesia, and in the marae of Aotearoa/New Zealand. These are places that represent the heart of the traditional social and spiritual life found on the islands of Papahānaumokuākea and throughout the vast Polynesian Triangle (Smith, 2016, p. 107).

Taputapuātea, in the 'Opoa Valley on the island of Ra'iatea in French Polynesia, was entered on the World Heritage List in 2016 as a cultural landscape and seascape, one whose transnational values illustrate the shared origins and histories of Polynesian communities in central Polynesia, Hawai'i, the Cook Islands, Rotuma, and Aotearoa/New Zealand. East Polynesian traditions identify the 'Opoa Valley as the place where Ta'aroa, father of the Polynesian gods and creator of all things, first entered the earth and thus created Havai'i (Ra'iatea). Polynesian communities across these islands acknowledge the island of Ra'iatea as their ancestral homeland, and Taputapuātea as the place from where navigators departed in their great voyaging canoes to find and settle these distant islands (Salmon, 2009; Smith, 2016). In 1975, the Polynesian voyaging canoe Hokule'a navigated the traditional route from Hawai'i to Taputapuātea, demonstrating for the first time that Polynesian ancestors had voyaged across the open seas using traditional navigation and seafaring techniques, and initiating a regional resurgence of voyaging using traditional wayfinding and navigation.

The third site we must consider here is a current nominee for Pacific World Heritage status, one whose nomination celebrates the international significance of the traditional knowledge of navigation and seafaring. It concerns the continuing cultural connections between communities on the islands of Yap and the nation of Palau, and is being developed by the Federated States of Micronesia (Smith, in press). The longevity of the cultural connections between communities on these islands is evidenced in Yapese 'Stone Money'—monumental stone discs, or *rai*, found throughout Yap. The *rai* were quarried in Palau, four hundred kilometres to the southwest of Yap, and then transported to that island across the open seas. Oral traditions of both Yap and Palau speak of this long cultural interaction, and of the kinships established and sustained by long-distance voyaging between the islands (Nero, 2011, pp. 140–150).

The potential for the UNESCO World Heritage Convention to provide protection for the cultural and natural heritage of the high seas, or for areas of the Pacific Ocean beyond national jurisdiction, has also been discussed in academic publications, and in documents produced by UNESCO (Freestone et al., 2016; Laffoley & Freestone, 2017). These papers primarily focus on the possibilities for the protection of marine biodiversity through the application of the Convention in areas of the Pacific Ocean beyond national jurisdiction, but note that 'the criteria for defining the OUV [outstanding universal value] of these areas goes beyond biodiversity to include... sites of historic, archaeological or cultural value' (Freestone et al., 2016, p. 55).

The increased awareness of the importance of ocean cultural knowledge in the Pacific, occurring through the documentation and practice of navigational expertise and the (associated) inscriptions of ocean based cultural sites, is an important step towards recognition of the integral place of culture in oceans. It has helped create better understandings of the value of traditional knowledge, enabling the latter to be acknowledged as an element worthy of consideration within the UNCLOS treaty on biodiversity, beyond negotiation between national jurisdictions. Culture as they key factor to consider in ocean policy and management, yet inclusion and authoritative participation of indigenous cultural knowledge holders remains marginal. For instance, the summary Implementation Action Plan of the UN Decade of Ocean Science for Sustainable Development merely states, timidly, that '[o]cean science recognises, respects and embraces local and indigenous knowledge' (United Nations, 2021).

What is needed now, if this forward momentum is to continue, is a focused effort to document the hugely rich living archive of marine knowledge held by Pacific communities, with the objective of shifting the current paradigm of ocean resource management. Some scholars and practitioners have begun to do this (Bambridge & D'Arcy, 2014; Mulalap et al., 2020; Torrente et al., 2018). The archive contains, but is not limited to, cosmological, ancestral, and totemic relationships of people with marine animals, and the linkages with other aspects of environmental knowledge (for example, meteorological or agricultural knowledge): undocumented sea routes across different sub-regions, and the natural and cultural phenomena found in these routes; Indigenous taxonomic and behavioural knowledge of marine creatures, as well as the values, practices, ethics, rituals, and ceremonies associated with the harvesting, conservation, and stewardship of marine nature.

The Pacific Blue Economy

There is a plethora of opportunities associated with the 'Blue Economy', the concept we use to denote the set of economic opportunities associated with the oceans: in the Pacific, however, the act of seizing those opportunities will not be without its challenges. In the case of the ocean and its resources, as with any other productive and lucrative natural resource, the primary focus has, in many instances, been on the economic and the financial benefits that could be attained from it, with little discussion of the 'human' component, or how the resources might be used to benefit everyone equally, especially at the community level. There is, therefore, an urgent need to ensure that the discourse on the Blue Economy embeds the rights of those who own the resources in the culture. This is particularly true in the case of those indigenous and local communities that depend on the sea for their economic livelihoods, and whose rights must be respected. In addition, it is imperative that all the socio-economic benefits derived from the 'Blue Economy' are equitably distributed amongst those at community level.

Beyond the current discourse of Blue Economy narratives at global and regional level, which have helped to put the spotlight on the need to balance sustainability with economic expansion, little work has been undertaken on the direct development impact of the Blue Economy, especially for vulnerable and ocean reliant communities. For regions such as the Pacific, reference to vulnerable and ocean reliant communities has been through the UN Sustainable Development Agenda 2030 and its associated Sustainable Development Goals (SDGs). Here, a central and transformative promise has been to 'leave no one behind' (LNOB) (UNSDG, 2021). LNOB not only entails reaching the poorest of the poor but requires combating discrimination and rising inequalities within and amongst countries, and their root causes (UNSDG, 2021).

For the Pacific, there appears to be (with a few noteworthy exceptions) an imbalance in the levels of attention paid to key components of the Blue Economy, with, consequently, missed opportunities for integration across scales, time and stakeholders; issues of power, agency, and gender remain weakly addressed even in recent Blue Economy initiatives in the region (Keen et al., 2018). The Blue Economy concept appears to have utility in the Pacific region as it brings to the fore the need to balance protection of ocean systems for the future with present, and pressing, development needs. It has thus acted some way as an engine of refinement of ocean management approaches, enabling them to better fit the regional context and heightening attention paid to place-specific issues of customary marine tenure, strong cultural ties to oceans, and the shifting power dynamics affecting ocean governance and ocean economies.

It is critical that we advocate for a Blue Economy discourse that strongly emphasises a 'justice' approach, one promoting people-centric policies that include indigenous and cultural dimensions. The estimated value of the Pacific Blue Economy is well in the billions of US dollars; for example, the Melanesia Ocean Economy alone is estimated to be worth US$548 billion (Hoegh-Guldberget & Ridgway, 2016). Yet despite 'owning' this vast lucrative resource, most Pacific Island countries remain underdeveloped, with 6 Pacific countries being listed in the top 40 poorest countries in the world. According to the Guardian, a newspaper, there are instances where, when it comes to natural resource use, the Pacific communities see less than 12% of the final value of the resources being extracted, with little compensation or reinvestment in the countries which own the resources: this means that despite their collective natural resource wealth, GDP per capita remains low for many Pacific island countries (*The Guardian*, 2021).

The Blue Economy concept must therefore ensure that the distributive injustice experienced by many Pacific Island countries surrounding the use of their natural resource is ended. Proper respect needs to be paid to those that rightfully own those places in which culture, identity, and resources are embedded. There needs to be an appreciation that the communities in the Pacific possess an almost globally unique right to manage the ocean resources that they own or once owned (Johannes, 2002). These ownership rights, in at least some cases in the region, are constitutionally assured (Lane, 2006) and need to be respected. This provides communities with the opportunity to ensure they are engaged with by Government, when formulating locally driven policies relating to the resources they have been entrusted with.

To ensure that a people centred Pacific Blue economy is not lost in the sea of competing actors and agendas, a global effort to protect the ocean and the ocean ecosystem is needed (Cokanasiga, 2020). There needs to be an improvement in engagement mechanisms, which includes all relevant actors to ensure proper stewardship of the ocean (Cokanasiga, 2020). For this to eventuate the Blue Economy discourse must meaningfully include the very people who depend on coastal areas and marine resources for their livelihoods and identity—thus prioritising indigenous peoples and local communities. They who live close to the ocean, and who have long depended on it, have spiritual, cultural, and traditional links to the ocean, and their understanding and leadership must guide future ocean discourse and governance (Cokanasiga, 2020).

Governing the Blue Pacific

Balancing Development

For centuries, Pacific people have had a deep and intrinsic connection to the ocean, one that was interwoven with their culture and society and which is, today, increasingly a means of driving their economies. Pacific people have been the custodians of the Pacific Ocean for millennia and have always recognised that a healthy, productive, and resilient marine ecosystem is the foundation of sustainable livelihoods. Even today the ocean remains central to the identity of Pacific people, as it guides their local, national, regional, and international policies. Core to this engagement is the recognition of the need to protect and manage ocean resources, and their accompanying culture and traditional knowledge. However, within the dimension of the blue economy there is an increasing mismatch between short-term economic gain and long-term prosperity, as competing uses, extreme events, and cumulative impacts take a toll on the goal of balancing protection with production, whilst also eroding the heritage, culture, and identities that make the Pacific unique.

The UN Decade of Ocean Science for sustainable development provides a significant opportunity for the Pacific region to take a holistic perspective on ocean science, ocean use, and management and the implementation of international, regional, and national measures. Any such initiative will require effective regional collaboration, as the challenges and aspirations involved will be best met collectively, and similar economic activities often occur in several jurisdictions, and have impacts that can be widespread and transboundary. These impacts include those from climate change, which is the single greatest threat to the livelihoods, security, and wellbeing of the people of the Pacific. The future of the ocean depends on our ability to address this issue properly.

Pacific Island Countries and Territories (PICTs) are increasingly recognising the need for a comprehensive perspective on the management of the marine environment and are turning to the development of National Ocean Policies (NOPs) to address these challenges in coordination with regional policies. This section focuses on regional ocean management, consideration for culture and the role of the UN Decade of Ocean Science in supporting these initiatives.

Regional Ocean Institutions

The Pacific Island Countries (PICs) are organised politically through the Pacific Island Forum (PIF). This group also includes Australia and New Zealand, although the PICs within PIF do not coordinate with these two more developed PIF member states. There is a distinct but related Pacific Small Island Developing States (PSIDS) group based at the United Nations in New York, and it is through this group that the PICs coordinate international negotiations and diplomatic efforts related to climate, oceans, and sustainable development.

The PICs are also supported by a range of regional organisations that serve various functions, from the promotion of cooperation, trade, and economic development, to fisheries management, and science. Regional organisations, groupings and entities that make significant contributions to the advancement of regional ocean priorities and the implementation of SDG14 include the following:

Pacific Islands Forum (PIF): Since its establishment in 1971, the PIF has been the region's principal political and economic grouping. Its membership comprises the 18 states of Australia, Cook Islands, Federated States of Micronesia, Fiji, French Polynesia, Kiribati, Nauru, New Caledonia, New Zealand, Niue, Palau, Papua New Guinea, the Republic of Marshall Islands, Samoa, Solomon Islands, Tonga, Tuvalu, and Vanuatu. The PIF is committed to regionalism which is defined as a 'a region of peace, harmony, security, social inclusion, and prosperity, so that all Pacific people can lead free, healthy, and productive lives' (PIF, 2014). The PIF is also committed to the principal objectives of sustainable development: these are inclusive and equitable economic growth, enhanced governance and strengthened legal, financial and administrative systems, and the promotion of peace and security for all, as set out in the Framework for Pacific Regionalism (PIF, 2014).

The PIF convenes an annual Leaders' Retreat to develop collective responses to regional issues. Decisions by the Leaders are outlined in a Forum Communiqué, from which policies are developed and a work programme is prepared.

Pacific Community (SPC): The Pacific Community (SPC) is the principal scientific and technical organisation in the Pacific region, and has been supporting development in the region since 1947. SPC is owned and governed by its 26 members, including Australia, France, New Zealand, and the United States of America. Its focus includes work across more than 20 sectors, including cross-cutting issues on climate change, disaster risk management, food security, and ocean management.

Pacific Islands Forum Fisheries Agency (FFA): The Pacific Islands Forum Fisheries Agency (FFA) was established in 1979 to help countries sustainably manage their offshore fishery resources. FFA is based in Honiara, Solomon Islands, and is an advisory body providing expertise, technical assistance, and other support to its 17 members who make sovereign decisions about their offshore fishery resources, and who participate in regional decision making on fisheries management through agencies such as the Western and Central Pacific Fisheries Commission (WCPFC).

Secretariat of the Pacific Regional Environment Programme (SPREP): The Secretariat of the Pacific Regional Environment Programme (SPREP) was established in 1993 to promote cooperation in the Pacific region, to assist in the protection and conservation of the environment and to ensure sustainable development for present and future generations. Based in Samoa, SPREP brings together 26 member governments, including 21 Pacific Island countries and territories.

Council of Regional Organisations in the Pacific (CROP): The Council of Regional Organisations of the Pacific (CROP) was established to improve cooperation, coordination, and collaboration amongst the various intergovernmental regional organisations to work towards achieving the common goal of sustainable development in the Pacific region. It comprises the above-mentioned regional organisations, amongst others, with the PIF Secretariat as the permanent chair of CROP as well as its main secretariat support.

The scope of CROP is to provide high-level policy advice, and a forum to collectively respond to and follow up on regional priorities which are prepared from the Forum Communiqué. The annual CROP strategic work agenda sets out how the organisation will work together to deliver on Forum Leaders' decisions. The 2021 CROP priorities include sustainable ocean and sea-level rise and maritime zones. The progress is monitored by the CROP Heads through quarterly meetings, which provide a progress report to the Forum Leaders at their annual meetings.

The CROP coordination arrangements include task forces and working groups. The four task forces are formal, result-oriented, and time-bound, and the current architecture includes a task force on international engagement and advocacy for ocean events. The existing eight working groups are informal and ongoing, and include the Marine Sector Working Group (MSWG). The MSWG was established in 1997 to facilitate the coordination of regional activities in the development of a regional strategy for the marine sector which resulted in the Pacific Islands Regional Oceans Policy (PIROP) in 2002 and the Framework for a Pacific Oceanscape (FPO) in 2010. Regional and institutional frameworks and cross-cutting issues touching on the marine sector continue to develop considerably, and current discussions are converging on the development of the 2050 Strategy for the Blue Pacific Continent.

Office of the Pacific Ocean Commissioner (OPOC): The OPOC was established under the Pacific Island Forum Secretariat, and is tasked with supporting the Pacific Ocean Commissioner in its role to strengthen ocean governance and coordination and collaboration between stakeholders. This role also includes tracking progress made with respect to the FPO and regional ocean priorities, the product of which can be found in the 2021 Blue Pacific Ocean Report The Pacific Ocean Alliance (POA) was established under the office of the Pacific Ocean Commissioner to further improve ocean governance by connecting all ocean players, coordinating the implementation of integrated ocean management and cross-sectoral ocean discussions, and the provision of high-level attention to ocean priorities and processes in the Pacific. *Pacific Community Centre for Ocean Science (PCCOS)*: SPC established the Pacific Community Centre for Ocean Science (PCCOS) in 2017 to provide members with excellence in ocean science, knowledge, and innovation and to support the conservation and sustainable use of the Blue Pacific Continent. The PCCOS mission is to be the platform for coordination of ocean science, knowledge and innovation and delivery of multi-disciplinary multi-sectoral integrated programming to assist the Pacific Community in achieving the Sustainable Development Goal 14 and contributing to other SDGs in the emerging framework of the Blue Pacific Continent. In particular, PCCOS aims to deliver integrated scientific services supporting ocean management, governance, and observations.

Pacific Youth Council: The Pacific Youth Council was established in 1996 as a regional voluntary, non-governmental organisation for and of national youth councils in the Pacific. Its mission is to effectively represent youth organisations throughout the Pacific on policies and issues that affect youth. The PYC aims to assist in the development of a collective voice and action on issues that concern Pacific young people, including through the promotion of a regional youth identity that is sensitive to the PYC member countries' spiritual, cultural, social, economic, and political diversity at all levels. PYC has an Executive Board that works on a voluntary basis and meets every three years to elect a new board.

Ocean Law and Policies

The development of ocean laws and policies have evolved over time with many conflicts ensuing over access rights and ownership. For example, between the fifteenth and seventeenth century there were many claims over entire swaths of ocean using the principle '*mare clausum*' or closed seas (Walker, 1945). These claims were largely political and related to exploration and the exploitation of new worlds. Spain, for example, claimed much of the Pacific Ocean in that period. Hugo Grotius soon proposed a new framework that made the case for the concept of '*mare liberum*', or free seas. This was a counter to the doctrine of *mare clausum*, and suggested that the ocean should be open to all. This principle is now enshrined in UNCLOS as 'freedom of the High Seas', and continues to be applied in the world's seas and oceans. Meanwhile, in the Pacific Ocean and the many island nations therein, this did not affect the way Pacific Islands and their people used and managed their oceans. There is a deep connection between Pacific Islands and their terrestrial and oceanic spaces, with many communities having complex customary marine tenure claims, which are often rooted in systems of communal ownership.

The overarching framework for oceans law and policies is primarily guided by the United Nations Law of the Sea Convention (UNCLOS). National legislatures in the region establish their own laws and policies to guide the use and management of oceanic resources. The regional ocean governance architecture is supported by international agreements and regional policies. Prior to the emergence of contemporary ocean policies, the use and management of oceanic resources was guided by the customary marine tenure systems which varied across the Pacific Island regions, and which often shared common features of communal ownership and access to resources within these zones. Today, customary marine tenure still exists, but the extent of its reach and acceptance varies and is typically limited to coastal zones. Below is an overview of relevant ocean laws and policies.

The United Nations Convention on the Law of the Sea: Passage of the United Nations Convention on the Law of the Sea (UNCLOS) brought about significant change in the way the Pacific approached the ocean. Prior to its entry into force the common practice held that state sovereignty extended to three nautical miles from the coastal line, in accordance with the cannon shot rule (Walker, 1945). UNCLOS introduced an Exclusive Economic Zone boundary that extends for two hundred nautical mines, extending sovereign rights and responsibilities of coastal states. Compared to previous claims, this represented a significant expansion. Fiji was the first country to ratify UNCLOS and accordingly played an important role in the negotiation of UNCLOS through its then Ambassador to the United Nations, Satya Nandan.

UNCLOS forms the backbone of global ocean policies today, by creating the framework for ocean governance, maritime jurisdiction, and management. As large ocean states, the securing of the maritime zones and limits of the Blue Pacific Continent underpins the realisation of sustainable development for PICs. Pacific leaders have consistently reaffirmed their collective commitment to the conclusion of outstanding maritime boundary claims and zones through the Forum Communiqués. Ultimately, the states of the Pacific region need to effectively manage their ocean resources, and establishing their exclusive economic zones and extended continental shelf areas with certainty provides the foundation for this.

Subsequent implementation agreements (such as the 1994 Agreement on the Implementation of Part XI of UNCLOS and the 1995 United Nations Fish Stocks Agreement) led to the expansion of UNCLOS' role to include regulation of deep-sea mining activity, and of fisheries management in areas beyond national jurisdiction.

The UN Fish Stocks Agreement helped establish Regional Fisheries Management Organizations (RFMOs), such as the Western Central Pacific Fisheries Commission (WCPFC) to which many PICs belong. RFMOs such as WCPFC help to coordinate the joint management of migratory and straddling fish stocks in the Western Central Pacific region. Technical agencies in the Pacific, such as the Pacific Island Forum Fisheries Agency, were created to support the technical management and implementation of regional fisheries programmes and plans.

The 1994 Agreement on the Implementation of Part XI of UNCLOS relates to deep-sea minerals beyond the jurisdiction of states and created an entity to regulate such activities, the International Seabed Authority.

Currently, negotiations are underway with a view to the creation of an international legally binding instrument under the United Nations Convention on the Law of Sea which will assure the conservation and sustainable use of marine biological diversity of areas beyond national jurisdiction (BBNJ). This represents a further expansion of the regulation of activities in areas beyond national jurisdiction. With BBNJ there is a focus on Area Based Management Tools, Environmental Impact Assessment, Marine Genetic Resources, and Capacity Building and Technology Transfer In the Pacific, Traditional Knowledge is an important aspect of these BBNJ negotiations. The existence of forms of traditional knowledge and associated custodianship that relate to the ocean and its resources is something that predates the establishment of current national borders. It continues to inform resource access and use of marine areas and resources. This custodianship and knowledge could also include locations outside of national jurisdiction, such as long-established open ocean voyaging routes over the high seas, traditional knowledge of navigation and seafaring (including knowledge of weather, environment, and marine biological diversity) and places with substantial sacred significance (Mulalap et al., 2020).

Sustainable Development Goal 14: The Pacific Islands via the Pacific Small Island Developing States were instrumental in the inclusion of *SDG 14 Life Below Water* in the *2030 Agenda for Sustainable Development* (Quirk & Hanich, 2016). SDG 14 relates to the conservation and sustainable use of the oceans, seas, and marine resources for sustainable development. Members of PIF have committed to the full implementation of the 2030 agenda and the SDGs, and it is implemented via *The Pacific Roadmap for Sustainable Development*. SDG 14 features prominently in Pacific regional and national policies.

The Pacific Island Countries have also, however, developed their own regional ocean policies. These include the Pacific Island Regional Ocean Policy (PIROP) and the Framework for a Pacific Oceanscape (FPO).

Pacific Islands Regional Ocean Policy (PIROP): This regional policy was endorsed by Pacific Island Forum Leaders in 2002. The focus of PIROP is the sustainable use of ocean resources: it was envisioned as a template for PICs to adopt and adapt for their national purposes and contexts, and to feed into their national ocean policies. There was a subsequent framework to assist with implementation, the Pacific Islands Regional Ocean Policy Framework for Integrated Strategic Action (PIROF-ISA); however, this was replaced with the *Framework for Pacific Oceanscape* (FPO) in 2010.

Framework for Pacific Oceanscape: The Framework for a Pacific Oceanscape (FPO), which effectively replaced PIROF-ISA in 2010, works to strengthen PIROP, particularly through stronger provisions in the areas of coordination, resourcing, and implementation. One of the FPO's three objectives is 'integrated ocean management that responds to nations aspirations and priorities'. The FPO does not, however, demonstrate how to implement integrated ocean management, and this chapter suggests some avenues to achieving this.

Parties to the Nauru Agreement: The group of states named Parties to the Nauru Agreement (PNA) are signatories to a sub-regional agreement regulating the terms and conditions for tuna purse seine fishing licences. PNA members include the Federated States of Micronesia, Kiribati, Marshall Islands, Nauru, Palau, Papua New Guinea, Solomon Islands, and Tuvalu. PNA members work together to sustainably manage their shared tuna stocks with the goal of increasing the resulting economic benefits.

National Ocean Policies

National Ocean Policies: In addition, to regional ocean-related policies, there are a number of National Ocean Policies, or similar ocean centred national plans, which have been developed by nation states. These include the following national policies which highlight integrated approaches:

Cook Islands Marae Moana Policy: Integrated management is noted as a principle. Principles in the Marae Moana Policy are stated to be applied to all decisions and actions relating to the Marae Moana. Principle 6 is Integrated Management: The integration of decision making by all relevant stakeholders (Government, non-government and external partners) should be pursued in the operationalisation of the Marae Moana.

Fiji's National Ocean Policy: Fiji is committed to the integrated management of its ocean and marine resources. The policy provides a holistic framework for integrated action and partnerships on all of Fiji's national, regional, and global commitments on the ocean. The policy frames a progression to the integrated management of all of Fiji's Ocean by 2030, to ensure the resilience and sustainability of marine ecosystems whilst maximising opportunities for socio-economic benefits. The country's 2021 Climate Change Act provides a legal underpinning for its National Ocean Policy.

National Ocean Policy of Papua New Guinea: The NOP framework is based on the principle of Integrated Oceans Management (IOM). There is now consensus amongst the mainline organisations in different government ministries for a paradigm shift to embrace a national oceans policy based on the IOM concept. Under the principles of good governance for an NOP policy, the implementation of oceans programmes and activities must also include the region's traditional ocean communities and strive for gender, inter-generational, and geographic diversity amongst its stakeholders. The NOP is a national strategic policy document that sets out a planned, system-wide, approach to ocean management, and which incorporates all sectors and levels of governments in an integrated mode of planning, developing, and implementation of policies.

Samoa Ocean Strategy: Integrated Management is built into the title of Samoa's ocean policy: 'Samoa Ocean Strategy 2020-2030 Integrated Management for a Healthy and Abundant Future of Samoa's Ocean'. This strategy is posited as an integrated policy framework, one that, noting how Samoa has developed a comprehensive strategy towards integrated management of their Exclusive

Economic Zone (EEZ), seeks to support economic growth, protect important ecological habitats, and safeguard important sources of protein and income for Samoans.

Solomon Islands National Ocean Policy: The Solomon Islands National Ocean Policy (SINOP) is, likewise, framed as a strategic roadmap for integrated management and governance of the ocean. The policy defines Integrated Ocean Governance as involving:

A cross sectoral commitment to promote and implement ocean management through an ***integrated framework*** that recognises the critical importance of goods and services provided by the ocean. (SINOP, 2018, p. 8)

Vanuatu National Ocean Policy: In this case, a novel policy notes that the existing policy framework is inadequate for supporting the integrated approach needed in ocean governance. Fundamental changes in the way the marine space is managed and regulated are required at the Central Government, Provincial Governments and the community level. Central to such a change is the need to transition to a more integrated governance approach that requires all uses, users, and values to be considered, whilst empowering traditional leaders and traditional marine resource managers. The policy also notes a need to embrace global and regional initiatives and local traditional-knowledge approaches that appeal for an integrated ecosystem-based approach to the management of the ocean.

As the development of ocean-centric policies have proliferated in the Pacific, and as this proliferation has coincided with emerging development agendas such as the Blue Economy, culture has become central to the success of both policies and of development agendas more broadly. Pacific culture includes the beliefs, customs, traditions, and practices of the range of Pacific peoples. An important reason for this is that culture and tradition exist, in the Pacific context, as the dominant norm. For example, customary land tenure, as well as customary marine tenure, is prevalent across the Pacific Islands, and this is bolstered by the fact that the indigenous peoples of the Pacific are not a minority, but form, rather, the majority of Pacific populations. Culture and traditions remain widely revered, and continue to persist even in the modern Pacific paradigm.

There is much to learn from traditional cultural ocean management practices. For example, Fiji's '*tabu*' and Palau's '*bul*' illustrate robust fisheries management and stewardship that existed prior to the introduction of modern marine protected areas. Many Pacific Island jurisdictions continue to recognise, in their national laws and policies, customary marine tenure and these traditional forms of marine conservation and sustainable use of marine resources.

In February 2021, Forum Leaders at the Special PIF Leaders Retreat endorsed the *Pacific Islands Forum Leaders Oceans Statement 2021*. This is a statement calling for great global commitment and action to protect and preserve our Ocean: it includes a commitment to strong regional action, that will harness shared stewardship of the Pacific Ocean, and allow its signatories to act as what the statement describes as one 'Blue Pacific Continent' (Pacific Islands Forum Special Leaders' Retreat, 2021). The statement reaffirms the duty to care for the planet and ocean and to manage the resources carefully as an endowment fund, one inherited from our ancestors and which we must care for, invest in, and nurture if we are to continue to benefit from it, and to share it with future generations.

This Leader's Statement was informed by the Blue Pacific Ocean Report, that consisted of 'a comprehensive, multi-facetted, cross-cutting and holistic review and stock take of the state of affairs of ocean governance in the region'. The Blue Pacific Ocean Report emphasised that adequately addressing the stressors and pressures on the Pacific Ocean cannot and should not be done in 'silos', and reinforces the call, made in the 2010 Framework for the Pacific Oceanscape, for effective support for the ocean to regain and maintain its health, productivity, and resilience through cross-sectoral cooperation, and an integrated approach in governance and implementation of plans, activities, and measures. The Blue Pacific Ocean Report, however, recognises that to date, integrated ocean management has not been adequately implemented, even if progress in some areas of the Framework for Pacific Oceanscape has been encouraging. The Blue Pacific Ocean Report further notes the opportunity for the region to build a plan for a decade of accelerated regional ocean action spanning the period 2021–2030 and will be in line with the ocean agenda that is a priority for the region requiring transformative action. This opportunity supports the 2030 Agenda for Sustainable Development, in particular SDG 14 on the ocean.

In August 2021, Pacific Island Forum Leaders issued the Declaration on Preserving Maritime Zones in the Face of Climate Change-Related Sea-Level Rise (PIF, 2021a). This landmark declaration has been many years in the making, since Leaders first committed to preserving the rights stemming from maritime zones in the Framework for a Pacific Oceanscape (2010). This latest effort started to gain momentum in 2019 when the International Law Commission included Sea-Level Rise in international law in its programme of work. With the Declaration on Preserving Maritime Zones in the Face of Climate Change-Related Sea-Level Rise, the Pacific Island Forum seeks to develop a solution that both preserves maritime zones in the face of sea-level rise and climate change, and also upholds the integrity of UNCLOS.

This technical, legal, and diplomatic effort has been relatively successful, and employs an approach that centres on fostered cooperation for the delivery on the Leaders' commitment. The Pacific Maritime Boundaries Programme coordinated by SPC, with support from its consortium partners, has held 20 regional maritime boundaries working sessions since the early 2000s. This long-term sustained approach has allowed the region to build a community of practice with strong technical and legal capacity, and has fostered the emergence of national ocean champions (Frost et al., 2018). This exemplifies the strong cooperation in the Pacific on ocean issues, and provides a model for the co-design and co-delivery of other regional efforts including the integrated ocean management programme under the UN Decade of Ocean Science for Sustainable Development, 2021–2030.

The UN Decade of Ocean Science

The UN Decade of Ocean Science for Sustainable Development, 2021–2030, provides a significant opportunity for the Pacific region to take an innovative approach to ocean science, ocean use, and management and the implementation of related international, regional, and national measures. Any such initiative will require effective regional collaboration as challenges and aspirations are best met collectively, and natural phenomena and economic aspirations often occur across several jurisdictions, and have impacts that can be widespread and transboundary. Such impacts include those from climate change, which is the single greatest threat to the livelihoods, security, and wellbeing of the people of the Pacific, and the future of the ocean depends on our ability to address this issue adequately.

Regional collaboration in the Pacific has always been strong, and a joined-up approach to integrated ocean management will provide a mechanism for sharing data, information, knowledge, legislative approaches, and experiences. An example of this is the integrated way in which the region has concluded maritime boundaries over the last two decades. This model is key to addressing large scale ocean issues, bridges the sectoral based management system that currently exists, and is a direct action that is going to be addressed by the UN Decade of Ocean Science endorsed Pacific Programme 'Pacific solutions for a healthy Blue Pacific Continent: Integrated Ocean Management to sustain livelihoods today and into the future'.

The Pacific Decade of Ocean Science Programme seeks to increase scientific capacity and create opportunities for ocean science to feed into decision making and bridge the gap between science and policy. This will be achieved by focusing on three major aspects: law and policy, decision support systems, and increased consideration of Pacific culture and context. As custodians of the world's largest marine ecosystems, termed the Blue Pacific Continent, this flagship Programme seeks to reverse the decline in ocean health and provide solutions to some of our existential challenges such as climate change.

The development of this Programme was achieved through several years of extensive consultation with major stakeholders involved in ocean governance, and commenced with the preparatory meeting of the UN Decade of Ocean Science in 2019 in Noumea, New Caledonia. This meeting provided the foundation for a determination of the needs and approaches required for an effective and holistic development of an integrated ocean programme in the Pacific (SPC, 2019). Representatives from scientific organisations, traditional and indigenous knowledge specialists, youth representatives, media partners, cultural experts, and representatives from agencies across the Pacific were consulted. Subsequently, after the Noumea consultation meeting, a mapping exercise of existing national ocean policies (NOPs) was undertaken by SPC in 2020. The analysis showed that a priority across all the NOPs was to include the principles of IOM. The policies commonly called for an understanding of the totality of ocean uses and pressures, as well as support for the long term, sustainable use of ocean resources. IOM effectively balances economic gain and development with preservation of the long-term health of the ocean—an issue the flagship programme seeks to directly address within the Pacific context.

To collectively achieve the outcomes and goal of this 10-year Programme, the following three thematic areas of work are proposed:

1. Develop robust legal and governance frameworks to implement national ocean policy.
2. Improve decision support systems so that policies are based on evidence-based science.
3. Integrate Pacific traditional knowledge, culture, and wisdom to support inclusive, consultative decision-making and improve decision support systems.

To establish the foundation of this ten-year programme and contribute to its results, several actions are proposed. Key activities include support for National Ocean Policies development and implementation, along with regional integration: improvement of knowledge platforms and development of science products; online and in-person capacity strengthening, and the fostering of 'champions' in ocean management; and the development of effective ocean management approaches in a community of practice which integrates traditional knowledge and science, communicated in a culturally sensitive and efficient way.

Moving Forward with Integrated Ocean Governance

The Pacific Region's ocean governance regime is robust, with links to global initiatives, whilst also maintaining a Pacific-centric focus through its emphasis on culture and the Pacific nations' connections to the ocean. UNCLOS provided a new mandate for the extension of maritime boundaries, and a slew of sovereign rights within the exclusive economic zones that extend for a distance of two hundred nautical miles from the shore.

Climate change and its impacts, environmental degradation, and development pressures are growing stressors in the Pacific, and are imposing direct and cumulative impacts on inherently vulnerable coastal communities. Whilst the FPO reflects the principles of integrated ocean management, the current practice in the region generally tends to manage the ocean domain on a largely sectoral basis. Recent National Ocean Policies in the Pacific demonstrate a growing move to begin integrating sectoral activities and the corresponding agencies.

Integrated ocean management aims to support a sustainable ocean economy that uses ocean resources in ways that preserve the health and resilience of marine ecosystems and improve livelihoods and jobs, balancing protection and production to achieve prosperity. To achieve this goal, it brings together relevant actors from government, business and civil society and across sectors of ocean industry. IOM is not a new concept in the Pacific, with several successful examples being deployed across the region (albeit on a smaller scale), from the Coral Triangle Initiative to the Reef to Ridge programme. These programmes used holistic approaches to resource management that engaged a broad range of stakeholders and exhibit a focus on co-management.

Whilst the 2010 FPO noted that there is little progress at the national level in an integrated approach to ocean governance, with not a single Pacific Island country considering development of a national ocean policy at that time, more than a decade later the region now boasts no less than 8 national ocean policies and plans. PICs are also setting up national institutions or agencies dedicated to coordinating ocean affairs, as is currently underway in Papua New Guinea and Vanuatu.

This change is beginning to highlight IOM as a key priority of Pacific national ocean policies. Ultimately, the task of addressing the multiple existential threats facing Pacific nations and their ocean requires a holistic and innovative approach that is difficult to achieve using current practices of isolated sector-based management. However, it is also important that those seeking to accelerate the uptake of IOM to avoid adopting imported concepts, as this manoeuvre would be plagued with pitfalls. Instead, they should continue the critical discourse on the key issues and, in so doing, develop an improved understanding of what IOM represents in the context of the Blue Pacific Continent.

The development of policy frameworks or programmes of action can be rife with unforeseen and negative issues, with activities failing because of unachievable goals or timelines, poor resourcing and design, or the absence of the kind of consultation required to achieve 'buy-in' from a broad base of stakeholders. The Pacific region can, however, already demonstrate positive outcomes of the investment in training, mentoring, and the establishment of strong regional networks to support ocean science in the maritime sectors. Experience has shown that, for a community of practice to emerge, it is essential to ensure a sustained momentum that couples a harmonised regional approach with technical and legal support to country teams. This will facilitate the emergence of ocean management champions and build the foundations for political buy-in.

A more recent development, and a potentially significant one, relates to the Ocean Decade, the UN's Decade of Ocean Science for Sustainable Development (Heymans et al., 2020). A crucial part of the Ocean Decade will be the engagement of youth and Early Career Ocean Professionals (ECOPs). This is crucial, as these youth and professionals can make vital contributions to the Decade by becoming active participants in, and advocates for, the decade, and, ultimately, by continuing the post-2030 legacy of the Ocean Decade. The Pacific Community Centre for Ocean Studies (PCCOS, see above) is working towards the formation of a Pacific ECOP network (ECOP Pacific) that will actively involve youth organisations and networks (including the Pacific Youth Council) in Ocean Decade activities. Pacific ECOPs are an integral part of our collective resolve to manage the Pacific Ocean and contribute to ocean science, ocean governance, marine conservation, and many more important (and often neglected) concerns. This network aims to empower Pacific youth to participate and to contribute solutions and their perceived priority actions to sustainably manage the Pacific Ocean to protect the people, places, and prospects of the Blue Pacific.

The Pacific Decade of Ocean Science Programme under the United Nations Decade of Ocean Science for Sustainable Development (2021–2030) seeks to increase scientific capacity and create opportunities for ocean science to feed into decision making and bridge the gap between science and policy. This will be achieved by focusing on three major aspects of the problem: law and policy, decision support systems, and the need for increased consideration of Pacific culture and context. As custodians of the world's largest marine ecosystems, 'the Blue Pacific Continent', this flagship Programme seeks to save our Ocean and provide solutions to some of the existential challenges facing us and the wider world, such as climate change. This Pacific Decade of Ocean Science Programme will involve actions enabling collective progress towards the achievement of the outcomes and goals of the 10-year programme.

Conclusion

This narrative has recognised that, due to the impacts of past colonisation and contemporary globalisation, Pacific indigenous and traditional knowledge has been endangered or even lost: it has further recognised this knowledge is in need of protection and revival, particularly in terms of reopening a huge number of inter-island routes and networks. This would unlock a rich reservoir of ancestral voyaging knowledge and culture that has not yet been recognised or connected with scientific research efforts and policy schemes. There is a need for a more sustained effort to include indigenous perspectives in policy and action, requiring the engagement of contemporary knowledge holders and there are some signs of revival and a renewed focus on blue heritage and culture. A focus on the Blue Economy has been emerging in the region, but any discourse on that Blue Economy must emphasise the need to ensure that there is respect for the rights of those who own the resources and are embedded in the culture. This is particularly so in the case of the indigenous communities, as well as those local communities that depend on the sea either for their economic livelihoods or for their subsistence.

In governance terms, the region is well provided for in regional institutions and regional policy, which is being further underpinned by National Ocean Polices. There appears to be increasing vigour in the recognition of the vitalness of the ocean for Pacific peoples and Pacific leaders. There is a progressive trend for the forging of further integration in ocean governance through regional and national level mechanisms, moving away from a physical and resource dominated view of the ocean. The tendency towards a more integrated approach opens up an opportunity in the region for further inclusion of blue heritage in policy and in decision making, and global efforts towards the formalisation of cultural values strengthen this process. Vehicles for promoting integration, such as the Pacific Decade of Ocean Science Programme (coordinated by SPC) and the Pacific 2050 plan (coordinated by PIFS), have been framed, setting out a path forward. To navigate this path, we may reflect on our ancestral Pacific voyagers, who set course over the blue horizon while trusting in their arrays of knowledge, and their honed abilities to integrate multiple knowledge systems.

References

Akimichi, T. (1980). Stars causing storms and the perception of natural phenomena: An ethno-meteorological study of Satawalese society. *Kikan Jinruigaku, 11*(4), 3–51.

Bambridge, T., & D'Arcy, P. (2014). Large scale marine areas in the Pacific: Cultural and social perspectives. In F. Féral & B. Salvat (Eds.), *Gouvernance, Enjeux et Mondialisation des grandes aires marines protégées: Recherches sur les politiques environnementales de zonage maritime, le challenge de la France de Méditerranée et d'Outre-Mer*. Ministère de l'Ecologie, du Développement Durable et de l'Energie.

Cokanasiga, L. (2020). *Blue economy, accelerating the industrializing of our ocean?* DAWNInforms. https://dawnnet.org/publication/dawn-informs-on-blue-economy/. Accessed 06/11/2021.

Freestone, D., Laffoley, D., Douvere, F., & Badman, T. (2016). *World heritage in the high seas: An idea whose time has come* (World Heritage Reports 44). UNESCO.

Frost, R., et al. (2018). Redrawing the map of the Pacific. *Marine Policy, 95*, 302–310.

Genz, J. H. (2008, May). *Marshallese navigation and voyaging: Relearning and reviving indigenous knowledge of the ocean* [Dissertation submitted to University Of Hawaii Anthropology].

George, M. (2012). Polynesian navigation and Te Lapa—"The flashing." *Time and Mind: The Journal of Archaeology, Consciousness, and Culture., 5*(2), 135–174.

George, M., & Wyeth, H. (2020a). *We, the voyagers: Our vaka*. Pacific Traditions Society.

George, M., & Wyeth, H. (2020b). *We, the voyagers: Our Moana*. Pacific Traditions Society.

Geraghty, P. (1993). Pulotu, Polynesian Homeland. *The Journal of the Polynesian Society, 102*(4), 343–384.

Gladwin, T. (1970). *East is a big bird: Navigation and logic on Puluwat Atoll.* Harvard University Press.

Gunn, M. J. (1980). Etak and the ghost islands of the Carolines. *The Journal of the Polynesian Society, 89*(4), 499–507.

Heymans, J. J., Bundy, A., Christensen, V., Coll, M., de Mutsert, K., Fulton, E. A., Piroddi, C., Shin, Y.-J., Steenbeek, J., & Travers-Trolet, M. (2020). The ocean decade: A true ecosystem modelling challenge. *Frontiers in Marine Science, 7.*

Hoegh-Guldberget, O., & Ridgway, T (2016). *Reviving Melanesia's ocean economy: The case for action: Summary.* WWF International.

Holton, G., Hachibmai, C., Haleyalur, A., Lipka, J., & Rubinstein, D. (2015, December). East is not a "big bird": The etymology of the Star Altair in the Carolinian Sidereal Compass. *Oceanic Linguistics, 54*(2).

Johannes, R. E. (2002). The renaissance of community-based marine resource management in Oceania. *Annual Review of Ecological Systems, 33,* 317–340.

Keen, M. R., Schwarzb, A.-M., & Wini-Simeon, L. (2018). Towards defining the Blue Economy: Practical lessons from pacific ocean governance. *Marine Policy, 88,* 333–341.

Kikiloi, K., Friedlander, A. M., Wilhelm, A., Lewis, N., Quiocho, K., 'Āila, W. Jr., & Kahoʻohalahala, S. (2017). Papahānaumokuākea: Integrating culture in the design and management of one of the world's largest marine protected areas. *Coastal Management, 45*(6), 436–451.

Laffoley, D., & Freestone, D. (2017). A world of difference—Opportunities for applying the 1972 world heritage convention to the high seas. *Aquatic Conservation: Marine Freshwater Ecosystems, 27*(1), 78–88.

Lane, M. B. (2006). The role of planning in achieving indigenous land justice and community goals. *Land Use Policy, 23*(4), 385–394.

Leblic, I. (1989). Notes sur les fonctions symboliques et rituelles de quelques animaux marins pour certains clans de Nouvelle-Calédonie. *Anthropozoologica,* 3è Numéro special, 187–196.

Lewis, D. H. (1972). *We, the navigators: The ancient art of landfinding in the Pacific.* University of Hawaii Press.

Low, S, (2013). *Hawaiki Rising: Hōkūleʻa, Nainoa Thompson and the Hawaiian renaissance.* Island Heritage Publishing.

Luomala, K. (1984). *Shark and shark fishing in the culture of Gilbert Islands.* Akademia Kiado.

Mulalap, C. Y., Frere, T., Huffer, E., Hviding, E., Paul, K., Smith, A., & Vierros, M. K. (2020). Traditional knowledge and the BBNJ instrument. *Marine Policy, 122*(1041), 13.

Nero, K. (2011). Paths to knowledge: Connecting experts in oral histories and archaeology. In J. Liston, G. Clark, & D. Alexander (Eds.), *Pacific island heritage, Terra Australis* (Vol. 35, pp. 127–154). ANU Press.

Oliver, D. (1974). *Ancient Tahitian society.* University of Hawai'i Press.

OPOC. (2021). *Blue Pacific ocean report: A report by the Pacific ocean commissioner to the Pacific islands forum leaders.* https://opocbluepacific.net/publications/#blue-pacific-ocean-report. Accessed 09/10/2021.

PIF. (2014). *The framework for Pacific regionalis.* https://www.forumsec.org/wp-content/uploads/2017/09/Framework-for-Pacific-Regionalism.pdf. Accessed 09/10/2021.

PIF. (2021a). *Declaration on preserving maritime zones in the face of climate change-related sea-level rise.* https://www.forumsec.org/2021a/08/11/declaration-on-preserving-maritime-zones-in-the-face-of-climate-change-related-sea-level-rise/. Accessed 09/10/2021.

PIF. (2021b). *Forum leaders ocean statement 2021b.* https://www.forumsec.org/wp-content/uploads/2021/03/Oceans-Satement_v8.pdf. Accessed 08/10/2021.

Quirk, G., & Hanich, Q. (2016). Ocean diplomacy: The Pacific island countries campaign to the UN for an ocean sustainable development goal. *Asia-Pacific Journal of Ocean Law and Policy, 1*(1), 68–95.

Ruddle, K. (2008). Introduction to the collected works of R.E. Johannes, publications on marine traditional knowledge and management. *Traditional Marine Resource Management and Knowledge Information Bulletin, 23*, 13–24.

Salmon, A. (2009). *Aphrodite's island: The European discovery of Tahiti.* University of California Press.

SINOP. (2018, November). *Solomon islands national ocean policy.* Government of the Solomon Islands, Ocean 12. SINOP

Smith, A. (in press). Mangyol village, Yap: A micronesian social landscape. In C. Goetcheus & S. Brown (Eds.), *Routledge handbook of cultural landscapes.* Routledge.

Smith, A. (2016). (Re)visioning the Ma'ohi landscape of Marae Taputapuatea, French Polynesia: World heritage and indigenous knowledge systems in the Pacific islands. In W. Logan, M. Nic Craith, & U. Kockel (Eds.), *A companion to heritage studies* (1st ed., pp. 101–114). Wiley.

Smith, A. (2017). *The cultural dimension of the Pacific ocean in Sustainable development in the Pacific.* Briefing Paper for the Office of the Pacific Ocean Commissioner, Pacific Islands Forum, Suva and the Pacific Ocean Alliance.

SPC. (n.d.). *Pacific maritime boundaries programme.* https://gem.spc.int/projects/pacific-maritime-boundaries-programme. Accessed 10 Oct 2021.

SPC. (2019). *Implementation plan summary.* https://www.oceandecade.org/wp-content/uploads//2021/09/337521-Ocean%20Decade%20Implementation%20Plan:%20Summary. Accessed 06/11/2021.

The Guardian. (2021). *Pacific plunder.* https://www.theguardian.com/world/ng-interactive/2021/may/31/pacific-plunder-this-is-who-profits-from-the-mass-extraction-of-the-regions-natural-resources-interactive. Accessed 28/09/2021.

Torrente, F., Bambridge, T., Planes, S., Guiart, J., & Clue, E. (2018). Sea swallowers and land devourers: Can Shark Lore facilitate conservation?. *Human Ecology, 46,* 717–726.

UNESCO. (2009). *Ocean declaration of Maupiti*. https://www.iucn.org/news/commission-environmental-economic-and-social-policy/201710/raising-and-integrating-cultural-values-ocean. Accessed 19/05/2021.

UNSDG. (2021). *Leave no one behind*. https://unsdg.un.org/2030-agenda/universal-values/leave-no-one-behind. Accessed 06/11/2021.

United Nations. (2021). *The United Nations decade of ocean science for sustainable development (2021–2030): Implementation plan, summary*. https://unesdoc.unesco.org/ark:/48223/pf0000376780. Accessed 29/04/2022.

Vierros, M. K., Harrison, A.-L, Sloat, M., Crespo, O. G. Moore, J. W., Dunn, D. C., Ota, Y., Cisneros-Montemayor, A. M., Shillinger, G. L., Watson, T. K., & Govan, H. (2020, September 1–13). Considering indigenous peoples and local communities in governance of the global ocean commons. *Marine Policy, 119*.

Walker, W. L. (1945). Territorial waters: The Cannon shot rule. *British Yearbook of International Law, 22*, 210–231.

Wilson, K. (2010). *Nā Moʻokū ʻauhau Holowaʻa: Native Hawaiʻian Women's stories of voyaging the canoe Hōkūleʻa* [Ph.D. Thesis, University of Otago].

17

Policy, Epistemology, and the Challenges of Inclusive Oceans Management in Sierra Leone

David O'Kane

Today, as Sierra Leone struggles for development and democracy, it contends with multiple challenges. These are more than just the obvious challenges of economic development, political stabilisation and so on: they include, also, those challenges represented by heritage and its preservation, the need for political inclusivity, and the ocean. The ocean is a medium, one which not only enables connectivity, but shapes the forms of connectivity that take place through it. It presents human communities with challenges of connectivity between individuals, coastal and inland cities, nation-states, environmental zones, and other entities, natural and artificial. The mediations required by oceans and the technologies required to travel through them are what allows resources, goods, commodities, ideas and information to pass and be exchanged between those entities. Some of those exchanges, of physical and intellectual objects, are short-term acts for practical purposes. Others are of long-term, inter-generational nature, and involve the assemblages of objects we know as 'heritage'. This chapter, therefore, begins an answer to the following question: how has the ocean as a mediating factor influenced the country of Sierra Leone ever since its first beginnings, when formerly

D. O'Kane (✉)
Max Planck Institute for Social Anthropology, Halle am der Saale, Germany
e-mail: okane@eth.mpg.de

enslaved persons were resettled on the site of what is now the Sierra Leonean capital, Freetown?

In history, ocean-mediated exchanges have allowed the flourishing of societies, or enabled their collapse: the Mediterranean as a sphere of cultural interaction, for example, provides multiple cases of such rises and falls, from before the time of the Roman empire to today (Borutta & Gekas, 2012). The Atlantic, as it evolved into a vast intercontinental social and cultural space after 1492, is another such case. Paul Gilroy's evocation of the 'Black Atlantic' is only one of many accounts of that space, and of the forms of violence that drove the rise, within it, of contemporary patterns of emerging nationhood, persistent economic dependence, and what Gilroy terms 'double consciousness' (Gilroy, 1993, p. 4). Gilroy's focus on the ship as a microcosm of the emerging Atlantic cultural sphere highlights a key factor in the mediating role of oceans, one already alluded to above: the ways in which transoceanic attempts to communicate, trade, or migrate (or force others to migrate) impose technological and other requirements on those who make them. The explosive surge of European powers into the Atlantic ocean after the Portuguese creation of the Atlantic slave trade was the result of maritime technologies first developed in the Mediterranean sea, and developed through both trade and military conflict. The slave trade and its social and political aftershocks is one example of oceanic mediations of human connectivity, one that took a grave toll in human lives. China's Belt and Road Initiative is another example of global connectivity mediated through oceanic infrastructure; it may not have the destructive consequences of the slave trade, but it has already perturbed some observers, who perceive in the rise of China, and its interventions in Africa, the coming of a new colonial power whose activities risk reproducing the patterns of domination and exploitation exhibited by the older colonialisms that once ruled the African continent. In this chapter I review some episodes of the old colonial era, and the new era of the China–Africa relationship: I situate those episodes within the mediation of political relationships via oceans and oceanic technologies; and I reflect on what these may mean for the politics of heritage in a world where 'inclusivity' is a key word, but also one that is rarely defined—and never defined, as far as I can see, in ways that would imply genuine participation of people in matters that affect them materially, and directly.

The denial of direct popular participation does not mean the exclusion of the masses from the making of history. To this day, the technologies of maritime communication in the era of slavery have left their mark in Sierra Leone's lingua franca, Krio. This language is a creation of the Krio ethnic group,[1] who make up around 3% of the population, but it is widely spoken across the country. In that tongue, a 'floor' is a 'deck', and what in other countries would be called a machete is a 'cutlass', after the weapons borne by naval officers and others who had resort to force and violence in the era of sail (Hancock, 1976). Krio is part of the national heritage of Sierra Leone, and it is not the only historical or contemporary element of that country's culture which bears the traces of oceanic mediation.

Such traces are often (if not in all cases) the result of political decisions, and they arise from such decisions both at the time they are carried out, and over longer periods also. Legacies of the British sea-borne empire are still identifiable in the political and legal system of Sierra Leone, for example, even as the economic dependencies which the west African country inherited from British rule are being challenged by the (arguably, equally imperialistic) movements of Chinese capital into Africa. Only the beginnings of an answer can be provided here, but this chapter will provide a new account of how today's mediation of Sierra Leone through oceanic connections has been formed by historical processes that date back to the late eighteenth century. Those processes, I argue, have been decisively shaped by political decision-making, and by those who made those decisions. There is a decisive contrast, I agree, between the colonial and imperial powers who made decisions over the heads of those who were affected by them in the nineteenth century, and the politically independent sovereign states that now rule the formerly colonised lands of what was once British West Africa. The hypothesis of this chapter is that the effects of working across oceanic distances will be seen in the ways in which the sea-borne British empire exercised its power in West Africa, especially during the crucial decades around 1860, when the question of further British colonial expansion into the region was decisively settled; and, further, that the mediating effects of oceans can also be detected today in the relationships of contemporary Sierra Leone to the outside world, as those relationships are conducted in and through external connections formed by maritime trade and related activities. This will have consequences for the management of heritage in twenty-first century Sierra Leone.

In this chapter, therefore, I describe the continuities and contrasts that exist between the world of the nineteenth century, in which Sierra Leone acquired both its land borders and territorial waters (or, to be precise, had those demarcated limits imposed upon it by the imperialist power of Britain), and the present, twenty-first-century situation. That latter situation is one in which Sierra Leone is no longer a colony, but an independent nation-state must confront a neo-colonial global political economy, while dealing with issues of continued nation-building. Its assertion of national rights is part of an effort to look outward to the world: and it, also, is connected to problems of law and power.

Oceans, in addition to the intra-human connections they enable, also connect humanity to nature, and to natural resources which humanity can exploit (sometimes with permanent costs). When we describe human societies as being connected to each other by oceanic mediations, the adjective is used advisedly: the distinction between land-based powers and sea-borne empires underlines the crucial role of oceanic mediation in the case of the latter, with consequences for historical and contemporary patterns of imperialism, colonialism and neo-colonialism. The concept of Thalassocracy highlights this distinction between land-based empires, which can be territorially contiguous, and those which use the sea to 'tie together scattered dominions' (Abulafia, 2014, p. 139). It was this act of tying together what would otherwise have been scattered that imposed certain obligations on the aspiring thalassocratic power: ports had to be built and maintained at key nodal points in the network, and this implied new populations whose ethnicity was not, and could not be identical to those of the 'mother country'[2] (Abulafia, 2014, pp. 149–150). In the context of Africa's Atlantic coastline—and not only in the Upper Guinea Coast—this means that today, as in the past, 'ports not only bestride the land-sea divide, but they also interface with institutions and interests' (Olukoju, 2020, p. 186; Banton's *West African City* [1969] gives an account of Freetown as a port where such interfaces arose in the immediate pre-independence context of Sierra Leone). The creation of geographical, social and political interfaces presents hegemonic powers, today as in the past, with new dilemmas. For the nineteenth-century British empire, for example, the maintenance of control required naval power, and investment in maritime technologies more generally. For a twenty-first-century China moving into new relationships with Africa, the establishment of those relationships requires forms of power that are more soft than hard: but this does not rule out adverse consequences for those affected by those relationships.

Such consequences have shaped the historical experience of the countries that were part of the sea-borne empires such as Portugal, Spain, and Britain and their transoceanic networks. This includes the countries of the Upper Guinea Coast, the arc of West African countries that stretches from Senegal and Gambia to Liberia and Ivory Coast, and which includes Guinea, Guinea-Bissau Cape Verde and Sierra Leone. The shores of the Upper Guinea Coast countries stand by the Guinea Coast Large Marine Environmental zone, 'characterized by a relatively wide continental shelf with low lying coastal topography and a large number of estuarine systems with strong river inflows'

(Belhabib et al., 2019, p. 81). One of those estuarine systems flows out into the harbour of the Sierra Leonean capital Freetown, the third-largest natural harbour in the world. For centuries, Sierra Leone has, like its neighbours and near-neighbours in the Upper Guinea Coast region, been enmeshed in relations of unequal exchange mediated by the Atlantic waters that connect it, as Atlantic slavery and its triangular trade did, to Europe and the Americas. Today, Sierra Leone, like its neighbouring countries, enjoys political independence. It has completed twenty years of post-civil war reconstruction, and has gone through a viral epidemic wrought by Ebola. The ocean as a medium is relevant to these concerns and that relevance takes the form of, for example, national assertion over territorial waters and the resources they contain. It also takes the form of participation in international legal regimes regulating the international use of oceanic waters.

These regimes rely on bodies of international law include the United Nations Convention on the Law of the Sea (UNCLOS), which specifies the rights and duties of all those who use the seas for commerce or other activities. This UN Convention was once seen as part of a drive for a 'New International Economic Order' that would help lift up the developmental status of countries such as those of the Upper Guinea Coast (Brown, 1984, p. 259). UNCLOS was a massive text, which took years to prepare, and which dealt with a vast array of matters relating to the oceans and their human users. When it was issued, in 1982, the world was on the brink of a new era in which matters of the environment and of heritage would become increasingly important in political debates at both national and the international levels.

UNCLOS is just one part of a wider body of international law that delimits territorial waters, sets out the responsibilities and freedoms of those who use the sea, and which has made innovations in the international laws that relate to cultural heritage (Frost, 2004). As such, this body of law is what allows for the organisation of oceanic connections between nation-states and other entities, and for the mediation of relations between those states and entities. Consideration of the history of these regimes indicates that any problems of organisation they may have are problems of policy, of the making and implementation of policy, and of the problems of policy for international orders, via the culturally and socially embedded contexts in which policy always exists.

This section of this chapter is concerned with the relationship between the British empire and Sierra Leone. For the purposes of this chapter, we need to understand that this sort of imperial decision-making and policy-making was also key to the evolution of British policy in Sierra Leone, including as it did the country's ocean-mediated connections to the world. The country began as a sea-borne colony whose construction was part of the wider movement for the abolition of the Atlantic slave trade. The earliest settlement of the Freetown colony was followed by the use of Freetown harbour as the headquarters of the African Squadron of the Royal Navy, which sought—where possible—to intercept and interdict the shipping fleets that remained involved in the abduction and enslavement of African people, and to rescue (where possible) enslaved persons and return them to Africa (Smith, 1848).

Those who implemented this policy experienced epistemological challenges arising out of the mediating role of the Atlantic ocean. The British parliamentary select committee of 1865, which considered the possible options for the extension or retrenchment of British-held enclaves in West Africa, took evidence from 'several Civil, Military, and Naval Officers, Merchants, and Missionaries, and from one Native Envoy' (Select Committee, 1865, p. x). The extensive body of evidence this produced required very careful and protracted analysis: the ultimate recommendation was that, with the exception of Sierra Leone, Britain should pull out of its West African enclaves.

Careful and protracted analysis was also the lot of the captains of the Royal Navy's anti-slavery squadron. Every time they sought to intercept a slaving ship, they were confronted with the problem of identifying a ship as both a slave ship, and as a ship sailing under the flag of one of the slaving powers. The problems of the counter-slavery squadron were determined not only by the oceanography of the Upper Guinea Coast littoral, but also by the symbolic boundaries and signs which were overlayed on the currents, deep sea shelves, *et cetera*, of that littoral area. The ocean, in the waters around Sierra Leone and its neighbouring countries, became a medium for the transmission and exchange of ideas, goods, (and, until the suppression of the trade, enslaved persons) and symbols. As such, that ocean became, in the perception of its travelers, not only a medium but a palimpsest, a text-bearing object on which several texts had been layered over each other. Over the course of the nineteenth century, that layering of symbols would become part of the process by which the British empire rediscovered its Sierra Leonean colony and its strategic importance.

By the end of that century, British naval superiority had been fully proven: Britain's military and political leaderships also knew that they would have to retain that superiority if they were to retain, also, their political and economic hegemony. The projection of power that this goal required entailed the maintenance of a network of 'coaling stations' where sea-going international patrols could refuel themselves: one of these was at Freetown (Hurd, 1898,

p. 721). It was the city's potential as a coaling station which had prevented it being included in the 1865 select committee's recommendation for British withdrawal (Ekoko, 1981, p. 138). In the 1860s, the British government, under Benjamin Disraeli, had given serious consideration to withdrawal from its African colonial possessions; changes in maritime technology (from sail to steam) and the emergence of new rivals (a resurgent French empire, and then a united Germany) produced a decision for the other option, the expansion of the empire into Africa. Sierra Leone would retain its strategic oceanic importance up until the early 1980s, when Freetown's harbour would be visited by the Task Force sent to eject Argentina's occupation forces from the Falkland Islands, thousands of miles away in the south Atlantic (Adams, 1984). The next time British forces would come to Freetown it would be to help keep the peace in the last days of the Sierra Leonean civil war. The restored Sierra Leonean state would have to restore its authority not only over its internal territory, but also over the territorial waters of the republic. There, the ocean would become a medium where the country was connected to the outside world: there would be visible, within that connection, both continuities and discontinuities with the past.

Those continuities remain mediated, in very different and contrasting ways, through oceanic connections. This was true of Sierra Leone under both British colonial rule, and in the twenty-first century as an independent African state. After the first freed slaves, and others, sailed into that harbour in 1787, they planted, or sheltered, under a cotton tree before founding the Freetown colony. In some accounts this tree had been used as a venue for the sale of enslaved persons, but it would grow to become the great national monument of Sierra Leone (Basu, 2016, p. 235). In 1810 the peninsula of Freetown became, officially, a colony of the British Empire,[3] and soon began to play a role in the regional trade that connected the Upper Guinea Coast with the European world, becoming a importer of firearms and cloth and an exporter of timber and other local commodities (McGowan, 1990, pp. 37–38). Participation in that trade was a harbinger of the dependent relationship of the colony with the world economy of the nineteenth century, a relationship that would also become a driver of integration with the interior of Sierra Leone. Trade with the interior grew over following decades, as did missionary activity by the Krio ethnic group who had emerged from the freed slaves, liberated Africans and Black Nova Scotians who had settled in the Freetown colony. At the end of the nineteenth century the Hut Tax war would bring the interior under full British control, as it was incorporated as a protectorate. Well before that, however, Sierra Leone had become a key node in the British Empire's global network of naval power. This would become visible in the late nineteenth century, as the international rivalries that exploded in the first world war grew more important: but it was already visible in the early decades of that century, after the slave trade was banned within the British Empire, and the Royal Navy became a vigilant pursuer of vessels still engaged in that traffic—or at least as vigilant a pursuer as it could be, under the circumstances.

This is one of the historical cases which underlines the importance of the ocean as a medium, and its relevance to the epistemological aspects of policy. Anthropologists of policy have always insisted on policy's cultural and socially embedded character, and the unique twists in the nature of the relationship between epistemological factors and policy decisions (Shore & Wright, 1997). For imperial powers and those who resisted them, the dangers and temporalities of ocean travel imposed certain policy choices on them, and deprived them of others. It was not a coincidence that the British empire was a seaborne empire, and that its sea-power allowed it to become the global political and economic hegemon at its height. From one point of view this might have been the inevitable outgrowth of a logic that began with the first colonisation of Britain's neighbour Ireland, in the twelfth century, and its later recolonization of that country from the sixteenth century onwards. From another point of view, there was no good reason to assume that the British empire would reach the heights it did, in Africa or anywhere else. It is tempting to retrospectively assume the inevitability of events like the Scramble for Africa, or even the incorporation of the Sierra Leonean protectorate into the British Empire; but that would be to engage in the logical fallacy of begging the question, of assuming that these events, and the decisions which caused them, were inevitable. We know that they were not inevitable because the British government of the 1860s seriously considered (as noted above) evacuating its personnel from Africa, and ending its rule over the handful of coastal enclaves which, at that time, composed its African colonies—though the one exception to this proposed withdrawal was the colony of Sierra Leone:

> The report of 1865 has often been criticised as shortsighted, but on the assumptions of the mid-Victorians it was perfectly sound. Its authors did not advocate a withdrawal from west Africa as such, but merely from those irritating hostages to fortune, the colonies of west Africa. At the same time the Committee genuflected towards the humanitarians by consenting to stay in Freetown. All that was proposed was that Britain should cut her colonial losses and concentrate on the more profitable regions where the gains of trade were not cancelled by the liabilities of rule. (Robinson & Gallagher, 1981, p. 30)

It is not only that the report was shortsighted, but that its drafters had difficulty obtaining reliable information of the kind that could form a 'knowledge base' for the drafting and implementation of policy. In the end, the epistemological challenges presented by the ocean did not prevent the British empire from remaining in Freetown and other enclaves, and neither did it prevent it from making deeper incursions inland, as it seized the protectorate in the interior of Sierra Leone in the 1890s.

One reason for this expansion of colonial rule was an imperial 'mission creep' in which the strategic demands implied in sustaining an imperial presence in one area or region would require imperialist expansion into neighbouring regions: but it was also part of a process whereby Sierra Leone became a strategically important node in the network of British sea-borne power. By then, sailing vessels had become obsolete, replaced by coal-fired steam ships in both civilian and naval shipping. This entailed the creation of a global network of coaling stations, among whose number Freetown would be ranked (De Thierry, 1899). At the same time as Sierra Leone was thus integrated into globalisation in the British imperial manner, however, its people were beginning to integrate themselves with the rest of West Africa, as a new generation of intellectuals and other leaders began to forge links with each other along the West African coast. This network of travel was accompanied by a network of print media that extended throughout the new English colonial possessions of the Upper Guinea Coast. The first Sierra Leonean newspaper was founded in 1801, and would later be followed by many successors (Parins, 1985, p. 26). These included the *New Era* newspaper of William Drape, whose politics led to attempts at its suppression by the local colonial governor. Campaigns in the imperial metropole led to enough controversy to induce the London government to rescind the governor's attempt at suppressing the newspaper, setting a precedent which made the West African press to enjoy a relative degree of freedom for the rest of the nineteenth century. During the rest of that century, that freedom was used by the nascent Freetown elites (business people and professionals, largely, at this time, Krio) to advance their interests and those (as they saw it) of the colony. The colony's newspapers, certainly, occupied a vital position in the networks through which trade and politics were carried out:

> The board of directors of the Sierra Leone Times, published every Saturday on Rawdon Street by the Sierra Leone Printing and Publishing Company read like a Who's Who of the colony in 1892. The chairman, J.H. Malamah Thomas, was a self-made man with extensive trading connections in the north of Sierra Leone who served several terms as mayor of Freetown. (Deveneaux, 1976, p. 49)

Such trading connections involved the building of links with the kind of economic and political networks that Howard and Skinner (1984) describe as being vital to the social evolution of what is now Sierra Leone's northern province (and it is also the kind of institution which would have functioned as a feedback loop of either the positive or negative kind, according to the ways in which Chairman Thomas would have chosen to deploy his economic and political powers). They also connected elites to growing numbers of literate people, in a context where, through 'public acts of reading', literacy came to be regarded as a 'public property', thus magnifying the effects of literacy on popular consciousness (Newell, 2011, p. 28). Ultimately, in the twentieth century, the Sierra Leonean press would help foster anticolonial leaders such as I.T.A. Wallace Johnson, who would go from being editor of the *African Standard* to an organiser of waged workers from the whole population (Denzer, 1982, p. 159).

They also forced both elites and masses to develop very significant (and ocean-mediated) connections with the wider populations of colonised West Africa (Newell, 2011). For Stephanie Newell, this helps illustrate a defect in Benedict Anderson's concept of 'print-capitalism' (1983), which she finds far too focussed on the nation-state to be able to grasp the highly heterogeneous reading publics of the late colonial era in Sierra Leone, Nigeria and other West African locales (2011, p. 28). She notes that 'until the mid-1930s in West Africa, the very word 'nation' was used to describe complex and diverse identities and affiliations' (Newell, 2011, p. 29). An 'Andersonian' analysis of this and later, related, cases can be retained, however, if we recognise that such print industries exhibited (as their present-day equivalents exhibit) all the crucial features of the system concept as it pertains to societies and cultures. Where such systems can maintain a state of equilibrium, they will likely continue to maintain their pre-existing condition: where positive feedback loops ensure that equilibrium cannot be maintained, crisis will ensue, as it did for European colonialism in West Africa after the Second World War, and as it did for Sierra Leone in the 1980s and 1990s. In the latter decade, crisis took the form of a descent into civil war (Gberie, 2005). After a decade of conflict, with severe consequences for the civilian population, the civil war ended with British intervention in its final weeks. Borne there by aircraft carriers, British forces helped local groups overcome the last remnants of the Revolutionary United Front, and restore some peace to the country. This may have been the last major intervention of its kind by the British state in Sierra Leone: since the end of the civil war, new influences and rising powers have appeared on the Sierra Leonean scene, as they have in African more generally. The most important of these rising powers is the People's Republic of China

(PRC), whose influence is (at present) driven more by soft power than the hard power of military action. In spite of this, the PRC's growing importance in Sierra Leonean political and economic life is increasingly controversial—and controversial in a way which highlights the relationship between oceans, power, sovereignty, and also (and indirectly) heritage.[4]

China and Sierra Leone on the Frontier Zone of Black Johnson Beach

The crisis of Sierra Leonean postcoloniality reached its climax with the civil war of 1991–2002 (Gberie, 2005). This was the conflict which saw Britain's final intervention in the country, and perhaps the 'last hurrah' of its ability to project sea-borne military power. The years that followed saw the reconstruction of the Sierra Leonean polity and its economy, and the resumption of an already-established relationship with a rising player in the geopolitical rivalries that form global imperialism—the People's Republic of China (PRC). There were many contrasts between the original relationship that was revived in those years, contrasts that derived from the changing nature of development strategy in the PRC and the ideological redefinitions required to legitimise those changes. Similarities and continuities were apparent too, however—similarities and continuities between the burgeoning Sierra Leone-China relationship (a part of the wider Africa–China relationship) and the older, colonial, relationship between Sierra Leone, its African neighbours, and the old, colonial powers of Europe. Chinese activity in Africa is widely seen in many countries—including Sierra Leone—as a thoroughly neo-colonial undertaking. In this section I describe some of the ways in which that neo-colonialism of Chinese activity in Africa can be seen in Sierra Leonean reactions to a fishmeal processing plant, which China and Sierra Leone's government have proposed be built at the site of Black Johnson beach on the Freetown peninsula. A community of artisanal fishers has been living there for a long time; the proposed Chinese plant may mean the end of their community, via the destruction of the local maritime environment on which it depends. It may also mean a new episode in the role of the ocean as a mediating factor in the evolution of Sierra Leone's culture, society, and politics.

This episode will be partially determined by trends occurring thousands of miles away, in the PRC. There, changing social and economic circumstances have led to major changes in local dietary patterns, leading to a situation where Chinese consumer demand for seafood is likely to outstrip supply by 2030 (Crona et al., 2020). This has already led, and will continue to lead to, a growing emphasis on aquaculture as a means of meeting that demand—aquaculture being the deliberate cultivation of fish and fish products in controlled or semi-controlled environments. This is a process that requires inputs, including fishmeal, a high-protein brown powder produced by the processing of several types of fish (Barlow, 2003, p. 2486). Demand for fishmeal products for aquaculture has driven the rise of new supply chains that connect its primary producers with its ultimate consumers: many of these supply chains ultimately end in, or connect with, China. The PRC government's goals of social stability and 'moderate prosperity' are strongly represented in its drive for the so-called 'Belt and Road Initiative' (BRI), the network of the infrastructural project that is intended to connect the country across both continental and oceanic mediations (Crona et al., 2020, p. 41). While the infrastructural developments the belt and road initiative involves are what has attracted the most international notice, it is important to remember that it has substantial ideological and policy dimensions as well. The initiative is, purportedly, one in which China will establish mutually beneficial relationships with those countries that are drawn into the long-term relationships which its infrastructural aspects imply. In addition to the infrastructural implications of the Belt and Road Initiative, however, there are other, less visible, aspects of the initiative, rooted in the BRI's wider context: the rise of China as a world power on all fronts.

China, today, has occupied the position once taken by the United Kingdom at the height of its imperial power: that of the 'workshop of the world'. The factories of the coastal provinces produce goods for sale all over the planet, and not only in the markets of Western Europe and North America. Less well known is the rise of China as an overseas investor (Liu and Dunford, 2016, p. 337). This is interpreted by some as meaning a 'win–win' relationship between those who choose to join the PRC in its BRI plans: but those who remember Africa's debt crisis will also recall that creditors create debtors, and that the inclusivity of indebtedness is not one that should be aspired to.

This, then, is the background to the controversy that has emerged around Black Johnson beach on the Freetown peninsula. The community of artisanal fishers who have lived there for generations could easily find their community, and the natural environment on which it depends, wiped away by the proposed building of a fishmeal processing plant.[5] Voices from the Sierra Leonean media are already sounding the alarm over the likely consequences of this proposal:

> A small, uninhabited island just off Black Johnson is called Barracuda Island by locals. It is no ordinary place; it is home to the prestigious barracuda, a ray-finned fish known for its fearsome looks and, of course, its taste. Along the beaches here are egg-laying homes of the endangered green turtle. The rainforest hills overlooking the Atlantic Ocean hold a variety of majestic animal species, including green monkeys, slow loris, porcupines, pangolins, and Gambian pouched rats.
>
> The peninsular forests are also home to a large collection of bird species, including the beautiful and rare rock fowl (Picathartes), the noisy plantain-eaters, wood doves, and the endangered Gola malimbe – the list is enviably long! Freetown and Greater Freetown surrounds a truly tropical rainforest in the peninsula – you can stand anywhere in Black Johnson and take photographs of an impressive collection of tropical birds. To sell this place for $55m is a heinous act. (Fallah-Williams, 2021)

This author notes the threat to local food security that the proposed fish factory represents, and the threats to local welfare that are likely to ensue when its construction displaces local people: mental illness and prostitution in informal settlements are listed among those threats, which may be described as 'transnational oceanic ecoviolence' (Stoett & Omrow, 2021). What, however, does this have to do with heritage? Intangible cultural heritage is something carried inter-generationally by communities: its preservation requires theirs. Tangible cultural heritage, of the kind listed by Basu and Sam (2016, p. 9) for Sierra Leone, may seem less dependent on the preservation of locally based social structures that can bear cultural capital. This is over-optimistic: loss of cultural heritage and patrimony is more likely to come in a context where short-term profits are seen as incompatible with the preservation of heritage and patrimony (this should be seen as the kind of narrow vision that Amartya Sen has criticised for its opposition to the deployment of development budgets on health care and education).

The Freetown peninsula is home to some of Sierra Leone's most important intangible and tangible cultural heritage. The latter takes the form of artefacts and monuments dating from the Atlantic slave trade and the wider geopolitics of its end: among those listed include shipwrecks, cannons, and a defensive fortification of the kind known as Martello towers[6] (Basu & Sam, 2016, p. 9). This is one point where the continuities between the imperialisms of the nineteenth and twenty-first centuries are most visible. Those continuities are those of a frontier zone, one which happens to be superimposed on the geographical zone where land meets the sea. It is particular to frontier zones that they produce policy challenges for both frontier communities and the hegemonic powers which rule them, or which propose to rule them. In the following section I begin to discuss this aspect of the problem.

Epistemology and Policy Through the Medium of the Oceans

This chapter considers two historical episodes that involved Sierra Leonean connections to the sea and the oceans, and compares the different challenges that were involved in both. The first is the period of British rule in the nineteenth century, when Sierra Leone was an integral part of a vast, ocean-spanning imperial network, but also the centre point of a new relationship between Africa and wider world. During that time, an ethnic group descended (partly) from formerly enslaved persons began to build a new identity and language that was part of the social and intellectual movements of colonised Africa that would one day blossom into the twentieth-century movements for African independence. The second period is that of the twenty-first century and its patterns of unequal globalisation. In this present era the Sierra Leone nation-state is trying to engage in nation-building while at the same time trying to re-establish itself in a context of post-imperial, globalised inequalities.

This is what brings into play the connection between policies, epistemologies and 'inclusivity'. I use the inverted commas on the last word advisedly. The principle of inclusion is, for many twenty-first-century people, the fundamental pillar of 'social justice'. These are both, however, more contestable concepts than they first appear. It may be right to be inclusive, and to implement inclusive policies; but we should first of all ask who is being included, and for what purpose.

Policy selection must rely on some theory of knowledge, implicit or explicit, to justify itself, even when conducted via the 'garbage can' model of policy selection (Cohen et al., 1972). Policy choice and implementation involves the mobilisation of 'epistemic communities' within the political or other apparatuses through which policy choices are chosen and implemented. I define epistemic community, here, as a set of agents engaged in knowledge work, agents for whom the conduct of that work requires certain shared epistemological assumptions, whatever their other shared (or not shared) traits may be. When such communities are engaged in policy affairs, there is a connection between power and knowledge, including the power to designate persons and their communities as included or excluded (we do not have to be Foucauldians to recognise this reality). There is nothing inherently wrong with inclusion or exclusion: what matters is who is being included or excluded, and why. Colonialism, in Sierra Leone, attracted moral opprobrium because its, involved a model of governance that rested on the exclusion of the colonised population from any say in the political direction of their communities or their country. As this became untenable over time, the movement to independence began to emerge. At the same time, exclusion can be a good thing, as the problem of the free-rider in collective action implies (the maintenance of coalitions requires, after all, the punishment of defectors).

Ever since the publication of Edward Said's *Orientalism* forty years ago, there has been an awareness of the relationship between power and knowledge (2003 [1978]). For many, however, awareness of this relationship has not involved any awareness of the nuances of that relationship, and there remains little on the exact relationship between systems of knowledge and the practical deployment of power through policy decisions. This should not be considered Said's fault: his book (one of the most important of the late twentieth century) is concerned with the nature and character of a paradigm that persisted in the western intellectual sphere from at least early modern times onward, and with the relationship between that paradigm and the political relationship between the west and the east, as that relationship existed from the late eighteenth century onward. When Said published his book, he could assume that his readership was aware (even in the most rudimentary sense) of the struggles which underlay that relationship, that they would be conscious that the collapse of the Ottoman empire was a slow process that lasted for most of the nineteenth century, that the British empire in India had evolved from private commercial enclaves in the eighteenth century to full conquest after 1857, and that its sea-borne empire contained not only its Eastern possessions but had penetrated far into Africa.

And the facts of that penetration highlights an unfortunate absence in Said's work, an absence that obscures the nuances of the historical episode he discusses: the absence of the policy choices that the Orientalist discipline was intended (at least in part) to support. For most European powers, the Ottoman empire was a culturally alien entity whose very existence had, in the historical past, been an existential threat. By the nineteenth century, that empire was also an entity with whom treaties could be signed and relied upon, and which could be brought into the set of peace-keeping alliances known, by then, as the 'concert of Europe' (Adanir, 2005, p. 408), and one which might be a military ally in one generation and an enemy in another. The Crimean war of 1854 involved British and French defence of the Ottoman empire against what was perceived as Russian aggression: the Balkan crisis of

1878 nearly led to war by Britain on an Ottoman empire that was engaging in religious persecution of Christian populations in the Balkans. That would have involved policy decisions by a British government, decisions informed by, but not exclusively determined, by Orientalist discourses or any other sort of discourse.

At roughly the same time as the British Empire was engaged in its diplomatic and military approaches to the Ottoman Empire in the east, it was running a parallel policy in West Africa. Like its Ottoman policy, this was one which was intended to advance its imperial interests. In the discourses at play in that case, representations of British interests were allied to discursive representations of Africa whose racialised stereotypes and assumptions need no introduction. The subsequent experience of British colonial rule in Sierra Leone, once it had expanded out from the Freetown peninsula and into the interior, involved the clash of those stereotypes with reality, and the rise of a new nationalist political consensus among the colonised population—a consensus enabled by, among other things, maritime connections forged along the Upper Guinea Coast, and across the Atlantic ocean (supporters of Marcus Garvey, for example, addressed meetings in Freetown in 1920: see Okonkwo, 1980, p. 107). When, in 1946, the Monuments and Relics Ordinance was passed by the colonial legislative council, it was directly intended to address the passage of West African relics and artworks into the global art market: but the need to conciliate a rising anti-colonial politics must have been involved in the passage of the ordinance also (Basu & Sam, 2016, p. 25).

In their review of the Monuments and Relics Act, Basu and Sam argue that it has become increasingly obsolete, and unable to meet the heritage needs of Sierra Leone in its current form (2016, p. 6). As Basu has noted elsewhere, those needs are urgent and pressing: and they are, as he notes, needs that imply temporal mediations between contemporary Sierra Leonean society and its recent and distant past (mass graves dating from the civil war of 1991–2002 are among the heritage sites which he sees as in need of protection; Basu, 2016, p. 247). For him, the accumulation of historical events which produce such sites requires the concept of the 'palimpsest', the document on which successive narratives have been inscribed, only to be later overwritten by new narrations. As various narratives compete for the attention of communities which contemplate the 'memoryscapes' in which their heritage is embedded, the outcomes are not predictable, and are not predictably positive. The potential of the mass grave to negatively influence future generations is the one cited in this argument (Basu, 2016, p. 254).

As the civil war recedes into the past, and as political democracy seems to have stabilised itself in Sierra Leone, the threat of negative potentials in the memoryscape may also diminish. The controversies over Black Johnson beach, however, imply the persistence of the abuse of power as a feature of Sierra Leonean politics, and this has implications for the ways in which the connection between heritage and the ocean should be conceptualised. What I have argued in this chapter is that the ocean generally, and the coastal zones where human populations meet the ocean in particular, is a palimpsest, one whose nature is shaped by the particular challenges which it sets those populations. So far, Sierra Leone has met those challenges. Will it continue to do so in the future?

Conclusion

When a nation-state adopts policies about heritage and its preservation, this indicates the intention of the nation and its political elites to take control of the discourses that affect the trajectories of its own social development. In such cases, the legacies of the past cannot be wished away or ignored. This chapter has dealt with the ways in legacies of the past have been transmitted into the present by Sierra Leone's continuing relationship with much more powerful external forces. The exploitation implied in the proposed construction of the fishmeal plant at Black Johnson beach will immediately evoke older histories of exploitation during the age of 'classic' colonialism, even if the explicit aspects of colonial tutelage (military occupation, usurpation of local self-government, and so on) are absent. One of the most important factors in this case is the ocean as mediator. During the several episodes that have formed the span of Sierra Leonean history since 1787, the ocean has imposed certain restraints on political actors in the country and the wider transcontinental networks into which the country was embedded. The British empire owed its power to the mastery of the seas: that mastery, in turn was shaped in particular ways by the struggles that grew up through the slave trade and the resistance it inspired. Since the end of the Sierra Leonean civil war, the question of heritage has returned as part of the general problems of nation-building through national culture that has been studied by writers such as Basu (2016). That question is also part of a wider set of policy choices, a set in which the option for the preservation of heritage must compete for support against other options, which often contradict, or if implemented would prevent, the goal of heritage conservation. Preservation of the coastal part of Sierra Leone's heritage, means preservation of that part of the national heritage which provides a marker of the relationship between land and sea, and of the historical, ocean-mediated relationships of Sierra Leone with the external world. The controversies over the threat to Black Johnson beach indicate that heritage preservation, here, will be especially problematic. In Sierra Leone, liberation has often come via the sea, but so too has slavery and other forms of exploitation. In the past, this has imposed policy choices on all Sierra Leone's people, and not only on its intellectuals and political leaders. Those choices are likely to continue today, and will continue to present challenges in all spheres, including that of heritage.

Acknowledgments The author gratefully acknowledges the assistance provided by the University of Durham during the period when this chapter was written, especially the access it provided to online archive resources.

Notes

1. The Krio are a key example of an ethnic group whose identity has been formed by creolisation processes, the same processes which led to the emergence of the Krio language: a large literature on this subject includes Knörr (2010) and Knörr and Kohl (2016). On the language policies of contemporary Sierra Leone, see O'Kane (2020).
2. The source cited here is discussing thalassocratic power as it was deployed in the Mediterranean Sea from antiquity to Early Modern times; I believe that the insight cited here is relevant to the African case as well.
3. The distinction between the original colony of Freetown and the protectorate that was, later, declared over the interior of Sierra Leone persists today. It is especially visible in the different land tenure regimes that govern property rights in the two parts of the country.
4. Chinese relations with Sierra Leone extend back to the 1970s (Brautigam, 1994, p. 326).
5. This is in a wider context in which the Freetown peninsula has become a zone of intense real estate speculation and a building boom that has helped intensify environmental degradation of the area, and exacerbated social inequalities in ways that have endangered the security of the poor, including artisanal fishing communities (Ménard, 2019).
6. Their full list of such heritage as part of the legacy of slavery is as follows:

    - Heddle's Farm (proclaimed 1948).
    - The original bastions of Fort Thornton (State House) (proclaimed 1949).
    - Gateway to the Old King's Yard, Wallace Johnson Street, Freetown (proclaimed 1949).
    - The old wharf steps and guard house, Wallace Johnson Street, Freetown (proclaimed 1953).

- Three old Freetown city boundary guns (Kissy Road; Leicester Road; Pademba Road) (proclaimed 1953).
- The original Fourah Bay College building, Cline Town (proclaimed 1955).
- St. John's Maroon Church, Siaka Stevens Street, Freetown (proclaimed 1956).
- St. Charles' Church and Remains of the King's Yard Wall, Regent (proclaimed 1959).
- Martello Tower, Tower Hill, Freetown (proclaimed 1961).
- Old military butts, Jomo Kenyatta Road, Freetown (proclaimed 1962).
- Grave of Captain Lendy and others, Waiima, Kono District (proclaimed 1965) (Basu & Sam, 2016, p. 9).

References

Abulafia, D. (2014). Thalassocracies. In P. Horden & S. Kinoshita (Eds.), *A companion to Mediterranean history* (pp. 137–153). Wiley-Blackwell.

Adams, V. (1984). Logistic support for the Falklands campaign. *The RUSI Journal, 129*(3), 43–49.

Adanir, F. (2005). Turkey's entry into the concert of Europe. *European Review, 13*(3), 395–417.

Anderson, B. (1983). *Imagined communities*. Verso.

Banton, M. (1969). *West African city: A study of tribal life in Freetown*. Repr. Oxford University Press.

Barlow, S. M. (2003). Fish meal. In B. Caballero (Ed.), *Encyclopedia of food sciences and nutrition* (2nd ed., pp. 2486–2491). Academic Press.

Basu, P. (2016). Palimpsest memoryscapes: Materializing and mediating war and peace in Sierra Leone. In F. de Jong & M. Rowlands (Eds.), *Reclaiming heritage* (1st ed., pp. 231–259). Routledge.

Basu, P., & Sam, M. A. (2016). *Review of the monuments and relics act and development of recommendations for new heritage legislation for Sierra Leone*. Consultation Report, Prepared for the Ministry of Tourism and Cultural Affairs and Monuments and Relics Commission.

Belhabib, D., Sumaila, U. R., & Le Billon, P. (2019). The fisheries of Africa: Exploitation, policy, and maritime security trends. *Marine Policy, 101*, 80–92.

Borutta, M., & Gekas, S. (2012, February). A colonial sea: The Mediterranean, 1798–1956. *European Review of History: Revue Europeenne d'histoire, 19*(1), 1–13.

Brautigam, D. A. (1994, November). Foreign assistance and the export of ideas: Chinese development aid in the Gambia and Sierra Leone. *The Journal of Commonwealth & Comparative Politics, 32*(3), 324–348.

Brown, E. D. (1984). The UN convention on the law of the sea 1982. *Journal of Energy & Natural Resources Law, 2*(4), 258–281.

Cohen, M. D., March, J. G., & Olsen, J. P. (1972). A garbage can model of organizational choice. *Administrative Science Quarterly, 17*(1), 1–25.

Crona, B., Wassénius, E., Troell, M., Barclay, K., Mallory, T., Fabinyi, M., Zhang, W., et al. (2020, July 24). China at a crossroads: An analysis of China's changing seafood production and consumption. *One Earth, 3*(1), 32–44.

De Thierry, C. (1899). The empire's coaling stations. *The English Illustrated Magazine, 194*, 115–125.

Denzer, L. (1982). Wallace-Johnson and the Sierra Leone labor crisis of 1939. *African Studies Review, 25*(2/3), 159–183.

Deveneaux, G. K. (1976). Public opinion and colonial policy in nineteenth-century Sierra Leone. *The International Journal of African Historical Studies, 9*(1), 45–67.

Ekoko, A. E. (1981 [1982]). The strategic-imperial factor in British expansion in Sierra-Leone, 1882–1889. *Journal of the Historical Society of Nigeria, 11*(1/2), 138–152.

Fallah-Williams, J. (2021, June 4). Sierra Leone's Black Johnson Chinese fishmeal factory deal exposed—Op ed. *Sierra Leone Telegraph* (blog). https://www.thesierraleonetelegraph.com/sierra-leones-black-johnson-chinese-fishmeal-factory-deal-exposed-op-ed/

Frost, R. (2004). Underwater cultural heritage protection. *Australian Year Book of International Law, 23*, 25–50.

Gberie, L. (2005). *A dirty war in West Africa: The RUF and the destruction of Sierra Leone*. Indiana University Press.

Gilroy, P. (1993). *The Black Atlantic: Modernity and double consciousness*. Verso.

Hancock, I. F. (1976). Nautical sources of Krio vocabulary. *International Journal of the Sociology of Language, 7*, 23–36.

Howard, A. M., & Skinner, D. E. (1984). Network building and political power in Northwestern Sierra Leone, 1800–65. *Africa: Journal of the International African Institute, 54*(2), 2–28.

Hurd, A. S. (1898). Coal, trade, and the empire. *The Nineteenth Century and After: A Monthly Review, 44*(261), 718–723.

Knörr, J. (2010). Contemporary creoleness; or, the world in pidginization? *Current Anthropology, 51*(6), 731–759.

Knörr, J., & Kohl, C. (2016). *The Upper Guinea Coast in global perspective*. Berghahn Books.

Liu, W., & Dunford, M. (2016, November 3). Inclusive globalization: Unpacking China's belt and road initiative. *Area Development and Policy, 1*(3), 323–340.

McGowan, W. (1990). The establishment of long-distance trade between Sierra Leone and its hinterland, 1787–1821. *The Journal of African History, 31*(1), 25–41.

Ménard, A. (2019, October 25). Land and tears: The political economy of deforestation on the freetown peninsula. *Integration and conflict along the Upper Guinea Coast/West Africa (IC_UGC)* (blog). https://upperguineacoast.wordpress.com/2019/10/25/land-and-tears-the-political-economy-of-deforestation-on-the-freetown-peninsula-by-anais-menard-2019/

Newell, S. (2011). Articulating empire: Newspaper readerships in colonial West Africa. *New Formations, 73*, 26–42.

O'Kane, D. (2020). *Language, nationhood, and systems: Insights from language policy at the University of Makeni, Sierra Leone* (Max Planck Institute for Social Anthropology Working Papers, Working Paper, no. 204).

Okonkwo, R. L. (1980). The Garvey movement in British West Africa. *The Journal of African History, 21*(1), 105–117.

Olukoju, A. (2020). African seaports and development in historical perspective. *International Journal of Maritime History, 32*(1), 185–200.

Parins, J. W. (1985). The English-language native press in the nineteenth century. *Victorian Periodicals Review, 18*(1), 17–33.

Robinson, R. E., & Gallagher, J. (1981). *Africa and the Victorians: The official mind of imperialism*. Macmillan.

Said, E. (2003 [1978]). *Orientalism*. Penguin.

Select Committee. (1865). *Report of the Select Committee on West African settlements*.

Shore, C., & Wright, S. (Eds.). (1997). *Anthropology of policy: Perspectives on governance and power*. Routledge.

Smith, G. (1848). *The case of our west-African cruisers and west-African settlements fairly considered*. William Watts.

Stoett, P., & Omrow, D. A. (2021). Transnational oceanic ecoviolence. In P. Stoett & D. A. Omrow (Eds.), *Spheres of transnational ecoviolence: Environmental crime, human security, and justice* (pp. 103–125). Springer International Publishing.

18

Navigating a Sea of Laws: Small-Scale Fishing Communities and Customary Rights in Ghana and South Africa

Anthea Christoffels-DuPlessis, Bola Erinosho, Laura Major, Elisa Morgera, Jackie Sunde, and Saskia Vermeylen

In the past two decades there has been growing recognition of and support for indigenous peoples' and other communities' customary law in international environmental and human rights law (Knox, 2018; Tobin, 2014; Trechera, 2010). In addition, the importance of customary systems of tenure in natural resource governance has been increasingly recognized, both for its local contributions to sustainability and culture (FAO, 2012, 2014; Jentoft & Bavinck, 2019; Sunde, 2017), as well as for its contributions to global objectives.

This paper has been prepared under the One Ocean Hub, a collaborative research for sustainable development project funded by UK Research and Innovation (UKRI) through the Global Challenges Research Fund (GCRF) (Grant Ref: NE/S008950/1). GCRF is a key component in delivering the UK AID strategy and puts UK-led research at the heart of efforts to tackle the United Nations Sustainable Development Goals. The authors and their affiliations are as follows: Anthea.Christoffels-DuPlessis, Nelson Mandela University, South Africa, Bola Erinosho, University of Cape Coast, Ghana, Laura Major, University of Strathclyde, UK, Elisa Morgera, University of Strathclyde, UK, Jackie Sunde, University of Cape Town, South Africa and Saskia Vermeylen, University of Strathclyde, UK.

A. Christoffels-DuPlessis (✉)
Faculty of Law, Nelson Mandela University, Gqeberha, South Africa
e-mail: anthea.christoffels-duplessis@mandela.ac.za

While these international developments have largely focused on land and terrestrial natural resources, the importance of legal pluralism for the governance of the ocean and marine resources is also receiving increasing attention internationally (Gupta & Bavinck, 2014; Jentoft & Bavinck, 2019; Parlee & Wiber, 2015). This attention has been catalyzed the struggles of indigenous peoples and small-scale fishing communities who have advocated for recognition of their customary systems of law and rights to marine resources (ICSF, 2008; Sunde, 2017). It has also come from within the arenas of both fisheries and marine conservation, as State and non-State actors seek policy reforms that will address conflicts over resources and promote resource sustainability in the face of past failures (Bavinck et al., 2014; Aswani & Cinner, 2007).

Yet, although customary law has gained formal recognition in many national jurisdictions (Cuskelly, 2010), many small-scale fishing communities' customary rights remain vulnerable due to a combination of factors: the complex legacy of colonialism in existing national legislation on natural resources, unduly restrictive approaches in statutory law, and continued prejudice toward all things customary (Jentoft & Bavinck, 2019; Sunde, 2017; Wilson, 2021). In some cases, it may also be due to the lack of awareness that the ocean-related rights of Indigenous Peoples and local communities have been the object of misrepresentation, misrecognition, or dispossession for decades and centuries (Wilson, 2021).

After reflecting on the current status of international law and insights from anthropology on the recognition of customary law, this chapter explores shared challenges and different approaches to ensuring the recognition of small-scale fishers' customary rights in Ghana and South Africa The chapter

---

B. Erinosho
Faculty of Law, University of Cape Coast, Cape Coast, Ghana
e-mail: berinosho@ucc.edu.gh

L. Major · S. Vermeylen
Strathclyde University, Glasgow, Scotland
e-mail: laura.major@strath.ac.uk

S. Vermeylen
e-mail: saskia.vermeylen@strath.ac.uk

E. Morgera
University of Strathclyde Law School, Glasgow, UK
e-mail: elisa.morgera@strath.ac.uk

J. Sunde
University of Cape Town, Cape Town, South Africa

concludes by drawing on international human rights law and international environmental law to identify procedural approaches to strengthen the protection for fishers' customary rights as a matter of implementation at the national level.

International Recognition of Customary Fishing Rights in the Context of International Biodiversity Law and International Human Rights Law

International human rights law requires States to recognize customary laws of indigenous peoples and other local communities as a way to protect their human rights to natural resources, as well as their rights to culture and livelihoods, property, and/or development. International biodiversity law imposes obligations on States to recognize and protect customary laws as a way to acknowledge and support indigenous peoples' and local communities' contributions to environmental sustainability. While these two areas of international law have evolved in separate ways, and for different purposes, they have come, over time, to converge and complement each other in determining the content of State obligations and the modalities for their implementation (Morgera, 2018).

In parallel, international recognition is growing of the global benefits of customary approaches: for instance, FAO has indicated that customary food systems are a "game-changing solution" for current global debates on sustainable food systems, the conservation of biodiversity, and the realization of the right to food (FAO, 2021, p. 42). FAO recognized that these food systems are "intimately connected to nature, promote the equitable distribution of resources and powers, while supporting indigenous identities and values" (FAO, 2021, p. xiv) although they "continue to be marginalized in policy" and escape characterization by dominant approaches to food systems as linear value chains (FAO, 2021, p. xiv).

On the side of international human rights law, the UN Declaration of the Rights of Indigenous Peoples (UNDRIP), calls upon States to give due consideration to customary laws and tenure systems in providing legal recognition and protection to indigenous peoples' territories and resources (Art. 26.3), "establish(ing) and implement(ing), in conjunction with indigenous peoples concerned, a fair, independent, impartial, open and transparent process(es)" (Art. 27), and providing effective remedies for all infringements of their rights (Art. 40). Along similar lines, ILO Convention 169 calls on

States to pay "due regard" to customary laws in applying national laws and regulations (Art. 8) so that indigenous peoples can retain their own customs and institutions that are compatible with fundamental rights defined by the national legal system and with internationally recognized human rights. In addition, States shall prevent third parties from taking advantage of customs to secure the ownership, possession, or use of land belonging to indigenous peoples (Art. 8). These obligations, while clear in scope and content, leave unclear which means of implementation should be used in the context of natural resource governance.

Under the Convention on Biological Diversity (CBD),[1] in turn, we have more open-ended obligations, but detailed guidance on means of implementation to protect indigenous and local custodianship embedded in customary rules and practices (including linguistic diversity).[2] A series of guidelines, adopted intergovernmentally and with inputs from indigenous peoples' representatives, explain how these obligations apply in the context of natural resource governance. For instance, CBD Parties defined, to the benefit of natural resource managers, customary laws as "law consisting of customs that are accepted as legal requirements or obligatory rules of conduct; practices and beliefs that are so vital and intrinsic a part of a social and economic system that they are treated as if they were laws" (Akwé: Kon Guidelines, para. c). CBD Parties are then expected to assess the potential impacts of development proposals affecting indigenous peoples' sacred and traditionally used resources on their customary laws, including with regard to tenure, distribution of resources and benefits, or their knowledge with a view to identifying the need to "codify certain parts of customary law, clarify matters of jurisdiction, and negotiate ways to minimize breaches of local laws" (Akwé: Kon Guidelines, paras. 29, 34 and 60). In addition, CBD Parties are expected to ensure that environmental and other laws are implemented according to customary laws, whereby seeking free, prior informed consent of indigenous peoples is understood as a "continual process of building mutually beneficial, ongoing arrangements ... in order to build trust, good relations, mutual understanding, intercultural spaces, knowledge exchanges, create new knowledge and reconciliation" (Mo'otz Kuxtal Voluntary Guidelines, para. 8).[3]

Further guidance, specifically targeted at fisheries managers has been developed, on the basis of international human rights and environmental law, under the aegis of the UN Food and Agriculture Organization (FAO). The Voluntary Guidelines for Securing Sustainable Small-Scale Fisheries in the context of Food Security and Poverty Eradication (SSF Guidelines) (ibid.) acknowledge that in many countries a range of tenure regimes may coexist, although some may not be recorded or legally protected. The Guidelines call on States to "recognize, respect and protect all forms of legitimate tenure rights, taking into account, where appropriate, customary rights, to aquatic resources and land and small-scale fishing areas enjoyed by small-scale fishing communities" (SSF Guidelines, para. 5.4; Morgera & Nakamura, 2022). To that end, States should take appropriate legal measures to identify, record, and respect legitimate tenure right holders and their rights, recognizing "local norms and practices, as well as customary or otherwise preferential access to fishery resources and land by small-scale fishing communities … in ways that are consistent with international human rights law" (SSF Guidelines, para. 5.4). In addition, formal planning systems should consider methods used by small-scale fishing, their customary tenure systems, and decision-making processes (SSF Guidelines, para. 10.2). At the same time, however, the SSF Guidelines recognize that "customary practices for the allocation and sharing of resource benefits in small-scale fisheries, which may have been in place for generations, have been changed as a result of non-participatory and often centralized fisheries management systems, rapid technology developments and demographic changes" (SSF Guidelines, Introduction). So processes of genuine engagement with and learning from indigenous peoples and small-scale fishers are necessary to understand the current status of customary laws, customary governance, and customary tenure of fisheries resources.

To that end, the FAO Voluntary Guidelines on the Responsible Governance of Tenure (FAO, Rome, 2012) provide a methodology and set of guiding steps for States to ensure that policy, legal, and organizational frameworks "reflect the social, cultural, economic and environmental significance" of fisheries and "the interconnected relationships between land, fisheries and forests and their uses, and establish an integrated approach to their administration" (para. 5.3) which may require adapting national laws to customary laws with full and effective participation of all members or representatives of affected communities, including vulnerable and marginalized members (paras. 9.6–9.7). A series of practical steps are then identified, whereby States are to:

> First identify all existing tenure rights and right holders, whether recorded or not, by consulting indigenous peoples and other communities with customary tenure systems, smallholders and anyone else who could be affected (para. 7.3);
> Provide support to tenure holders so that they can enjoy their rights and fulfil their duties (para. 7.5); ensure coordination between implementing agencies, as well as with local governments, and indigenous peoples and other communities with customary tenure systems (para. 5.6);

Publicize categories of legitimate tenure rights through a transparent process, recording indigenous peoples' and other communities' tenure rights and private sector's rights in a single recording system, or systems linked by a common framework (paras. 8.4 and 9.4);

Establish safeguards to avoid infringing on or extinguishing customary tenure, including legitimate tenure rights that are not currently protected by law, including for women and the vulnerable who hold subsidiary tenure rights (para. 5.6);

Where it is not possible to provide legal recognition of tenure rights, prevent forced evictions that are inconsistent with their existing obligations under national and international law (para. 7.6); and

Provide access to justice if people believe their tenure rights are not recognized (para. 7.3).

On the whole, while all these international instruments have been developed at different times, and have different formal legal status, they are generally recognized as providing a coherent interpretation of relevant international binding obligations on States.[4] Even where States may argue that relevant international treaties do not apply to them, these international legal materials must be understood as "best practices" should be "adopt(ed) as expeditiously as possible" (Knox, 2018, paras. 7–9). In other words, it would be extremely difficult for a State to defend an approach that goes against an internationally recognized best practice, particularly when it has agreed upon the international guidance after intensely participating in its negotiations (Morgera, 2018).

Perspectives on Legal Pluralism

The recognition of customary laws and the role they can play in fisheries management is part of a wider acknowledgment that resource management is regulated through different legal regimes, including: international, regional, national, and customary laws. One way to engage with these multiple normative layers is to employ the concept of legal pluralism, which can be defined as "a situation in which two or more legal systems coexist in the same social field" (Merry, 1988, p. 871).

Legal pluralism raises interesting questions on how these different legal scales operate together and how they are able to influence each other in order to create an inclusive best-practice approach, without distorting or creating tensions and/or contradictions between the different norms that are underpinning the legal scales. Based on the discussion of international law above, the recognition of customary law is not a discreet legal issue, but part of a wider inquiry that touches upon other human rights issues, such as the right to health, food, education, and so forth. Therefore, legal pluralism can be considered as part of a wider debate about global legal systems and how to manage legal hybridity. The global legal system consists of "an interlocking web of jurisdictional assertions by state, international and non-state normative communities" (Berman, 2007, p. 1159). Increasingly States must share jurisdiction over the same activity or they share their legal authority with multiple courts (e.g., at international, national, and regional level) or indeed may share authority with non-State legal or quasi-legal norms. In these sites of hybrid legality, the overlapping legal authorities may clash and cause conflict and confusion. Different options exist to resolve such conflicts, the most common ones are imposing or reinstating the primacy of territorial or State governed legal systems or seeking universal harmonization of legal systems and norms (Berman, 2007). An alternative option that is endorsed by global legal pluralists is to embrace legal pluralism and actively "seek to create or preserve spaces for conflict among multiple, overlapping legal systems" (Berman, 2007, p. 1164). It studies the interaction between the different actors without suggesting a hierarchical ordering between them. Instead, it focuses on the plural procedural mechanisms, institutions, and practices that provide a platform for the communication between plural voices (Berman, 2007).

In this area of thinking, Teubner calls for moving away from distinguishing what is law and what is not law, and rather focus on what is legal and illegal (Teubner, 1983, p. 1415). For Teubner and Fischer-Lescano, the major problem does not lie in the incompatibility between customary law and modern international or national law, but in how highly specialized action centers, may have, torn customary law through legal fragmentation "out of its context on which it has been embedded and transform it in their own metabolism" (Teubner & Fischer-Lescano, 2008, p. 8). The end result is that the wider surviving strategies of indigenous peoples and local communities have become incorporated into mainstream society and therefore the whole process of knowledge production and recognition of customs and norms should be included in basic rights protection (Teubner & Fischer-Lescano, 2008, p. 6). This supports reliance on international human rights law for recognizing customary law at the national level, with a view to avoiding infringements of other internationally protected human rights.

Similarly to international biodiversity law, legal pluralism also points to procedural pluralism as a way forward. One way of achieving this is through focusing on a technique that is widely used in anthropology and that is thick descriptions of the ways in which various procedural mechanisms, institutions, and practices actually operate as sites of contestation and creative innovation. This opens up the opportunity in implementing the procedural steps identified in international legal instruments for the recognition of customary laws to employ different methods to support genuine and potentially transformative processes of recognizing traditional fishing rights. One of these could indeed be story-telling, as illustrated in the Chapter by Erwin, McGarry, Perreira, and Coppen in this book.

Customary Fishing Rights and Wrongs: Reflections from Ghana and South Africa

The sections that follow will set out details of the legal and material environment in which day-to-day fishing activities and their governance plays out in Ghana and in South Africa. Both are characterized by the kind of complex plural "sea" of legal discourse and overlapping regimes that has been discussed in a section further on. We will discuss some of the differences and similarities between these two legally plural contexts, noting the implications for the actual or potential protection of small-scale fishers' access to the ocean, use of customary fishing techniques, and their preferred harvest. We will point out situations in which national legal mediation and decision-making processes appear to disadvantage or misunderstand both the situation and wishes of small-scale fishers.

Ghana

Ghana, with a 500 km coastline, has a thriving and important trade in fisheries to which small-scale and semi-industrial fishers are a key contributor.[5] Although robust stock assessment data is limited, the catastrophic decline in fish stocks, particularly affecting the sardinella species that are the mainstay of smaller vessels and local fishing trade (Cook et al., 2021) has increased the risk of economic and social destitution for many of the coastal communities who are extremely economically poor and are not able to compensate for the attendant loss of livelihood. The decline is linked to many factors, including global warming, increased pollutants from terrestrial sources, and over-fishing (Atta-Mills et al., 2004). The threat presented by a loss of fish and fishing practice is about more than just economics. Fishing in Ghana also holds great significance for community, cultural, and spiritual life. Individual and group identities are closely linked to values and meanings that stem from fishing heritage and ongoing practice of fishing and its trade (Odotei, 2002).

Research is exploring where customary governance systems become relevant in the move to secure the rights of small-scale fishing communities to fish stocks and to continue to practice long established fishing techniques, trade, and associated community activities in view of a declining resource base. This is not a straightforward issue in a context in which traditional authorities are currently positioned in national legal and policy reform efforts as interlocutors, tasked with integrating and even enforcing State-led fisheries policy and legal reform—a positioning that is often challenged by traditional authorities and their respective communities. Such reforms have very particular framing that may not easily align with indigenous and local community customary practice, these efforts are also hampered by a complex political environment in which fisheries across scales, including small-scale fishers, have effectively carried on fishing practice, often under customary management arrangements, in the gap between regulation as written and a relative lack of implementation of that regulation.

Customary Law Within the Legally Plural System of Ghana

The roots of present-day legal pluralism in Ghana lie with British colonization, via indirect rule, of the Gold Coast area from the 1800s. From the 1400s Portuguese, then Dutch and British companies battled to set up and control profitable trading outposts in the region. The British ultimately gained the upper hand among competing colonial authorities. A series of treaties including the Bond of 1844, which were initially targeted at ensuring trading routes were undisturbed and military protection for the Fanti coastal communities, created the opportunity for the British to assume control of parts of the Gold Coast. The Bond of 1844 had required the chiefs to submit serious crimes such as murder and robberies to British jurisdiction. This laid the legal foundation for subsequent colonization of parts of the Gold Coast. Earlier in 1843, the British Settlements Acts were passed to empower colonial administrators on behalf of the British crown to "establish such laws, institutions and ordinances, and to constitute such courts and offices as may be necessary for the peace, order and good government" of the territories concerned. This declaration laid the groundwork for Ghana's contemporary legally plural system. Victor Essien (2020, p. 2) helpfully summarizes this original setting and its references to the continued recognition of customary law as part of this legal framework:

In 1876, the Gold Coast Supreme Court Ordinance (No. 4 of 1876) was passed. Section 14 of the Ordinance stipulated that: "The Common law, the doctrines of equity, and the statutes of general application which were in force in England at the date when the colony obtained the local legislature, that is to say, on the 24th day of July 1874, shall be in force within the jurisdiction of the Court."

Section 19 of this Act … reads in part as follows:

> Nothing in this Ordinance shall deprive the Supreme Court of the right to observe and enforce the observance, or shall deprive any person of the benefit of any law or custom existing in the said colony and territories subject to its jurisdiction, such law or custom not being repugnant to natural justice, equity and good conscience, not incompatible either directly or by necessary implication with any enactment of the colonial legislature.

The effect of these enactments was not only to create the foundations of a legally pluralistic system, but also one steeped in hierarchy and value judgments based on western philosophical ideologies of "natural justice equity and good conscience" (repugnancy test) weighed against the customary law rules in these communities. At independence, the "repugnancy test" was abolished along with the requirement to prove the existence of a customary law as facts before court. Nevertheless, customary law continues to fall under broader common law as a source of law in Ghana, recognized under the 1992 Constitution. Specifically, defined under Article 11(3) of the 1992 Constitution as the "rules of law, which by custom are applicable to particular communities" Customary law is however still subject by the courts to a compatibility test and in Article 26(2) of the 1992 Constitution, "customary practices which dehumanise or are injurious to the physical and mental well-being of a person are prohibited." In ascertaining the rules of customary law, the Courts Act of 1993 (Act 459) notes that if there is doubt as to the existence or content of a rule of customary law relevant in any proceedings before a court, it may consider testimonies and or depositions of persons held to be knowledgeable in a particular custom, judicial precedents, textbooks, and commentaries by scholars on particular customary law rules. It further notes that the court may request a House of Chiefs, Divisional or Traditional Council, or other body with knowledge of the customary law in question to state its opinion which may be laid before the inquiry in written form (S. 55). As again, Essien (2020, p. 10) summarizes "under Sections 42 and 43 of the Ghana Chieftaincy Act, 1971 (with some later amendments) the National House of Chiefs and/or Regional House of Chiefs, can draft their declaration of customary law for approval and publication as legislative instrument by the President after consultation with the Chief Justice." In recent years, the House of Chiefs has developed a program to research and systematically record customary law, efforts that could perhaps assist with efforts to secure co-management and community tenure rights for certain sections of the fisheries.

The Constitution of Ghana therefore recognizes the institution of Chieftaincy as the custodians and an arbiter of disputes that can concern customary laws, that is the system of customary values and norms of the nation. As noted by *Amegatcher JSC in Republic v National House of Chiefs & Ors (2019)* "it is very clear from the intention of the framers of the Constitution and the lawmakers that the responsibility given to the National and Regional Houses of Chiefs is to do everything within its power to preserve the customary practices of this revered institution in our culture. This is to be done by advising the individuals, bodies, and groups vested with the authority of state...."

The Chieftaincy Act, 2008 (Act 759) sets out the definition of the chief, the need for candidates to be eligible according to lineage, and the hierarchical structure of chiefs to be recognized at national level. Although customary law and the Chieftaincy is enshrined in formal law within the legally plural system, there is considerable debate about where and how the system could be reformed (Marfo, 2019; Ubink, 2007). Certainly, the colonial era created a rupture in the system of Chieftaincy in part because those who remained in office during the colonial administration may have been able to do so because they acquiesced to the demands of colonial authorities, those who were not, were deposed and sometimes deported. This undermined faith in the system among communities that Chiefs were selected to represent and positioned traditional authorities, sometimes awkwardly, between customary and State authority, an ambiguity that remains an issue for those seeking to exercise customary authority. Despite its uneasy position within Ghana's legally plural system, traditional authorities are held in high regard by communities as custodians of customary law.[6] This is not to say that customary law has not been subject to critique, with some arguing that such systems are sometimes inequitable and may not ensure gender equity (Britwum, 2009; Tendayi et al., 2016).

This uncertainty has been further complicated by the lack of codification of customary law, and by the formal court system being vested with the power to decide the validity of customary laws. Judges in the Supreme Court, who are in many cases outsiders to the communities and customary practices which they have been called to adjudicate upon, have the final decision-making powers to the detriment of Chiefs. This has often resulted in litigants/complainants overriding the chief's decisions in superior courts. This practice, which started in colonial times, continues to persist today. In *Akwufio & Ors v Mensah & Ors* (1898), the courts overruled the decision of the chief of Winneba on the ban of "Ali nets" which the chief and some of the fishermen believed was detrimental to the sustainability of the fishing industry. The court not only ignored existing marine tenure regimes and reinforced the then European perceptions of the sea as an unlimited resource which should be utilized more efficiently (Endemano Walker, 2002). It also effectively disrupted the power of the chiefs to make decisions on matters of marine resources. More recently the Supreme Court has often had to make pronouncements on matters decided at the National House of Chiefs (c.f., Andakwei [Substituted by Kow Atteh] & Ors. v Abaka & Ors, 2018).

Customary Practice, Traditional Authorities, and the Marine Space

There are specific customary laws and traditions which relate to fisheries practice along the coast of Ghana, and these are managed by traditional authorities arranged in a hierarchy and associated with particular ethnic groups of fishers. These are loosely tied to areas of marine space although there is a strong history and contemporary practice of seasonal migration among fishers in the region which routinely extends across the national borders of present-day Ghana and into Nigeria, Liberia, and further afield (Overå, 2001). This movement is in keeping with the origins of commercial sea fishing in the region in which Ghanaian seafaring companies were a powerful force in the West African region particularly in the nineteenth and twentieth centuries (Atta-Mills et al., 2004). Innovation in artisanal and industrial fishing methods is an important part of fishing culture in the region, and it was the modification of river boats for the sea in the nineteenth century which allowed the commercial fishing sector to emerge (Agbodeka, 1992 in Atta-Mills et al., 2004). There is a continuity in efforts to adapt to changing conditions, as new technologies have become available.

The traditional authorities for small-scale marine fisheries run parallel to and in some communities in conjunction with traditional political authorities. Small-scale fishers are represented by a Chief Fisherman Apofohene (Fanti), Woleiatse (Ga-Adangme), Dotorwofia (Anlo) who is assisted by a council of men. Reflecting the gendered nature of fisheries in Ghana, the women are led by a Chief Fishmonger/Processor Konkohene (Fanti) and her council Besonfo in coastal communities. These positions, which may be hereditary or elected on merit, are responsible for upholding customary practices, fishing taboos in coastal communities as well as representing the interests of small-scale fishers.

This traditional governance model is however not acknowledged in the existing legal framework represented by the Fisheries Act (2002) Act 625 and related legislative instruments. The Act vests power for fisheries management in the National Fisheries Commission, recently housed under the Ministry of Fisheries and Aquaculture Development, created under the 1992 Constitution in order to manage fisheries resources and coordinate policy. There is broadly speaking, an acknowldgement of a problematic lack of clarity over the responsibilities and powers of the Commission and of fisheries management across scales (Coastal Resources Center, 2021, p. 22). In response, the national Fisheries Management Plan (2015–2020) proposes as solutions reducing "excessive fishing effort" and reviewing and enforcing existing regulations. For the fishing sector that uses canoes (which may be motorized or non-motorized), solutions proposed included the survey and registration of canoes, "education of fishers in collaboration with traditional authorities and local assemblies," and a heavy emphasis on the development of co-management systems in communities (Fisheries Commission, 2015, p. 22).

The 2020 Co-Management Plan for the Fisheries Sector adopted as part of the current wave of reforms adopted the FAO definition of co-management as "a partnership arrangement between the government and the local community of resource users, sometimes also connected with agents such as NGOs and research institutions, and other resource stakeholders, to share the responsibility and authority for the management of a resource." The intention of the policy is that it will delegate fisheries management responsibilities and authority to resource users and other stakeholders through a "pluralistic management approach" to co-management units or areas.

Traditional authorities are thus to play specific roles in the co-management process and be involved in management committees. The specifics of this involvement are not set out in the policy, but it is noted that "the process of empowerment of traditional authorities must be accompanied by careful integration of conventional co-management approaches with traditional beliefs and practices" and that "co-management processes and institutions should build on existing forms of functional and traditional management where they exist and have some degree of efficiency." The Policy references "Traditional Institutions and women" in one specific section noting that "The roles of traditional authorities in fishing communities varies from place to place and do not usually appear to promote good fishing practice. However, overall, they represent a positive force for good management even though their influence in some circumstances has been compromised over time" …. "Chief Fishermen have over the years been a fulcrum for resource management due to their widespread respect and social influence in fishing communities … This Policy will encourage the involvement of traditional institutions such as the Chief Fisherman (in maritime areas), and river priests or the fisher folk headman (in inland fishing communities), in the co-management committees … To ensure their effective supervision by the Traditional Area or community chieftaincy, the Ministry will liaise with the Ministry of Chieftaincy and Traditional Affairs to recognize the position of the Chief Fisherman, whether by inheritance or by appointment by fishers, under the Chieftaincy arrangement in Ghana" (Government of Ghana, Ministry of Fisheries and Aquaculture, 2020). A similar arrangement is proposed for the traditional institution of the "fish market queen" and her elders, though there is no reference to recognizing the position under Chieftaincy arrangements.

This reference to traditional authorities "not usually appear(ing) to promote good fishing" picks up on a general set of tensions within the current regulatory environment. On the one hand the view that small-scale fishers represented by traditional authorities such as the Chief Fishermen are to blame for the challenges within the fishing industry including the IUU fishing.[7] On the other hand, traditional fishing authorities, who are not constitutionally precluded from participating as political party members, can also sometimes be seen to be partisan in their dealings deeply entangled in the culture of "political influence" which is a limiting factor in the successful development and implementation of legal and policy reforms (Stark et al., 2019). Traditional authorities are part of the political nexus that is concerned with securing electoral votes and the goodwill of fisheries communities that have a relatively large sway in terms of political influence. These complexities in the positioning of traditional authorities raise questions about whether these custodians of customary law and practice will be able to integrate these systems into conventional co-management approaches in the way that the policy directs. There may be particular difficulties for Chief Fishermen who already struggle to arbitrate disputes, and maintain the trust in their authority by the communities.[8]

Where these efforts have been successful, and where the co-management policy seems to make the most of traditional and customary systems of practice is in the case of relatively bounded areas of marine space and relatively specific small-scale fishing practice.[9] Where the co-management policy may face particular difficulty is in the areas in which there is significant conflict over both space and resources across fisheries scales. The most critical of these areas will be in relation to Ghana's inshore coastal region where access to fishing of pelagic fish stocks (particularly sardinella species) has been the source of clashes between fisheries groups of a similar size as well as across scales. It is in these areas that the desired shift from "open access" fisheries to managed units,[10] will be particularly challenging and in which the issue of rights-based approaches and international law more broadly may be best placed to mediate in ensuring that small-scale fishers have equitable access to resources and that customary management systems, values, and practices are protected and effectively integrated into new arrangements.

Furthermore, fishers in this area, although almost always in agreement that fish stocks are in catastrophic decline, have thus far benefitted in some cases from the freedom that came from a lack of regulation of their activities, allowing them to follow customary practice without the need to justify this to authorities. This along with overcrowding in the sector and competition with industrial trawlers has led to considerable tension which policy and legal reforms have so far failed to sufficiently resolve. This competition is further heightened by other ocean-based activities such as oil prospecting which has excluded fishers from resource-rich waters in and around the buffer zone introduced around the oil rigs, sea defense projects as well as pressures from the tourism industry which has meant that small-scale fishers find themselves potentially being squeezed out in the access to the resources of the ocean.[11]

On the whole, it remains to be seen if the 2020 co-management fisheries policy, and follow-up regulations and plans will sufficiently protect small-scale fishers' customary rights. Research is needed to establish where and how small-scale fishing communities continue to draw on customary law and both "traditional" and other formal legal processes in efforts to govern fishing and associated activities. In addition, an investigation is needed of whether there is a route to protecting and elevating the rights of small-scale fishers by harnessing international law to further limit the size and influence of industrial fishing fleets.

South Africa

South Africa's history of legal pluralism extends back to the first days of the Dutch colonial occupation of the country in 1652 when the Dutch arrived at the Cape. Historically, South African waters and shores have been a rich source of marine resources upon which its earliest indigenous coastal communities have depended for their food (Sowman, 2006). Many of these indigenous peoples had their own set of customs and practices that constituted their local systems of customary law that governed their societies (Rautenbach, 2010). Their customary law was unwritten and orally handed down from one generation to the next generation (Sunde, 2014). As in many other places in the world, the impact of colonialism by the Dutch and British Settlers invariably affected the indigenous communities of South Africa (*Alexkor Ltd and Another v Richtersveld Community and Others* 2003. Hereinafter the *Richtersveld Community* judgment).[12] South Africa's common law is thus comprised of both Roman-Dutch law and English law (Rautenbach, 2010).

The colonial regimes sought to regulate customary law by means of legislation as a way of controlling the indigenous communities. This codification of customary law rules did not appreciate the fluid nature of customs, thus rendering it fixed. Further, the customary law rules could only be applied to disputes among members of customary communities and on condition that they did not offend public policy or the principles of natural justice (Hamukuaya & Christoffels-Du Plessis, 2018).[13]

Development of the Law in South Africa Since 1994

The marginalization of and discrimination against traditional fishing communities in South Africa persisted pre-constitutional era, as result of racial segregation and the treatment of customary law as inferior to statutory law by the Apartheid government (Sunde, 2014). At the time of the first democratic elections in 1994, it was estimated that 50% of African residents lived according to some form of customary law (Mnisi, 2009). Many of the coastal communities on the eastern seaboard of the country, particularly those living in areas designated by the apartheid regime as "bantustan homelands," lived according to an African customary system of law. The systems of marine tenure differed considerably from region to region, on account of the different histories of the peoples of the coast and the distinctive ways in which their customary systems and practices have interfaced with colonial and apartheid governance (Sunde, 2014).

The legal reforms post 1994 provided important recognition for customary law, as an independent source of law, in terms of a quartet of provisions in the Constitution, 1996 (*Gongqose and Others v Minister of Agriculture, Forestry and Fisheries and Others 2018*, para. 22). Section 211(3) empowers "the courts to apply customary law when that law is applicable, subject to the Constitution and any legislation that specifically deals with customary law." The right to participate in a chosen cultural life is also protected in section 30. Section 31 ensures that indigenous communities have the right to enjoy their culture and to form cultural associations. Furthermore, courts are obliged to promote the spirit, purport, and objects of the Bill of Rights when developing customary law in terms of section 39(2) of the Constitution. Under section 39(3) the Bill of Rights does not deny the existence of any rights and freedoms arising from customary law provided they are consistent with the Bill of Rights. The South African legal system is indeed a plural system of law; and customary law is a defining and independent feature thereof (*Bhe and others v Khayelitsha Magistrate & Others*, 2005, para. 235; *Gumede v President of the Republic of South Africa*, 2009, para. 22). The Constitution, as the supreme[14] law of the land, recognizes both customary law and common law as equal and independent sources of law in the South African legal system and both should be equally applied and developed in a manner consistent with the Constitution and its values.[15]

The Recognition of Customary Law Through the Constitutional Lens

In order to understand the evolution of the recognition of customary rights in South Africa, it is necessary to have regard to the landmark rulings by the Constitutional Court, which set the stage for the recognition of customary fishing rights discussed in the following subsections.

The constitutional protection and recognition of customary law includes both living and official customary law (Section 211 of the 1996 Constitution). Although there is no express hierarchy between the two forms of customary law, the Constitutional Court expressed its preference to living customary law due to its authenticity as it is the law which is practiced by the communities to whom it applies. In the *Richtersveld*[16] judgment, the Constitutional Court found that the customary law does not comprise of a fixed set of rules that are easily ascertainable. Instead its evolving nature makes it subject to change, as the lives of the people who live by customary norms change and this evolution now continues within the context of the Constitution and its values (paras. 52–53). The court determined that the content of the land rights claimed by the community had to be ascertained by taking into account the community's history and usage of the land and its natural resources (para. 60).

In *Bhe*[17] judgment (*Bhe and Others v Khayelitsha Magistrate & Others*, 2005), the Constitutional Court added that customary law should be interpreted in its own setting and not through the prism of the common law or other system of law (para. 43). In this case, the court acknowledged the importance of adopting a flexible approach in developing customary law owing to its fluidity and adapting to the needs of community practicing (paras. 110–113).

In *Shilubana* judgment (*Shilubana and Others v Nwamitwa*, 2008),[18] the Constitutional Court outlined that customary laws norms must be determined with reference to certain factors: the traditions of the community concerned in order to establish the position under customary law by enquiring into the past practices of the community; the community's right to be able to develop its own customary laws; the fact that customary law regulates the lives of people, so the court must carefully balance the flexibility required to recognize customary law against the need for legal certainty and respect for vested rights, as well the protection of constitutional rights (para. 44–47); the need to promote the spirit, purport, and objects of the Bill of Rights in order to give effect to its duties under Constitution section 39(2) (paras. 44–48).

On the whole, the journey of customary law in South Africa has been truly remarkable: its historically relegated position could no longer subsist in the new constitutional order and the Constitutional Court has breathed life into customary law as an indispensable and integral part of the South African plural legal order. The Constitutional Court has also offered important words of caution in this connection: official customary law found in legislation, case law and academic writings may not necessarily reflect the contemporary practices of the customary law of its people and is often susceptible to distortion (*Richtersveld Community* judgment, para. 54; *Shilubana* judgment, para. 44). What the aforementioned cases illustrate is that the South African courts have provided guidance on a procedural approach to recognizing customary law in a given dispute, while giving deference to the living practices in a culturally conscious manner.

Legal Reforms in the Marine and Coastal Governance Sector

The Marine Living Resources Act, 1998 as amended (MLRA)[19] was the first legislative instrument by which the new government sought to address the inadequacies inherited from its predecessors in order to achieve *inter alia* equality in the fishing industry while maintaining and sustaining marine resources use (Witbooi, 2005). It included three categories of fishing—namely commercial, recreational, and subsistence fishing. Although the above-cited constitutional provisions recognized customary law and rights arising in terms of this law, the MLRA did not recognize *customary* fishing rights.

Traditional fishers in the Western Cape, where the commercially orientated fishing industry was located, argued that the definition of subsistence was too restrictive and did not recognize their occupation as traditional, small-scale, artisanal fishers (Jaffer & Sunde, 2006). They also argued that the individual, property rights-orientated regime introduced in South Africa did not accommodate their collective, community-based rights system. In 2005, these traditional fishers, represented by the Legal Resources Centre, a public interest legal organization, brought a challenge in the Cape Town High Court and simultaneously in the Equality Court (*George and Others V Minister of Environmental Affairs and Tourism [EqC]*, 2005). The Ministry responsible for the regulation of the fishing industry and under whom the authority to execute the mandate of the MLRA sits, unsuccessfully attempted to resist the legal action (*Minister of Environmental Affairs and Tourism v George*, 2007 [SCA]). An out-of-court settlement was then reached between the erstwhile Department of Environmental Affairs and Tourism and the fishing communities, which required the government to properly and effectively make provision for the socio-economic rights of traditional fishing communities in legislation and rights allocation process by creating a special policy fit for this purpose (Equality Court order issued under File No: Ec 1 on 2 May 2007, paras. 8–10).

Following an extensive nation-wide public participation process, which included the small-scale fishers requesting a community-based rights allocation, and recognition of customary fishing rights, the Policy for the Small-Scale Fisheries Sector was gazetted in 2012. The Policy was welcomed for adopting a human-rights-based approach (Coastal Links-Masifundise, 2013), which was based on Constitution Section 39(3) recognizing rights arising from common law, customary law, or legislation, at the request of traditional fishing communities' legal experts.[20] This is a significant development as during the negotiation phase, the Fisheries Department and its legal team had initially denied that communities had placed evidence of customary rights before the policy task team (DAFF, 2011).[21]

In 2014, an amendment to the MLRA inserted a new category of fishing rights namely, small-scale fishing rights into the statute. The section mentions that the Minister, in trying to achieve the objectives of Constitution Sections 39(3),[22] makes provision for communities who "have a history of shared small-scale fishing and who are, but for the impact of forced removals, tied to particular waters or geographic area, and were or still are operating where they previously enjoyed access to fish, or continue to exercise their rights in a communal manner in terms of an agreement, custom or law" (Section 1 of 2014 Marine Living Resources Amendment Act). This broad definition of small-scale fishing communities appears apt to include customary fishing communities and the term is used interchangeably here.

In addition, the Regulations relating to Small-scale Fishing were gazetted in 2016 (Government Notice 227, Government Gazette No. 39790), to operationalize the Policy. But unlike the Policy, these regulations do not refer to customary fishing rights, despite activists' detailed submissions (Legal Resources Centre 2015; Sunde, 2015), and this has created legal uncertainty.

Small-Scale Fisheries and Multiple Legal Currents: The Case of David Gongqose and the Communities of Dwesa-Cwebe

As the Policy developed through the national policy process, a subsequent articulation of customary fishing rights was taking place along the Eastern Cape coastline. Dwesa-Cwebe is a coastal forest reserve and a marine protected area. The land comprising the reserve was settled by the ancestors of the current residents several centuries ago. The seven communities who lived on this land regard the land and associated natural resources as their common property, upon which they depend for their livelihoods. Their relationship with this land and the coast, derived through their relationship with their ancestors, is reflected in their culture and their customary governance system. Authority to access marine resources is vested among elders in the community and the sub-headman, not with the Chief or Traditional Authority (Sunde, 2014).

Over the course of the past century, commencing in the 1890s, these communities faced widespread and continuous loss of their land, their forest and marine resources in the name of nature conservation, with devastating consequences for their basic food security and livelihoods (Sunde, 2014). In 1994, the communities embarked on an advocacy campaign to reclaim their tenure rights. Ultimately the communities accepted a compromise settlement: their land ownership was recognized, but the reserve was to remain under conservation status owing to the involvement of a powerful conservation lobby. Importantly, the Settlement Agreement made provision for their sustainable use of resources, for them to co-manage the reserve as equal partners with the State and to benefit from tourism within the reserve. Yet despite *de jure* recognition of their tenure rights in 2001, implementation of the obligations to recognize their rights to resources and to co-manage their territories was not honored and no co-management or beneficiation was implemented (Sunde, 2014). Instead, the department responsible for fisheries management and marine conservation declared that the area would be a "no-take marine protected area and all use of resources was prohibited" (Government Gazette No. 21948 of 2000). On 29 December 2000 the then Minister of Environmental Affairs and Tourism, acting under section 43 of the MLRA, effectively prohibited surrounding communities from exercising any form of access to the marine resources, making it an offense for anyone to fish or attempt to fish in a marine protected area without the permission of the Minister responsible.

Despite this prohibition, the communities continued to access marine resources as this was the only source of protein for many households and they considered it their customary right (Sunde, 2014). Considerable conflict with the conservation authorities ensued and many residents were arrested, prosecuted, and imprisoned as a result (Sunde, 2014). In 2010, three members of this community were arrested and criminally charged for contravening MLRA section 43, read with the Criminal Procedure Act, 1977. The resulting *Gongqose* case (*Gongqose and Others v Minister of Agriculture, Forestry and Fisheries and Others; Gongqose & Others v State*, 2018) is the first judicial decision on a criminal defense of customary fishing rights in South Africa. During the case, the State Prosecutor repeatedly asserted that there was only one law, "the right law" of the State. In response, the accused fisherman, David Gongqose, repeatedly attempted to inform the court that while he respected "the government's law," he wished the government to also take notice of "our law," the "community's law" (Gongqose 2014 in Sunde, 2014). David Gongqose testified that he learnt about these rules from his father, and a traditional healer (*sangoma*) testified that she is called to the sea by her ancestors, thereby providing evidence of how access to these natural resources and ongoing relationship to ancestors is the material basis of their culture (Sunde, 2014) and of how knowledge of local resources, coupled with local laws about how to use these resources, including rules about fishing, are passed down from elders to children (Fig. 18.1).

Having regard to the numerous constitutional and international human rights instruments and jurisprudence cited by the communities' legal representatives and witnesses in their defense, the Magistrate, in his judgment, noted that "South Africa's new constitutional dispensation began not only a political but also a legal revolution. With the inclusion of a justiciable Bill of Rights in the Constitution, the validity of a wide range of laws, whether public or private, could now be tested against the standards of fundamental human rights" (*State versus Gongqose and others* E382/10). He expressed strong criticism of the conservation authorities for their failure to recognize the livelihood needs of this community. He confirmed that this community had a customary system of law, however he did not have the authority to find

Fig. 18.1 Community on the beach describing their customary rights and territory to the conservation officials at Hobeni, Dwesa-Cwebe (Photo source Jackie Sunde [co-author])

a national statute (i.e., the MLRA) unconstitutional. But he indicated that it was doubtful that the MLRA would survive constitutional scrutiny, urging the parties to appeal this matter in a higher court.

The Supreme Court of Appeal found that the law as it stood at the time the fishermen were arrested, neither recognized nor extinguished their customary right to fish (*Gongqose and Others v Minister of Agriculture, Forestry and Fisheries and Others 2018*, para. 59). It held that the existence of this customary right negated the unlawfulness required to hold the small-scale fishers criminally liable in terms of the now defunct provision of the MLRA read with the Criminal Procedure Act.[23] The court held that the fishermen had demonstrated and exercised their customary right to fish and that their customary right was duly recognized and protected under the provisions of section 211 of the Constitution, 1996 (ibid., paras. 56–57). The court emphasized this constitutional interpretation to customary law and customary fishing rights and found that:

An interpretation that the appellants' customary rights survived the enactment of the MLRA not only grants them the fullest protection of their customary system guaranteed by section 211 of the Constitution, but also accords with the position in international law—which a court is enjoined to consider when interpreting the Bill of Rights—that indigenous peoples have the right to their lands and resources traditionally owned (ibid., para. 57).

Further, in the absence of laws dealing with customary fishing rights in South Africa, the Court, in line with section 39(1)(c) of the Constitution identified foreign authorities as guidance to determine whether the customary rights of the accused members of the fishing community had been extinguished under the Constitutional dispensation (ibid., para. 41). In consideration of the suitability of the Canadian test of extinguishment and applying it in the South African context, the Court held that the customary rights are equally protected in terms of both the Constitution first and second any legislation specifically dealing with the particular customary rights. Therefore, the Court held that customary rights could only be extinguished if done so expressly by specific legislation dealing with customary rights (ibid., para. 50). The court determined that the MLRA prior to its amendment did not constitute such legislation as it did not deal with customary fishing rights and held: "Applying these principles, there is nothing in the language of the MLRA that specifically deals with customary rights. At most, it provided a right of access to marine resources by 'subsistence fishers…'." (ibid., para. 52).

Additionally, the court, in endorsing this constitutional interpretation, found that it accorded with both the African Charter on Human and People's Rights Charter and the United Nations Declaration on Indigenous People's Rights, which recognizes the rights of indigenous people to, *inter alia*, access their natural resources in manner that promotes economic, social, and cultural development and to be granted legal recognition to land, territories, and resources with due respect to their customs, traditions, and land tenure systems (ibid., para. 58).

The Implementation of *Gongqose* Decision

Despite extensive correspondence initiated by the Legal Resource Centre and offers to assist authorities in understanding the substance of the *Gongqose* judgment and in developing a procedural approach in order to give effect to it (LRC, 2021a, 2021b), there has been no public acknowledgment of customary fishing rights by the government and arrests have continued, which are considered unlawful (LRC 2020). The DEFF is treating communities holding customary fishing rights in the same manner as other small-scale fishing communities. When challenged about their failure to institute a process to understand local customary fishing system pertaining to particular

Fig. 18.2 Co-accused in the Gongqose customary rights court case—Co-accused fishers and their legal team and expert witnesses (Photo source Jackie Sunde [co-author])

communities, the State merely says that it has "recognised and provided for the exercise of these communities' customary fishing rights" (DFFE, 2021).[24]

The evidence from the response of the fisheries and conservation authorities since the 2018 Supreme Court Judgment suggests that, unless fisheries governance is transformed so as to truly "engage with customary law in its own setting" and "on its own terms," as required by the Constitutional Court (*Shilubana* 2009 in Sunde, 2014), real recognition of customary fishing rights will not be possible. There remains the need to develop "procedural mechanisms, institutional designs and discursive practices" "to "manage" hybridity" in the fisheries sector (Berman in Jentoft & Bavinck, 2019, pp. 283–284) (Fig. 18.2).

Conclusions

As a consequence of colonialism and its legacies that linger in post-colonial governance of marine and coastal resources, many fishing communities in Africa operate in a sea of both State and customary laws, with complex expressions of this pluralism at different scales of governance (Sunde, 2014; Wilson, 2021). Despite recognition of customary law in many constitutions and national legislation on the continent, the experiences of Ghana and South Africa illustrate the difficulties that small-scale fishing communities experience as a result of the way in which the recognition of customary law at national level translates into day-to-day governance of fisheries at local level.

In both countries, mainstream fisheries reforms have tended to ignore critical texture and specific context underpinning customary systems of tenure and customary fishing rights. More specifically, although fisheries reforms happened earlier in South Africa than in Ghana, in both countries they have been focused on promoting sustainability and equity without adequately integrating and ensuring respect for customary fishing rights.

In Ghana, State fisheries governance reforms are or propose to engage with traditional authorities, with a focus on the Chieftaincy and an aspiration that Chiefs will play a central role in directing and enforcing change. Evidence thus far suggests that these proposals are mismatched with the realities of customary governance structures. There is also a problematic gap between the aspiration to refer to customary law and the availability of records and systems to refer to that law. In addition, courts have not yet engaged with customary fisheries rights although they have recognized customary rights generally.

In South Africa, the judiciary has moved away from the approach focused on chiefs and in all events chiefs have not played a significant role in customary governance of marine and coastal resources (Sunde, 2014). Courts have been increasingly called upon to recognize customary laws as living sources of law in their own setting, through a procedural approach that echoes international law and policy instruments. However, so far administrative implementation of judicial decisions is severely lacking. Further developing a procedural approach at the administrative level has been supported by human rights lawyers and activists in South Africa in their interactions with fisheries and conservation authorities.

Implementation issues in both cases point to the need to develop more detailed procedural approaches to match the changing reality of customary laws, customary fishing rights, and the broader cultural relationships of indigenous and other communities with the ocean and its resources. While international law and guidance provide a series of steps to be followed to that end, our case studies show that tackling complexity at the local level requires understanding and skills that go beyond traditional legal education and rather rely on insights from anthropology. Story-telling and other arts-based approaches could be usefully embedded in the needed procedures for recognizing customary fishing rights as part of genuine efforts to protect the internationally recognized human rights of those engaging in customary fishing activities and supporting their sustainable practices that contribute to the global goals of biodiversity conservation and food security. Fundamentally, recognition of customary law also requires the recognition of multiple sources of law and law making as part of living, daily practices, and rituals of fishing; stories that represent these daily rhythms and customs are indeed the law laying the foundation of what is legal and illegal.

Notes

1. Convention on Biological Diversity (CBD), Rio de Janeiro, 5 June 1993, in force 29 December 1993.
2. CBD Art. 8(j) and 10(c), as interpreted in Tkarihwaié:ri Code of Ethical Conduct to Ensure Respect for the Cultural and Intellectual Heritage of Indigenous and Local Communities Relevant to the Conservation and Sustainable Use of Biological Diversity, paras. 18 and 20; and Mo'otz Kuxtal Voluntary Guidelines.
3. CBD Mo'otz Kuxtal Voluntary Guidelines for the development of mechanisms, legislation or other appropriate initiatives to ensure the "prior and informed consent," "free, prior and informed consent," or "approval and involvement," depending on national circumstances, of indigenous peoples and local communities for accessing their knowledge, innovations, and practices, for fair and equitable sharing of benefits arising from the use of their knowledge, innovations, and practices relevant for the conservation and sustainable use of biological diversity, and for reporting and preventing unlawful appropriation of traditional knowledge. Available at: https://www.cbd.int/doc/publications/8j-cbd-mootz-kuxtal-en.pdf.
4. For a detailed discussion of the legal value of CBD guidelines from an international human rights law perspective, see E. Morgera (2020), 'Biodiversity as a Human Right and its Implications for the EU's External Action', Report to the European Parliament, available at: www.europarl.europa.eu/RegData/etudes/STUD/2020/603491/EXPO_STU(2020)603491_EN.pdf; and for a discussion of the legal strength of international human rights instruments, see 'Under the radar: fair and equitable benefit-sharing and the human rights of indigenous peoples and local communities connected to natural resources' (2019) 23 *International Journal of Human Rights* 1098–1139.
5. The Ghanaian Fisheries Commission classifies actors in the industry as:small scale/artisanal fishers who operate in canoes (made of wood, motorized and non-motorized); semi-industrial fishers (mainly wooden); and industrial vessels which are generally over 25 m Length Overall (LOA) made of steel hull and with capacity to operate in areas beyond national jurisdiction (Government of Ghana, Fisheries Commission, 2015, p. 12).

6. In terrestrial natural resource management, for example, customary law and customary practice have been cited in efforts to negotiate equitable access and benefits for communities amidst natural resource extraction efforts, indigenous knowledge systems, particularly with reference to the association between land, natural entities, and the sacred, have been referenced in efforts to ensure that local communities are able to practice traditional stewardship of natural resources.
7. Chief Fishermen have often been criticized, either because they did not want to be involved in enforcing regulation, or because fisherfolk reported that Chief Fishermen were sometimes themselves involved in IUU fishing, and illegal transshipment known as saiko fishing. See http://www.graphic.com.gh/news/general-news/chief-fishermen-indicted-for-illegal-fishing.html.
8. For instance, between artisanal/small-scale canoe fishers and those engaging in the "saiko" trade.
9. For example, the Volta Estuary Clam Fisheries Management Unit which does employ VGGT principles to secure tenure, culture, and access. Available at: https://henmpoano.org/wp-content/uploads/2021/01/Tranversal_Issue-Brief_VGGT_HM.pdf.
10. As the co-management policy states: "entrust control and access to the fisheries resources to both central government and local communities as well as associations for various management units" (Government of Ghana, MOFAD, 2021, Section 1.2).
11. At present, the only attempt toward a marine spatial plan is the Abidjan Convention funded pilot project based on an ecosystem-based approach to integrated marine and coastal management which commenced in 2020. Stakeholder consultations are currently ongoing on a plan for a section of the coast along the western region of Ghana. The pilot is part of the larger marine spatial plan for the West African coast, a regional plan which can potentially coordinate and holistically benefit marine biodiversity. See Sagoe et al. (2021) for one of the first reports of activities toward setting-up the MPA.
12. *Alexkor Ltd and Another v Richtersveld Community and Others* (2003) ZACC 18; 2004 (5) SA 460.

    (CC) (Hereinafter the *Richtersveld* Community judgment) (53); South African Law Commission Project.
13. Peart 'Section 11(1) of the Black Administration Act No. 38 of 1927: The Application of the Repugnancy Clause'1982 *Acta Juridica*, 99–116.
14. Section 2 of the Constitution, 1996.
15. Section 1 of the Constitution, 1996.
16. In this case the role of customary law and the claim of customary rights to land and natural resources by an indigenous community was considered for the first time in this case by the Constitutional Court of South Africa. The indigenous Richtersveld community successfully instituted a claim for the restoration of land and the rights of access to the natural resources produced on the land.

17. The *Bhe* case considered the rules of intestate succession in the context of customary law in particular the constitutional validity of the principle of primogeniture in the context of the customary law of succession.
18. The *Shilubana* case dealt with the issue of whether a woman could succeed as the Chief of a traditional community and whether that community could develop their customs and traditions in order to promote gender equality in the succession of traditional leadership, as envisaged in the Constitution.
19. Act 18 of 1998.
20. Community response to DEAT response to community submission to NEDLAC dated 23/11/2011.
21. At the time of the Equality Court Ruling, responsibility for fisheries management fell under the Department of Environmental Affairs and Tourism (DEAT). In 2009, fisheries management was moved to the Department of Agriculture, Forestry and Fisheries (DAFF). In June 2019, the Department of Environment, Forestry and Fisheries (DEFF) was established, and forestry and fisheries management functions were incorporated into the Department of Environmental Affairs.
22. The provision holds that: "The Bill of Rights does not deny the existence of any other rights or freedoms that are recognised or conferred by common law, customary law or legislation, to the extent that they are consistent with the Bill.".
23. Act 51 of 1977.
24. DFFE 21 June 2021 Letter from Minister Barbara Creecy to Ms W. Wicomb, Legal Resources Centre.

References

Agbodeka, F. (1992). *An economic history of Ghana from the earliest times*. Ghana University Press.

Aswani, S., & Cinner, E. J. (2007). Integrating customary management into marine conservation. *Biological Conservation, 140*, 201.

Atta-Mills, J., Alder, J., & Sumailia, U. S. (2004). The decline of a regional fishing nation: The case of Ghana and West Africa. *Natural Resources Forum, 28*, 13.

Bavinck, M., Sowman, M., & Menon, A. (2014). Theorizing participatory governance in contexts of legal pluralism—A conceptual reconnaissance of fishing conflicts and their resolution. In M. Bavinck, L. Pellegrini & E. Mostert (Eds.), *Conflict over natural resources in the global south—Conceptual approaches*. CRC Press/Taylor & Francis.

Berman, P. S. (2007). Global legal Pluralism. *Southern California Law Review, 80*, 1155.

Britwum, A. O. (2009). The gendered dynamics of production relations in Ghanaian coastal fishing. *Feminist Africa, 12,* 69.

Coastal Links-Masifundise. (2013). Coastal Links and Masifundise *Visser's Net* (p. 13). Coastal Links and Masifundise. www.masifundise.org/fishersnet/

Coastal Resources Center. (2021). Final report, 2014–2021. USAID/Ghana Sustainable Fisheries Management Project. Narragansett, RI: Coastal Resources Center, Graduate School of Oceanography, University of Rhode Island. https://www.crc.uri.edu/download/GH2014_PGM357_CRC_FIN508.pdf

Cook, R., Acheampong, E., Aggrey Fynn, J., & Heath, M. (2021). A fleet based surplus production model that accounts for increases in fishing power with application to two West African pelagic stocks. *Fisheries Research, 243,* 1.

Cuskelly, K. (2010). *Customs and Constitutions: State recognition of customary law around the world.* IUCN.

DAFF. (2011). Unpublished letter to Masifundise entitled Response to Community Proposal dated 23 November 2011. Fisheries Branch, DAFF (On file with authors).

DFFE. (2021). 21 June 2021 Letter from Minister Barbara Creecy to Ms W. Wicomb, Legal Resources Centre (On file with authors).

Endemano Walker, B. (2002). Endangering Ghana's seascape: Fanti fishtraders and marine property in colonial history. *Society & Natural Resources, 15,* 389.

Essien, V. (2020). *UPDATE: Researching Ghanaian Law.* GlobaLex. https://www.nyulawglobal.org/globalex/Ghana1.html

FAO. (2012). *The voluntary guidelines on the responsible governance of tenure to land, fisheries and forests in the context of national food security.* FAO.

FAO. (2014). *Voluntary guidelines for securing sustainable small-scale fisheries* in the context of Food Security and Poverty Alleviation. FAO.

FAO. (2021). *Indigenous peoples' food systems: Insights on sustainability and resilience from the frontline of climate change.*

Government of Ghana, Fisheries Commission. (2015). *Fisheries management plan of Ghana: A national policy for the management of the marine fisheries sector 2015–2019.* www.crc.uri.edu/download/GH2014_POL005_FC_Fisheries MgtPlan2016.pdf

Government of Ghana, Ministry of Fisheries and Aquaculture Development (MOFAD). (2020). *Co-management policy for the fisheries sector.* https://www.crc.uri.edu/download/GH2014_POL112_MOFAD_FIN508.pdf

Gupta, J., & Bavinck, M. (2014). Towards an elaborated theory of legal pluralism and aquatic resources. *Current Opinion in Environmental Sustainability, 11,* 86.

Hamukuaya, H., & Christoffels-Du Plessis, A. (2018). The impact of customary rights on marine spatial planning. *Journal of Ocean Law and Governance in Africa, 1,* 78.

ICSF (International Collective in Support of Fish Workers). (2008, November). A human rights approach to fisheries. *Samudra Report No 51.* ICSF.

Jaffer, N. & Sunde, J. (2006, July). *Fishing rights versus human rights?* (SAMUDRA Report, No. 44). ICSF.

Jentoft, S., & Bavinck, M. (2019). Reconciling human rights and customary law: Legal pluralism in the governance of small-scale *fisheries*. *The Journal of Legal Pluralism and Unofficial Law, 51*, 271.

Knox, J. (2018). *Report of the Special Rapporteur on the issue of human rights obligations relating to the enjoyment of a safe, clean, healthy and sustainable environment.* UN Framework Principles on Human Rights and the Environment, UN Doc. A/HRC/37/59.

Legal Resources Centre. (2015, May 28). *Submissions to the minister of agriculture, forestry and fisheries on the proposed regulations relating to small-scale fishing.* Legal Resources Centre.

Legal Resources Centre. (2020). *Letter to Colonel Mbena 20 July 2020 from Ms Wilmien Wicomb.* Legal Resources Centre (On file with authors).

Legal Resources Centre. (2021a). *26 January 2021a Letter to Mr Abongile Ngqongwa from Ms Wilmien Wicomb.* Legal Resources Centre (On file with authors).

Legal Resources Centre. (2021b). Letter to Minister Barbara Creecy from Ms Wilmien Wicomb. Legal Resources Centre.

Marfo, S. (2019). Chiefs as judges in Modern Ghana: Exploring the judicial role and challenges confronting the Ashanti regional house of chiefs UDS. *International Journal of Development, 6*, 160.

Merry, A. (1988). Legal Pluralism. *Law and Society Review, 22*, 869.

Mnisi, S. (2009). *(Post)-Colonial culture and its influence on the South African legal system—Exploring the relationship between living customary law and state law.* Unpublished PhD dissertation, University of Oxford.

Morgera, E. (2018). Dawn of a new day? The evolving relationship between the convention on biological diversity and international human rights law. *Wake Forest Law Review, 54*, 691–712.

Morgera, E., & Nakamura, J. (2022). Shedding a light on the human rights of small-scale fisherfolk: Complementarities and contrasts between the UN declaration on peasants' rights and the small-scale fisheries guidelines. In M. Alabrese, A. Bessa, M. Brunori, & P. F. Giuggioli (Eds.), *Commentary on the declaration on the rights of peasants*. Routledge.

Odotei, I. (2002). *The artisanal marine fishing industry in Ghana: A historical overview.* Institute of African Studies University of Ghana.

Overå, R. (2001). *Institutions, mobility and resilience in the Fante migratory fisheries of West Africa* (CMI Working Paper 2001). Christian Michelsen Institute.

Parlee, C. E., & Wiber, M. (2015). Whose audit is it? Harnessing the power of audit culture in conditions of legal pluralism. *Journal of Legal Pluralism and Unofficial Law, 47*, 96.

Rautenbach, C. (2010). Deep legal pluralism in South Africa: Judicial accommodation of non-state law. *The Journal of Legal Pluralism and Unofficial Law, 42*, 143.

Sagoe, A. A., Aheto, D. W., Okyere, I., Adade, R., & Odoi, J. (2021). Community participation in assessment of fisheries related ecosystem services towards the establishment of marine protected area in the Greater Three Points area in Ghana. *Marine Policy, 124*, 104336.

Sowman, M. (2006). Subsistence and small-scale fisheries in South Africa: A ten-year review. *Marine Policy, 30*, 60.

Stark, H. Z., Schuttenberg, M., Newton, S., Edminster, G., Asiedu, E., Ekekpi, & Torrens-Spence, G. J. (2019). *Advancing reforms to promote sustainable management of Ghana's small pelagic fisheries*. USAID. https://biodiversitylinks.org/projects/completed-projects/bridge/bridge-resources/advancing-reforms-to-promote-sustainable-management-of-ghana2019s-small-pelagic-fisheries

Sunde, J. (2014). *Customary governance and expressions of living customary law at Dwesa-Cwebe: Contributions to small-scale fisheries governance in South Africa* (Doctoral thesis). https://open.uct.ac.za/handle/11427/13275

Sunde, J. (2015). *Submission of comments DAFF draft regulations on small-scale fisheries Gazette No. 38536*. Coastal and Small-scale Fisheries Research Group, Department of Environmental and Geographical Science, University of Cape Town, Cape Town.

Sunde, J. (2017). Expressions of tenure in South Africa in the context of the small-scale fisheries guidelines. In S. Jentoft, R. Chuenpagdee, M. Barragan-Paladines, & N. Franz (Eds.), *The small-scale fisheries guidelines global implementation* (p. 139). Springer.

Tendayi M. M., Mills, D. J., Asare, C., & Asieddu, G. A. (2016). Enhancing women's participation in decision-making in artisanal fisheries in the Anlo Beach fishing community, Ghana. *Water Resources and Rural Development*. www.researchgate.net/publication/303026727_Enhancing_women%27s_participation_in_decisionmaking_in_artisanal_fisheries_in_the_Anlo_Beach_fishing_community_Ghana

Teubner, G. (1983). Substantive and reflexive elements in modern law. *Law and Society Review, 17*, 239.

Teubner, G., & Fischer-Lescano, A. (2008). Cannibalising epistemes: Will modern law protect traditional cultural expressions? In C. Graber (Ed.), *Intellectual property and traditional cultural expressions in a digital environment* (p. 17). Edward Elgar.

Tobin, B. (2014). *Indigenous peoples, Customary law and human rights—Why living law matters* (1st ed.). Milton Park.

Trechera, E. (2010). Legal pluralism, customary law and environmental management: The role of international law for the south pacific. *The Journal of Legal Pluralism and Unoffical Law, 42*, 171.

Ubink, J. (2007). Traditional authority revisited: Popular perceptions of chiefs and chieftaincy in peri-urban Kumasi, Ghana. *The Journal of Legal Pluralism and Unofficial Law, 39*, 123.

Wilson, D. (2021). European colonisation, law, and indigenous marine dispossession: Historical perspectives on the construction and entrenchment of unequal marine governance. *Maritime Studies, 20*(4), 387–407.

Witbooi, E. (2005). Fishing rights: A new dawn for South Africa's marine subsistence fishers. *Ocean Yearbook Online, 19*, 74.

Case Law

Andakwei ((Subsituted By Kow Atteh) and Others. Vrs ABaka and Others (J2/03/2017) (2018) GHASC 47 (18 July 2018).

Akwufio & Ors v Mensah & Ors. (1898). https://ghalii.org/gh/judgment/supreme-court/2019/6

Alexkor Ltd and Another v Richtersveld Community and Others. (2003). ZACC 18; 2004 (5) SA 460.

Bhe and others v Khayelitsha Magistrate & Others 2005 (1) SA 580 (CC); (2004) ZACC 17.

George and Others V Minister of Environmental Affairs and Tourism 2005 (6) SA 297 (Equality Court).

Gongqose and Others v Minister of Agriculture, Forestry and Fisheries and Others (1340/16, 287/17) (2018) ZASCA 87; (2018) 3 All SA 307 (SCA); 2018 (5) SA 104 (SCA); 2018 (2) SACR 367 (SCA) (1 June 2018).

Gongqose and Others v State 2018 (5) SA 104 (SCA), 2018 (2) SACR 367 (SCA), 2018 (3) All SA 307 (SCA) 39.

Gumede (born Shange) v President of the Republic of South Africa and Others (CCT 50/08) [2008] ZACC 23; 2009 (3) BCLR 243 (CC); 2009 (3) SA 152 (CC) (8 December 2008).

Minister of Environmental Affairs and Tourism v George and Others (437/05, 437/05) (2006) ZASCA 57; 2007 (3) SA 62 (SCA) (18 May 2006).

Republic VRS National house of chiefs And Others (J4/32/2018) (2019) GHASC 6 (30 January 2019).

Shilubana and Others v Nwamitwa (CCT 03/07) (2008) ZACC 9; 2008 (9) BCLR 914 (CC); 2009 (2) SA 66 (CC) (4 June 2008).

State v Gongqose and Others Willowvale Magistrates' Court, held at Elliotdale (Eastern Cape) Case no. E382/10 (date of judgment unknown).

19

Narratives of Non-Compliance in "Tuesday Non-Fishing Day" in Ghana

John Windie Ansah, Georgina Yaa Oduro, and David Wilson

Introduction

The ocean is an integral part of Earth's ecosystem and one strand of the many social and economic spheres of human existence (Allison, 2011, p. 2; Campbell et al., 2016, p. 4). The ocean is a ritual space, an historical place, a commercial zone, and a hub for the production and reproduction of culture and livelihood. Scholarly analyses of the ocean clearly demonstrate close

---

This research was sponsored by the One Ocean Hub. The One Ocean Hub is a collaborative research for sustainable development project funded by UK Research and Innovation (UKRI) through the Global Challenges Research Fund (GCRF) (Grant Ref: NE/S008950/1). GCRF is a key component in delivering the UK AID strategy and puts UK-led research at the heart of efforts to tackle the United Nations Sustainable Development Goals.

J. W. Ansah (✉) · G. Y. Oduro
Department of Sociology and Anthropology,
University of Cape Coast, Cape Coast, Ghana
e-mail: john.ansah@ucc.edu.gh

G. Y. Oduro
e-mail: gyoduro@ucc.edu.gh

D. Wilson
University of Strathclyde, Glasgow, UK
e-mail: david.wilson.101@strath.ac.uk

links between culture, livelihood, and ritual in the ocean space (Ford, 2011; Khakzad & Griffith, 2016; Nuttall, 2000; Ransley, 2011). This is true also in coastal Ghana, where different beliefs surrounding the sea inform human interactions with marine spaces and resources. This includes Akan belief that certain sea creatures are physical manifestations of supernatural beings and should therefore not be harvested or consumed. Rituals are performed to respect their presence and funerals are organized for such species at their death (Boamah, 2015, p. 7). Another example is the *agbodedefu* (translated literally as "sending a ram into the sea") ceremony traditionally performed by the Yewe cult among the Anlo of the Volta Region. This is a sacrifice to the gods of the sea in the hopes of encouraging a bumper fishing season (Akyeampong, 2002, p. 121). An equally notable representation of this argument is found in the creation and observance of non-fishing days in Ghanaian coastal communities.

Depending on the community and its beliefs, by custom, fishing activities are abhorred on Tuesdays, Sundays, and Thursdays. Such a custom has been visibly acknowledged in Ghana's modern policy arrangement, as expressed in the Sustainable Fisheries Management Project (SFMP), which proposed the implementation of additional non-fishing days in line with existing practices (Coastal Resources Center, 2018, p. 9). This chapter focuses specifically on the custom of non-fishing on a Tuesday, as this is the practice of the study sites in the Central Region of Ghana. This custom is anchored in the belief that the sea is a supernatural being and requires days of rest (Adjei & Sika-Bright, 2019, p. 2). While the idea of rest justifies observation of this custom, there are multiple outlays of economic value and, by extension, livelihood value, in this custom, in which its breach constitutes a socio-cultural taboo (Adjei & Sika-Bright, 2019, p. 11; Dosu, 2017, p. 47). In an effort to sustain the observance of the custom, violators of these norms face spiritual and societal sanctions, including fines, suspension from using the sea, confiscation of harvests, and obligations to pacify the relevant gods or spirits with items such as fowls, goats, and alcoholic beverages (Chief Fisherman, Cape Coast, interview 25/06/2021).

Artisanal sea-based fisheries in Ghana consist of dugout canoes—including canoes with outboard motors and use of a variety of fishing gear. Currently, there are approximately 12,000 canoes operating along the coast. About 6405 of the canoes use outboard motors of diverse horsepower capacities (Afoakwah et al., 2018, p. 3). Artisanal sea-based fisheries in Ghana are carried out in the four coastal regions—Greater Accra, the Central Region, the Western Region, and the Volta Region—each of which has unique cultural and social practices. Despite the diverse cultural milieus, there is a unified plethora of beliefs, customary laws, and norms that shape the fishing activities from harvesting to distribution. One of them is the observation of non-fishing days. As mentioned, these are diversities around the observation of non-fishing days, especially regarding the specific days observed. For example, Fante and Ga communities predominantly observe Tuesday non-fishing days, while some communities in the Western Region either observe Thursdays or Sundays, the latter predominantly by communities residing close to the border with the Ivory Coast where communities also observe Sundays (National Secretary of the Canoe and Fishing Gear Owners Association of Ghana, Abandze, interview 12/09/2021). Meanwhile, Anlo communities generally observe Wednesday as a day of rest and fishing rituals, although some Anlo communities prefer Tuesday too (Akyeampong, 2002, p. 37). Asafo companies—patrilineal and predominantly Akan political and military organizations responsible for the protection and defense of their respective societies (Chukwekere, 1970; Aggrey, 1978; Edusei, 1981; Acquah et al., 2014, p. 20; Nti, 2011, p. 3; Sam, 2014, p. 22)—include membership of fisherfolk and play a critical role of in the observance of the Tuesday non-fishing day across many coastal communities.

There are also visible gender dimensions in the performance of the rituals associated with the custom. Though the Akan/Fante Asafo companies of Ghana have traditionally been a masculine-dominated organization traced through one's patrilineal line (Aidoo, 2011, p. xi; De Graft Johnson, 1932, p. 308), it also has place for women who hold certain gender roles. The effect of social change culminating in a shift from the primary warfare or militancy role of Asafo companies has created more space for both men and women. De Graft Johnson (1932, p. 308) opines that both men and women occupy stool holding positions such as *Asafobaatan* (Mother of the Asafo group) and/or advisor, *Asafohen* (leader or chief of the Asafo group), or *Asafoakyere* in case of female traditional leaders, *Supi* among others. Aidoo (2011, p. xi) added that religious roles such as that of the *Asafokomfo* (traditional priest or priestess) are another key feminine role often occupied by women with significance for fishing in coastal Ghana. Since the traditional priest(ess) perform rituals during festivals such as the *Bakatue* (opening of the Lagoon) among the people of Elmina to determine whether the impending fishing season will be a bumper or lean season, the priest(ess) is also involved in cleansing ceremonies for fishers who break the taboo of fishing on sacred days. On a typical non-fishing day, women who are part of the *Asafo* group are usually asked to cook. They prepare white and yellow cocoyam with boiled eggs (*eto*). The males then pray and sprinkle the food to the gods. The males could also eat some of the food and ask them to bring more when it is consumed. These rituals are performed for the gods in the sea to request their support for a bumper fish harvest. These cultural components are used as reference points for routinized daily fishing activities, the usage of the sea, recruitment, and socialization as well as the allocation of fishing-related responsibilities among gender and age groups (Adom et al., 2019, p. 97; Boamah, 2015, p. 10; Salm & Falola, 2002, p. 6). Some of the shrines or *pusiban* of the Asafo groups also carry gendered and ocean-related images including the proverbial *mamiwata*, mermaid (Ross, 2007, p. 12). Another gendered dimension of fishing in coastal Ghana is that women are not encouraged/allowed to go to sea because of beliefs surrounding menstruation. However, upon the return of men from the sea, women play dominant roles in the processing and marketing of the fish.

Recent scholarly works have identified that customary laws restricting fishing on Tuesdays are still adhered to in many fishing communities in Ghana, but that this has also been disrupted across various fishing communities. Two key reasons have been offered as the cause for this disruption: (i) the expansion of Christianity and (ii) the deterioration of or disputes surrounding chiefly authority (Adjei & Sika-Bright, 2019, p. 2; Dosu, 2017, p. 59; Kalanda-Sabola et al., 2007, p. 26). Evaluating these reasons through an historical and contemporary lens, this chapter considers the entanglement of religion and authority in shaping disputes surrounding intangible heritages of the coast and sea. Following a methodological overview, this discussion is provided across two main sections. The first offers an historical study of non-fishing days in coastal Ghana through analysis of written documentation from the seventeenth century onwards, before contextualizing perceptions surrounding the impact of the spread of Christianity as one reason for the deterioration of non-fishing day customs. The second section then focuses on the role of traditional authorities in facilitating this deterioration through examination of the reasons behind recent communal violations of Tuesday non-fishing days in the Central Region. In particular, the section focuses on *Abandze* as a case study to understand these violations, outlining the politico-legal and economic realities that have shaped communal violations of the previously long-standing Tuesday non-fishing custom and the ways in which this has diffused to the surrounding communities of Ekumfi-Narkwa, Mumford, Otuam, and Moree.

Methodology

Two components comprise the methodology of this chapter: (i) analysis of written historical documentation and (ii) interviews and focus group discussions (FGD) conducted with the National Secretary of the Canoe and Fishing Gear Owners Association of Ghana and five communities in the Central Region: Abandze, Ekumfi-Narkwa, Mumford, Otuam and Moree over a period of five months in 2021. Historical analysis is focused on a variety of written documentation, particularly accounts by Europeans who travelled to, and resided in coastal Ghana in the seventeenth and eighteenth centuries (Cruickshank, 1853), British colonial reports in the nineteenth and twentieth centuries, and African-owned English-language newspapers in the early twentieth century. These provide various descriptions and accounts of the Tuesday non-fishing day custom across four centuries. It is important to stress that although much of this information was recorded following observation and dialogue with communities, the information contained therein was filtered through a lens removed from community perspectives and deeply influenced by the racialized and/or religious perceptions of the authors alongside the specific localized contexts and circumstances in which they were writing. These accounts reflect fishing customs as they were viewed through predominantly European, colonial, and/or Christian perspectives. Although this requires caution, scrutinizing how these perspectives influenced individual observations or descriptions, this written documentation is a valuable resource in gaining an understanding of the prevalence and practices of fishing customs in coastal Ghana in the precolonial and colonial periods when taken collectively. It is important to stress from the offset, however, that the appearance of descriptions of these practices from the seventeenth century onwards is *not* evidencing that these practices only *began* in the seventeenth century. Instead, this is when these practices started to be recorded in the surviving written records of Europeans. With these caveats in mind, this written documentation can be carefully employed to analyse these practices within the specific contexts in which they were discussed.

To understand contemporary developments and perceptions surrounding the violation of Tuesday non-fishing day in the Central Region, interviews and FGDs were conducted from May to September 2021 with the National Secretary of the Canoe and Fishing Gear Owners Association of Ghana, Chief Fishermen, Asafo company leaders, and fishermen of the five identified communities. The Canoe and Fishing Gear Owners Association of Ghana was officially established in 2021. It is composed of canoe and fishing gear owners. Until its formation, state institutions overseeing fishing activities were engaging the Ghana National Canoe Fishermen Council formed in 1982. This council was, however, found to have leaders who were neither fishermen nor canoe owners and were therefore seen not to be promoting the interest of fishermen and canoe owners. The council could not serve as any useful intermediary between the government and fishermen, thereby rendering current fishing policies and laws inappropriate, unimplementable, and unenforceable. Consequently, the Canoe and Fishing Gear Owners Association of Ghana was formed as a breakaway group to make up for the deficiencies in the artisanal fishing industry created by the Ghana National Canoe Fishermen Council. The Association's main objectives are as follows: first, to serve as a mouthpiece of canoe owners; and second, to help the Ministry of Fisheries and Aquaculture as well as the Fisheries Commission to develop strategies to address fishing-related illegalities such as light fishing, use of detergents and dynamites, and the use of unapproved fishing nets. The Association also collaborates with the Ministry and Commission to enforce the strategies. The Secretary's role is to collect the biographical data of the members of the Association and ensure their registration. The Secretary also gathers and provides all the relevant information needed to promote the welfare concerns of the members. The interview with the National Secretary of the Canoe and Fishing Gear Owners Association of Ghana focused on the general and overarching realities associated with the contexts and conditions surrounding the violation of the Tuesday non-fishing custom. The interview also helped to identify the five key communities involved in violating Tuesday non-fishing days in the Central Region. Trips were then made to the five communities where ten FGDs with fishermen were conducted—two per community in the five selected communities—alongside five interviews with Chief Fishermen and two interviews with Asafo Company leaders. Each FGD contained six discussants. In all, 68 research participants were engaged in the study. In FGDs with fishermen and interviews with Chief Fishermen, open and direct questions were asked in the Fante dialect about the reasons, motivations, causes contexts, conditions, and covariance associated with their engagement in Tuesday fishing.[1]

The interviews and FGDs were recorded to allow the free flow of conversation between the interviewer and moderators on the one side and interviewees and discussants on the other side. The recorded data was then translated, transcribed, and analysed. Comparative analysis of the narratives surrounding Tuesday non-fishing day across the five sites provided by diverse participants, who were of different socio-demographic, geographical, and experiential backgrounds, allowed for investigation and extrication of the similarities and differences between the various groups who had been exposed to or had exhibited the violations of the Tuesday non-fishing custom. Deploying the phenomenological analytical technique (Groenewald, 2004, p. 1), which is employed to analyse the encounters and perspectives of individuals and groups on social phenomena, the study evaluated responses from the interviews and FGDs, in which participants were key informants who have had direct encounters and experiences of the unique social structures and economic, political, and cultural exigencies associated with ocean usage. In particular, the data was used to analyse the unique lived experiences of the people relative to the processes, contexts, and conditions leading to the violations of the Tuesday non-fishing custom.

Through this analysis, two major themes emerged, which are closely aligned with central issues identified in previous studies concerning the breakdown of customary fishing practices. These are: (1) the impact of Christianity and (2) chieftaincy-related disputes or activities, which were also paired with considerations of economic realities. Across these discussions, it was clear that the violation or reconfiguration of Tuesday non-fishing customs had been shaped by the social forces that set the contexts for people living in coastal communities who use the ocean for survival and livelihoods.

Fishing Restrictions and Religion in Precolonial and Colonial Ghana

Already globalizing through trans-Saharan trading networks before sustained European voyages to coast, the region comprising modern-day coastal Ghana has been shaped and influenced through its active participation in Atlantic trading networks since the fifteenth century (Green, 2019, p. 123). These new ocean-based networks and connections had profound impacts on coastal communities and practices through their direct and indirect engagement and integration into networks surrounding the gold and transatlantic slave trades (c. 1471–1807), which was followed by the gradual encroachment of British authority following abolition and the era of "legitimate trade" (c. 1807–1874). This then provided the foundations for the establishment of the British Gold Coast Colony (1874–1957). Across these five centuries of change and transformation in Ghana, the centrality of fishing to the livelihoods and sustenance of coastal communities was one pillar of continuity, but one that was also representative of adaptation and resistance (Cruickshank, 1853).

Surviving accounts of coastal Ghana by Europeans from the fifteenth century onwards are littered with descriptions of the canoes and fishing activities of coastal inhabitants, particularly among the communities in what would become known as the Fante territory, as well as the Ga peoples in the coastal region surrounding Accra. For example, describing coastal Ghana in 1482, Duarte Pacheco Perreira wrote about "the many nets that were found here when this land was discovered" and the "great fishermen who go fishing two or three leagues at sea in some canoes resembling a weaver's shuttle" (quoted in Kraan, 2009, p. 14). The Fante were generally regarded and praised as the group with the most sea-based fishing expertise, knowledge which they spread to other groups on the coast. Oral traditions of the Ga, recorded in the early twentieth century, recount that it was the Fante who were the original group to teach them how to fish on the sea using canoes. Prior to this, the Ga and other coastal groups such as the Anlo in the Volta region had predominantly concentrated on fishing in the calmer lagoons close to the coast (Kraan, 2009, p. 14; Nti, 2011, p. 137; Parker, 2000, p. 4).

It was these fishing settlements and communities that Europeans first encountered and engaged with, and it was here that their commercial activities came to be concentrated. The written records date from fifteenth century but, of course, these fishing communities long pre-dated European arrival. These communities were already well connected to complex trading infrastructures, linking surrounding villages and interior markets (Kea, 1982, p. 48). With regard to fish markets, women predominantly controlled the processing and marketing of the catch on land (Adu-Boahen, 2018, pp. 179–180). This was commented on in the early 1600s by Dutch trader Pieter de Marees who wrote "These women and Peasants' wives, very often buy fish and carry it to towns in other Countries [in the interior] in order to make profit: thus the Fish caught in the Sea is carried well over 100 or 200 miles into the Interior" (quoted in Adu-Boahen, 2018, p. 179). Alongside communities specialised in fish and salt production, there were also other specialized craft villages linked to interior markets, including villages specializing in canoe building (Dawson, 2018, pp. 105–107; Kea, 1982, p. 48).

With the steady increase in trade with Europeans on the Atlantic coast, these towns developed into larger urban centres with a gradual increase in the regularity and number of markets being held on the coast as opposed to in interior centres. The majority of these towns did not have natural harbours but were "surf-ports" where goods had to be landed through surf-battered beaches offering little protection from the sea and no docks. As a result, European ocean-going vessels had no choice but to anchor far offshore as there were no harbour or port facilities to accommodate the depth of their vessels. This meant that goods and peoples had to be transported between shore and ship on smaller vessels. This required substantial navigation skills and local knowledge as the rough pounding surf along Ghana's coastline meant that this was a dangerous and difficult crossing. As a result, the canoes and canoe men of Ghana quickly became the essential link between shore and ship. Adapting and employing these canoes for Atlantic-based trade, canoe men were able to quickly monopolize the transport between ships and shore. Their employment included transporting goods and peoples—both free merchants, mariners, and officials as well as enslaved peoples—between ship and shore, but also included carrying letters between coastal centres, peddling provisions to vessels anchored offshore, and carrying on their own coasting trade (Dawson, 2018, pp. 100–103, 111–113; Dickson, 1965, pp. 98–111; Gutkind, 1985, pp. 29–33; Kea, 1982, pp. 220–221; Sparks, 2014, pp. 143–144). At first, this provided additional employment for fishermen who utilized their skill and expertise navigating these waters to increase their incomes. As Atlantic trade expanded in the seventeenth and eighteenth centuries, at least in some surf-ports canoe men became a distinctive class whose livelihood was derived directly and completely from international and interregional trade (Dawson, 2018, pp. 121, 126–127; Gutkind, 1985, pp. 29–37; Sparks, 2014, pp. 143–144). This employment continued after the British abolition of the slave trade with canoe men central to coastal transportation throughout the colonial period too with the role of surf canoes in shore-to-ship transport only declining in the twentieth century following moves to construct harbours improvements and, particularly, the development of artificial harbours at Takoradi in 1928 and Tema in 1962 (Dickson, 1965, pp. 98–111).

While canoe men utilized their expertise to carve out a profitable and influential niche within Atlantic trade,[2] fishing continued to form the primary employment for most coastal dwellers throughout the precolonial and colonial period. This is reflected in European and later British colonial reports, which provide written documentation concerning the long-standing powers of chiefs and religious authorities to restrict fishing activities. This included religious ceremonies controlling closed seasons for fishing in rivers or lagoons, such as the Edina Bakatue festival that ends the closed seasons in the Benya Lagoon in Elmina, and specific rights to fishing grounds as well as beach, river, or lagoon access held by certain chiefs, religious authorities, or communities who could restrict fishing activities and practices (Akyeampong, 2007, 177; *Gold Coast Leader*, 1913, 14 July; 1920, 31 July; Walker, 2002, pp. 397–398). These rights were acknowledged in a report by M. J. Field (1935, pp. 5–6) who, when discussing the opening of a fishery station in Greater Accra, warned:

> These sites [the beaches surrounding Greater Accra], or rather rights of fishing from these sites, are hereditarily owned. Certain fishing grounds belong by immemorial custom to certain individuals. Any violation of these and other customs would bring bitter opposition and possibly the wrecking of the whole scheme.

Field (1935, p. 5) also acknowledged the role of the Wulomai (the Ga high priest) in fishing restrictions in Greater Accra, writing "certain kinds of fishing are closed at certain seasons by order of the Nai Wulamo who by ancient custom is 'owner of the sea'." Unlike many of his peers, who often disregarded the fisheries knowledge of coastal dwellers, Field recognized that these prohibitions "have a sound basis of knowledge and fishing wisdom (for instance the closing of the lagoon fishing enables fish to spawn in peace)." He declared that "in spite of their 'heathen' trappings," these customs were "worth respecting" (Field, 1935, p. 5). Alongside chiefs and religious authorities, it appears that Asafo companies also played a role in upholding the fishing rights and restrictions of their communities.

Asafo companies were indigenous Akan political and military organizations responsible for the protection and defence of their respective societies. As a patrilineal organization, every man, woman, or youth joined their father's Asafo company. Among the coastal Akan the base of the Asafo organization was made up of fishermen and women (Nti, 2011, p. 3). The origins of Asafo companies have been traced to pre-colonial Akan coastal towns, particularly in the sixteenth and seventeenth centuries as a result of the growth and development of coastal townships (Anderson, 2016, pp. 245–247; Kea, 1982, pp. 131–168). Asafo companies typically functioned as the town militia. De Graft Johnson (1932), for example, identifies them as essentially a warrior organization while Datta and Porter (1971) describe Asafo companies as organized military bands and a well-structured unified force that served as the backbone of Akan polities. There were often multiple Asafo companies in each town, in which larger towns were divided into wards with each ward containing its own Asafo company with officers and rank-and-file member. A typical Asafo contingent comprised a commander called Supi, captains called Asafohenfo, a standard-bearer, an Odomankoma Kyerema or a drummer or a bugle, and a priest called Asafokomfo. The Supi, heads of individual Asafo companies, also reported to the Tufuhen, who was the supreme head of all Asafo companies in a town and who was responsible for their general conduct, reporting to the Omanhin. Leaders were frequently—although not universally—appointed on the basis of personal merit, overruling the ascriptive kinship structuring which characterizes Akan social and political institutions (Hernaes, 1998, p. 2; Nti, 2011, pp. 111–112). While the military role of Asafo companies has received the most emphasis, it is clear that they were also a social force that performed various social, political, religious, educative, entertainment, vocational, and environmental roles (Acquah et al., 2014, p. 50; Anderson, 2016, p. 237; Bentum, 2006, p. 102; Sam, 2014, p. 22). Asafo companies played an important political role in pre-colonial coast society, often achieving an autonomous position in relation to chiefs and elders, with the Tufuhen sitting on the Oman Council. The consent of Asafo—"commoners"—was required in policy decisions and lawmaking. The Asafo, therefore, may be taken to have articulated "commoner" interests. In this respect, the social basis of the Asafo was "the people", or the social groups included in the broad term "commoners" (Hernaes, 1998, p. 2; Nti, 2011, p. 111). During the colonial period, Asafo institutions also functioned as a medium of popular protest against escalating exploitation by chiefs and, indirectly, against the colonial administration (Nti, 2011, pp. 20–25). Equally, there are indications that the Asafo movement became an important mobilizing force in the political struggle leading up to Independence (Hernaes, 1998). While much more research is needed in this area, Asafo companies also seem to have played an important role in enforcing fishing restrictions, advancing community claims to fishing rights, or responding to infringements of these restrictions or rights by neighbouring communities. Nti, for example, discusses disputes in 1931 between the Asafo companies of Narkwa and Ekumpuano in the Saltpond District surrounding fishing rights in the Narkwa lagoon. There had been a history of disturbances between the two villages over these fishing rights, which came to a head in September 1931 following a small conflict between boys from Narkwa and Ekumpuano who were fishing in the lagoon. When one of the boys from Narkwa insulted one of the Asafo company captains of Ekumpuano, the boys from Ekumpuano tied him up and beat him. The men of Ekumpuano then gathered at a village near Narkwa and started beating war drums and singing war songs. They also fired one shot into the lagoon and another into the sea as a challenge. In response, the people of Narkwa gathered their weapons and charged the men of Ekumpuano. Fighting then ensued, in which three men died (Nti, 2011, pp. 294–296). In this case, ongoing disputes surrounding claimed lagoon fishing rights prompted conflict between the Asafo companies of Narkwa and Ekumpuano, who both claimed the rights to fish these waters and who responded to the infringements and insults of the other side.

Alongside broader rights to restrict fishing practices and declare closed seasons, and the resulting disputes between Asafo companies, written records also provide frequent mention of the observance of non-fishing days by fishing communities in coastal Ghana. Accounts of non-fishing days appear in European accounts from the seventeenth century onwards. As discussed, it is important to stress that this is not a suggestion that these practices only *began* in the seventeenth century. By the following accounts, it would not be too speculative to suggest that these practices were already well established by the time of European observation in the seventeenth century. In published accounts from the second half of the seventeenth century, Nicolas Villault de Bellefond (1670, pp. 166–167)—a Frenchman whose observations are based on his voyage to the coast in 1666–1667—and Willem Bosman (1705, p. 160)—who resided on the coast for more than ten years as an agent of Dutch West India Company—both referred to Tuesday as a "Sabbath" among coastal communities in Ghana. While de Bellefond suggests that "every man rests that day from his labour" and "the Peasants bring nothing to market," Bosman writes that this "differs from other Days no otherwise" except that "no Person … is permitted to Fish." Bosman, who spent much longer on the coast in comparison to de Bellefond, appears to provide a more accurate account of this practice, which aligns with the majority of other written records suggesting restrictions on Tuesdays focused specifically on fishing practices. In the 1660s, Wilhelm Johann Muller, a German chaplain at the Danish Fort Frederiksborg in the vicinity of Cape Coast, wrote that Tuesdays were held so sacred by the fishermen that "they believe a great disaster would befall them if they went to sea to fish with hooks on that day" (Jones, 1983, p. 167). This strict adherence was also noted by Thomas Thompson (1758, p. 36), an eighteenth-century missionary, who wrote that this was a custom "they keep so strictly, that even the Necessity of Hunger, in the scarcest Times, makes no Exception to their established Rule in this Particular." While these earlier accounts do not describe the punishment for non-compliance, aside from the displeasure of the gods, Alfred Burton Ellis (1887, pp. 220–221)—a British major of the West India Regiment and colonial ethnographer stationed in the Gold Coast Colony writing in the second half of the nineteenth century—describes that Akan fishermen who violated this rule were fined and their fish were cast into the sea.

Considering the origins of the Tuesday no-fishing day, Ellis (1887, pp. 220–221) recounted the tradition that the first fishermen, named "Kwegia," chose Tuesday for a day of rest. This tradition was also discussed by Carl Christian Reindorf (1895, p. 5), a Euro-African missionary born in Prampram with Ga and Danish heritage, who described that the Farnyi

Kwegya was one of two giants, alongside Asebu Amanfi, who emerged from the sea with a great number of followers; Amanfi travelled to the interior and founded Asebu, whereas Kwegya travelled to the beachside and founded Moree as "the place well suited for fishing" where he established a fishing industry. As a result, Kwegya was considered "the first fisherman" from whom "all the rest of the people of the Gold Coast acquired the knowledge of fishing in the sea" (Reindorf, 1895, p. 270). However, Reindorf does not suggest that Kwegya implemented the day of rest but, instead, relates the origins of the Tuesday non-fishing day to the Genesis creation narrative. Reindorf (1985, p. 270) writes, "Fishermen keep Tuesday as their holiday, and as our holidays always fall on the day of the week on which one was born, so our fisherman had known by tradition that the sea came into existence on the third day of creation, which was Tuesday." An editorial in the *Gold Coast Leader* (1905, 4 March)—an African-owned, English-language newspaper based out of Cape Coast—also suggested this as the reason fisherfolk did not fish on a Tuesday, describing "the Sea was created on a Tuesday and that day must be counted sacred to it." While couched in colonial perceptions and Christian narratives, these accounts provide clear evidence of the long-standing nature of the Tuesday non-fishing day, which was intrinsically linked to coastal beliefs surrounding the gods of the sea.

Signs of tension surrounding non-fishing days started to arise following the spread of Christianity. In the nineteenth century, the presence and activities of mission societies, particularly the Basel and Wesleyan Missionary Societies in Ghana who had been present in Ghana since 1828 and 1835, respectively, had led to a growing Christian population among coastal communities where missions were established (Jedwab et al., 2021, pp. 576–577). As early as the 1880s, Ellis (1887, p. 221) wrote that "the spread of Christianity and scepticism has caused this observation of Tuesday as a day of rest to fall into disuse." Ellis was clearly exaggerating as the majority of evidence suggests that Tuesday non-fishing day continued to be adhered to by the majority of fisherfolk in the nineteenth and twentieth centuries, but his observation speaks to the fact that some Christian fishers had started to violate Tuesday non-fishing day by the late nineteenth and early twentieth century. This was discussed in a report in the *Gold Coast Leader* (1908, 23 May) from Appam on 20 April 1908 concerning a meeting between the fishing community and Church authorities to discuss contributions for building a new Wesleyan chapel. During the meeting, the non-Christian members of the fishing community asked what measures had been taken in the matter of Church members—the Christian members of the fishing community—who were fishing on Tuesdays, which the non-Christian fishers regarded as "a desecration." The Wesleyan minister,

James Reynolds, responded by stating that just as non-Christian fishers do not observe Sundays as a day of no work, Christian fishers do not observe Tuesdays (*Gold Coast Leader*, 1908, 20 June). Other reports, however, suggest that a majority of Christian fishers refrained from fishing on both Tuesday and Sunday, thereby observing Sabbath while avoiding conflict with the broader fishing community and, conceivably, continuing to hold belief in the importance of Tuesday as a sacred day whether due to the creation of the sea in the Genesis creation narrative or due to continued belief in local narratives (Brown, 1947, p. 41; *Gold Coast Leader*, 1905, 4 March). Field (1935, p. 5), for example, writes that Christian fishers (who he estimates at "probably less than 2% of the fishing community") would instead spend Tuesday repairing nets and undertaking other related tasks on land.

The fact that Christian fishers observed both Tuesday and Sunday as no-fishing days also led to some calls for non-Christian fishers to observe the same. Such a call appeared in the *Gold Coast Leader* (1908, 7 March) from Elmina on 25 February 1908, which also suggested that this was the practice in some communities already:

> We call attention of the Omanhene and Tufuhene to stop fishing on Sundays by the non-christians. The christians are not fishing on Tuesdays and it is only but fair that the non-christians should pay some respect too by putting a stop to fishing on Sundays as is done in some places.

A follow-up report from Elmina on 5 December 1910 suggests that the Omanhen of Elmina had consented to such requests as a gong-gong had been beaten signifying the halting of fishing on Sundays. Celebrating this development, the reporter only regretted that "there is no remedy for the travelling of women folk into the bush for food on the Sacred Day" (*Gold Coast Leader*, 1910, 5 December). This prohibition seems to have continued until at least March 1911 when it was described as "an annoyance to the fishing folk" (*Gold Coast Leader*, 1911, 11 March) and there is also mention in October 1921 that there was no fishing permitted on Sundays in Elmina (*Gold Coast Leader*, 1921, 1 October), but it is unclear how long this prohibition lasted in practice. These disputes surrounding Tuesday and Sunday non-fishing days at the beginning of the twentieth century coincided with various other complaints being raised in newspapers surrounding activities occurring on Sundays, including the dragging and repairing of nets, the beating of Asafo company drums, and parades of women fish traders to ask the gods to provide fish in times of scarcity (*Gold Coast Leader*, 1908, 24 October; 1909, 1 May). There were also complaints that European shipping companies encouraged these practices by requiring canoe men and other labourers to work on Sundays (*Gold Coast Leader*, 1906, 29 December; 1915, 3 April). While the evidence presented here is fragmentary, what is clear is that such disputes differed across diverse communities at different times and had varied outcomes. More research is needed on the impact of missions on fishing practices, but these disputes would clearly have been more prevalent in communities, such as Appam, where Christian missions had an established presence. Whatever the case, despite some limited tensions and violations rising as a result of Christian conversion, as well as some local arrangements and negotiations, it appears that the Tuesday non-fishing day continued to be adhered to by the majority of the fishing population in the colonial period.

Since the mid-twentieth century and following Independence in 1957, Christianity has spread throughout Ghana with Christians comprising approximately 80% of the population as of 2017 (Jedwab et al., 2021, p. 594). As discussed, the surge of Christianity has been identified in previous studies as one of the central reasons behind changing perceptions and non-compliance of Tuesday non-fishing day in modern-day Ghana. This was also voiced by two Asafo company leaders in interviews conducted in Cape Coast in 2021. One leader, for example, stated:

> For about 30 to 35 years now we have stopped all these things… everyone says he or she goes to church and all the rituals we perform are fetish and pagan. Today being Tuesday for instance, we have to play the Asafo drums, but if I go and call my son and my brother, he will ask me to take the lead because he sees it to be a disgrace to be part of the Asafo drumming… (52-year-old male, Asafo Company Leader, Cape Coast, interview 10/08/2021)

The "disgrace" alluded to here aligns to perceptions that the introduction of new forms of religious belief alongside modernization has led to a reluctance of community members to uphold traditional beliefs and practices, including Tuesday non-fishing day and the activities of Asafo companies. This includes a high level of non-compliance of Asafo traditions and a lack of respect for Asafo deities. Yet, the functions of the Asafo companies have not become extinct but rather they continue to play several roles in Akan communities (Anderson, 2016, p. 248). Christianity has also transformed the relevance of other long-standing practices, such as pouring libation and sprinkling food to the gods.

If not fully erasing the fear of punishment in the event of any violation, the undermining of belief systems undergirding such practices has led to assertions that Christianity has played a significant role in eroding traditional beliefs in fishing communities (Adjei & Sika-Bright, 2019, p. 2;

Dosu, 2017, p. 57; Kalanda-Sabola et al., 2007, pp. 26–27). Such arguments surrounding Christianity and its contribution to the violation of the Tuesday non-fishing custom cannot be sidestepped. Yet, despite the expansion of Christianity, Adjei and Sika-Bright (2019, p. 2) have recently found in their study of traditional beliefs among sea fishers in the Western Region that fishing communities continue to transmit, imbibe, and practice traditional beliefs about the sea, and most fishers continue to perform practices, particularly the pouring of libation to invoke the protective powers of sea gods and spirits, that are frowned upon by Christianity. However, it has also been observed that some Christian fishers have replaced the traditional libation with Christian prayer (Abane et al., 2013, p. 69). Just as the compliance or violation of non-fishing days in the early twentieth century was determined by specific groups within communities rather than collective communal disregard, contemporary violations of the Tuesday non-fishing custom on religious grounds continue to be undertaken by specific individuals or groups rather than through broader communal or collective disregard for the custom altogether. Instead, as the following case study surrounding Abandze suggests, the reasons for widespread communal violations of Tuesday non-fishing days have to be understood with the context of specific politico-legal issues which, in conjunction with broader economic realities, has led to a broader diffusion of non-compliance among fishing communities in the Central Region.

Politico-Legal Stressors, Redefinition of Customs, and Communal Violation of Tuesday Non-Fishing Day in Abandze

The recent history of violations of the Tuesday non-fishing custom as a collectively driven communal venture in the Central Region can be traced to a small fishing community called Abandze. These events surrounding this community provide opportunity to understand the ways in which political and legal conditions can influence behaviours, which may alter the essence of heritage-based practices. Information gathered from the National Secretary of the Canoe and Fishing Gear Owners Association of Ghana, corroborated by the Chief Fisherman of Abandze, indicated how the violation of the Tuesday non-fishing custom became a political-cum-legal-influenced communal activity in Abandze. The people of Abandze, as the interview disclosed, were faced with a chieftaincy court dispute. This dispute started in 1972 and lasted in court for 31 years. The dispute involved the entire Abandze community who accused the then Chief, Nana Kwesi III, of usurping the powers of some traditional authorities. In reaction to the Chief's behavior, the traditional authorities and community members started disregarding his authority thereby setting the grounds for an acrimonious relationship that would end up in the courtroom. This dispute required the mobilization of funds to enable the community to pay for the legal fees. Led by some traditional authorities, it was decided that fishermen—who constitute the majority of the population in the community—could fish on Tuesdays in order to raise enough money to pay the legal fees. With this arrangement, fishermen would deduct their cost of operations from the sale of the harvested fish and present the proceeds to the newly installed Chief.

The act of fishing on Tuesdays was, therefore, supposed to be temporary. Yet, the success of the measure led to the discovery of two social realities that would eventually render the violation of the Tuesday non-fishing custom a normality in that community and beyond. Of scholarly relevance is how this observation adds to the heritage change discourse. First, the fishermen and the community, generally, discovered that the punishment that was expected to incur as a result of violating the Tuesday non-fishing custom was never experienced. This was manifest in the words of a fisherman during a FGD in Abandze: "When we went to fishing on Tuesday, we thought the gods will punish, but when we went and came back nothing happened so we felt we could go again" (37-year-old fisherman, Abandze, FGD 25/05/2021). This rendered them to question the spiritual significance of the custom and interpret the custom as lacking any repercussions to restrain their behaviour. Secondly, the Tuesday fish harvest was, by the estimation of the fishermen, more productive than the other days:

> We even realized that when we go for fishing on Tuesdays, we get more fish than the other days. At first we gave the money to the chief, but now we make more money because we don't give the money to the chief again. (45 year-old fisherman, Abandze, FGD, 13/07/2021)

These responses show how the initial quest to address immediate community needs became the fulcrum around which the violation of the Tuesday non-fishing custom was first exhibited. These experiences then led to a continuation of violations, which were viewed by fishermen and community members as a means to improve their livelihoods. Thus we discover plain subjective discourses in the violations to the Tuesday non-fishing custom. The loss of the supernatural's coercive power, as perceived following initial violations without punishment, seems to have dominated their motivations towards making Tuesday a normal fishing day for livelihood purposes. This perception has then encouraged the deterioration of a practice that, like other maritime cultural practices based on social values and norms among coastal communities, had formerly alleviated pressure on fish stocks on at least one day per week in regions where fish stocks have already been assessed as overexploited and near to collapse (Brown, 2004, p. 7; Cook et al., 2021, p. 255; Khakzad & Griffith, 2016, p. 96). This, then, has the potential to further compromise the balance between economy, environment, and social lives of the people over the long term (Okafor-Yarwood et al., 2020, p. 2). It is equally notable from the responses that the way the people interpreted the violations based on the outcomes has led their disposition of overlooking the symbolic and developmental essence of the custom. It is thus apparent that, for long, the developmental and ecological essence of the Tuesday non-fishing is not factored into the matrix of the observance of the custom. This, among others, is largely attributed to the fact that the existing changes in belief systems have not been associated with education on fishing practices (Tilley & Roscher, 2020, p. 16).

The trend of fishing on Tuesdays then continued even after the chieftaincy court dispute had been settled. This now occurs with a reconfigured purpose—as a tool to promote community development. An FGD with the fishermen in Abandze, for example, revealed that two sheds for fishermen, culverts which connect the Abandze to the seashore, and an 18-seater toilet facility were constructed using the proceeds from the Tuesday fishing. This change in the purpose of the violation has resulted in some changes in the distribution of the proceeds from the fish harvest on Tuesdays. Currently, as mentioned by the Chief Fisherman in Otuam (interview 22/06/2021), the distribution of Tuesday non-fishing day harvest is as follows: one-third of the proceeds goes to the Chief Fisherman, one-third goes to the Chief Fisherman's Council of Elders and the final third goes to the Chief. Further evidence from the interviews and FGDs suggests that, over the years, this violation, which has currently assumed regularity, has diffused from Abandze into nearby coastal communities including Otuam, Moree, Mumford, and Narkwa. This was confirmed by the National Secretary of the Canoe and Fishing Gear Owners Association of Ghana, and corroborated by the fishermen in the said communities:

> The people of Ekumfi-Narkwa started going to sea on Tuesdays about five years ago but, for those in Moree, they started about three years ago. (National Secretary of the Canoe and Fishing Gear Owners Association of Ghana, Abandze, interview 12/09/2021)

This could be largely attributed to the frequent interactions between fishermen along the coast of Ghana spawned by the migratory nature of the fishing activities in the country, and by extension, the West African coast (Haakonsen, 1991, p. 4). These interactions among fishermen have shaped their lived experiences with the ocean, which in turn have influenced their interpretations about the Tuesday non-fishing custom. As fishers from Abandze spread their perceptions about the positive, rather than negative, consequences arising from fishing on a Tuesday, this has encouraged other communities to replicate their behaviour. When these communities do not face immediate repercussions for their actions, the pattern continues. It is important to recognize that this is taking place at a time of recognized over-exploitation, in which continuing depletion of fish stocks means that increasingly smaller yields require increasing greater fishing effort (*Fisheries Act*, 2002; Cook et al., 2021, p. 255). Rather than preserve the no-fishing policy, communities whose livelihoods depend on fishing have increased their efforts for short-term benefit but with potential for drastic long-term consequences.

It is important to emphasize the role that traditional authorities play in facilitating these developments. Chief Fishermen occupy great importance in a traditional community fisheries structure. They serve as intermediaries between the fishers and state institutions, and represent fishing communities on consultative engagements for policy and law purposes. Crucially, Chief Fishermen play important roles in enforcing customary laws. The Asafo companies also work to preserve the sea and lagoons in communities in the Central Region of Ghana. An Asafo leader specified his responsibility as follows.

> When I see someone catch fingerlings from the lagoon and the sea, I call the Chief Fisherman to come and see what is happening. When the people who fish with the seine nets catch the juvenile fishes, we talk about it… I have caused the arrest of some people recently. I think it happened in February. I caused the arrest of eight people… and they were put behind bars for five days because they caught juvenile fishes… with small nets. (47 years old Asafo Company leader, Cape Coast, interview 25/06/2021)

It is thus clear that the Chief Fishermen pursue their customary law enforcement responsibilities in collaboration with the Asafo companies. However, an erosion in their roles is visible. Their role of punishing offenders has been taken over by modern law enforcers such as the Police Service. As Asafo company leader in Cape Coast recounted, "when we found them dragging the seine net in the lagoon I called the Police who got them arrested and they had stayed behind bars for five days" (47 years old Asafo Company leader, Cape Coast, interview 25/06/2021).

Under the current arrangement in the communities examined, the Chief Fisherman benefits from the fish harvest taking place on previously non-fishing days, which has impacted on their willingness to reintroduce and enforce these practices. The National Secretary outlined that:

> On a normal day the Chief Fisherman is given two buckets of fish when a fisherman is able to harvest ten pans. (National Secretary of the Canoe and Fishing Gear Owners Association of Ghana, Abandze, interview 13/07/2021)

However, this practice differs on a Tuesday as recounted by a fisherman in Abandze during an FGD: "On Tuesdays, the Chief Fisherman is given a pan which is equal to 10 buckets after every 10 pans of fish harvest by a fisherman or group of fishermen" (43 years old, fisherman, Abandze, FGD 13/07/2021). The distributive pattern with specific reference to the quantity received by the Chief Fisherman varies among communities. This came out during an FGD in Otuam, "In our community the Chief Fisherman receives a pan after every seven pans of fish harvest" (52 years old, fisherman, Otuam, FGD 22/06/2021). Despite the variations, there is a clear indication of a cultural heritage which has symbolic and survival significance being sacrificed due to financial interest. In the light of the collection of fish by the Chief Fisherman his law-enforcing role has not just been compromised but he has also lost recognition and veneration among the fishermen. An Asafo company leader recounted his experience in this narrative:

> Nowadays, the boys at the beach do not respect the Chief Fisherman… It is because they have seen that he is not the one who is supposed to occupy the stool. (45 years old, male Asafo Company Leader, Cape Coast, interview 25/06/2021)

In this regard, Chief Fishermen have not just reneged on his responsibility as a customary law enforcer but have become an active participant and a beneficiary of the violation of the Tuesday non-fishing custom. This speaks to the central role that customary authority plays in shaping the resiliency of fishing customs. As these practices are deeply embedded in customary authorities, if this authority is corrupted in favour of individual gain (i.e. "elite capture") or disputed by community members, then there is a direct impact on the integrity of and compliance towards the customs associated with that authority (Njaya & Donda, 2012, p. 9; Russell et al., 2018, p. 87). Moreover, as this authority is directly linked to spiritual beliefs surrounding the sea, which underpins customary fishing practices, then the initial undermining of these beliefs—in this case to meet legal costs related to a chieftaincy dispute—can inevitably lead to the collapse of effective traditional authority too. In the communities discussed, the two pillars upholding this custom—spiritual belief and chiefly authority—are being disrupted as a result of a combination of local economic interest, communal necessity, erosion of beliefs, and the loss of traditional political influence. In the process, a long-standing and highly symbolic cultural heritage, which has the potential to inform and encourage more sustainable fishing practices at a time when fish stocks are being over-exploited and near to collapse, is being gradually disputed and disrupted after centuries of practice.

Conclusion

Narratives of non-compliance surrounding Tuesday no-fishing days are deeply entangled with expressions of authority, both spiritual and chiefly. Historical examination of customs demonstrated that no-fishing days continued to be observed by fishing communities in coastal Ghana despite substantial transformations of coastal spaces as a result of active participation in Atlantic trading networks and the sustained coastal presence of European merchants between the fifteenth and eighteenth centuries. Although increased missionary activity and eighty years of British colonization in the nineteenth and twentieth centuries led to an increasing number of Christian fisherfolk and some disputes surrounding the observance of no-fishing days, this appears to have been upheld in the majority of communities with only certain groups violating or reconfiguring the practices (i.e. observing non-fishing days on Sunday instead or observing two non-fishing days). This meant that violation of Tuesday non-fishing days on religious grounds was largely undertaken by specific individuals or groups, rather than whole communities. This only took on a communal form, at least in the communities examined in the Central Region, when specific political and legal conditions led to the violation of customs by the community as facilitated by traditional authorities. The initial violation of these customs then led to sustained violation as community perceptions surrounding the custom transformed when there were no immediate repercussions for their activities. On the contrary, fishers observed positive short-term impacts from their increased fishing effort. This practice, originating in one community, was then diffused to surrounding communities as word spread that the punitive power of the supernatural was either non-existent or less powerful. This was directly linked to livelihood issues as well as financial and community development issues, in which fisherfolk choose to sacrifice heritage-based customs for the achievement of immediate livelihood and community needs. Hence, these issues cannot be disconnected from the violation of an ocean-related customary practice. However, such violations also have to be understood within the context of the interactive engagements between the violators and the enforcers as well as the meanings that fisherfolk derive from the actions and inactions of the enforcers and communities close by.

Despite displays of violations by some groups and by some communities, the blue heritage practice of Tuesday non-fishing day is not fully lost. There are some groups and some communities that continue to observe the custom. As it stands now, there is only a dialectical display of observance and the non-observance of the Tuesday non-fishing day. At the same time, given continuing over-exploitation of fish stocks in the waters surrounding Ghana and the resulting impact on yields, which is encouraging greater fishing effort rather than less, it is important to be cautious about the ease with which such violations could be reversed. The diffusive nature of these violations, alongside their intrinsic ties to the erosion of beliefs and customary authority surrounding fishing practices, means that reconstruction of these customs relies on the willingness of fisherfolk across different communities to reengage with these or similar practices on the grounds of their potential long-term benefits. Still, given the long-standing and symbolic nature of these customs, their basis in the rich blue heritage of Ghana's coastal communities, and their continuing legitimacy across many groups and communities, these are practices that could prove vital to broader programmes aimed at the realization of sustainable fisheries management across the fishing sector in Ghana.

Notes

1. The interview with the National Secretary of the Canoe and Fishing Gear Owners Association of Ghana was conducted in English.
2. This does not mean, however, that they were treated well by Europeans who often viewed canoe men with contempt, describing them as 'thieves', withholding their wages, and also seizing canoe men and selling them into slavery (Gutkind, 1985, pp. 30, 36–37).

References

Abane, H., Akonor, E., Ekumah, E., & Adjei, J. (2013). *Four governance case studies and their implications for Ghana fisheries sector.* USAID-URI Integrated Coastal and Fisheries Governance (ICFG) Initiative. Coastal Resources Center, Graduate School of Oceanography, University of Rhode Island.

Acquah, E. O., Amuah, J. A., & Annan, J. F. (2014). The contextual and performance dimensions of asafo music from the perspective of annual Akwambo festival. *International Journal of African Society Cultures and Traditions, 2*(2), 50–60.

Adjei, J. K., & Sika-Bright, S. (2019). Traditional beliefs and sea fishing in selected coastal communities in the Western Region of Ghana. *Ghana Journal of Geography, 11*(1), 1–19.

Adom, D., Sekyere, P. A., & Yarney, L. (2019). A return to the ghanaian cultural values of closed fishing season in ghana's artisanal marine fishing: An essential means of restoring small pelagic fish stocks. *Transylvanian Review of Systematical and Ecological Research, 21*(3), 95–110.

Adu-Boahen, K. (2018). Female agency in a cultural confluence: Women, trade and politics in seventeenth- and eighteenth-century gold coast. In J. K. Osei-Tutu & V. E. Smith (Eds.), *Shadows of empire in West Africa: New perspectives on European fortifications* (pp. 169–199). Palgrave Macmillan.

Afoakwah, R., Osei, M. B. D., & Effah, E. (2018). *A guide on illegal fishing activities in Ghana.* USAID/Ghana Sustainable Fisheries Management Project. Coastal Resources Center, Graduate School of Oceanography, University of Rhode Island. Prepared by the University of Cape Coast, Ghana.

Aggrey, J. E. S. (1978). *Asafo.* Ghana Publishing Corporation.

Aidoo, E. K. (2011). *Documentation of the Fante Asafo flags since the year 2000 and their socio-cultural significance.* Master's thesis, Kwame Nkrumah University of Science and Technology.

Akyeampong, E. K. (2002) *Between the sea & the lagoon: An eco-social history of the Anlo of southeastern Ghana c.1850 to recent times.* James Currey.

Akyeampong, E. (2007). Indigenous knowledge and maritime fishing in West Africa: The case of Ghana. *Tribes and Tribals, Special Volume,* 173–182.

Allison, E. H. (2011). *Aquaculture, fisheries, poverty and food security* (WorldFish Center Working Paper 65). https://aquadocs.org/handle/1834/24445

Anderson, G. (2016). Akan Asafo Company: A practical model for achieving true African liberation and sustainable development. In M. Mawere & T. R. Mubaya (Eds.), *Colonial heritage, memory and sustainability in Africa: Challenges, opportunities and prospects* (pp. 237–57). Langaa RPCIG.

Bentum, S. A. (2006). *Cultural significance of Edina Asafo Company posts.* Doctoral dissertation, Kwame Nkrumah University of Science and Technology.

Boamah, A. D. (2015). *Akan indigenous religio-cultural beliefs and environmental preservation: The role of taboos.* Master's thesis, Queen's University, Canada.

Bosman, W. (1705). *A new and accurate description of the coast of Guinea*. J. Knapton, A. bell, R. Smith, D. Midwinter, W. Haws, W. Davis, G. Strahan, B. Lintott, J. Round, and J. Wale.

Brown, A. P. (1947). The fishing industry of the Labadi district. In F. R. Irvine (Ed.), *The fishes and fisheries of the Gold Coast* (pp. 23–44). Government of the Gold Coast.

Brown, J. E. (2004). *Economic values and cultural heritage conservation: Assessing the use of stated preference techniques for measuring changes in visitor welfare*. Doctoral dissertation, Imperial College London.

Campbell, L. M., Gray, N. J., Fairbanks, L., Silver, J. J., Gruby, R. L., Dubik, B. A., & Basurto, X. (2016). Global oceans governance: New and emerging issues. *Annual Review of Environment and Resources, 41*, 517–543.

Chukwekere, B. I. (1970). *Cultural residence: The Asafo company system of the Fanti*. Research Report Series, No. 3. Social Studies Project. Cape Coast: University of Cape Coast.

Cruickshank, B. (1853). *Eighteen years on the Gold Coast of Africa including an account of the native tribes, and their intercourse with Europeans*. Cass & Co.

Coastal Resources Center. (2018). *Technical brief: The rationale and impact of a proposed second fishing holiday for the artisanal fisheries of Ghana*. Submitted to The Ministry of Fisheries and Aquaculture Development. Ghana Sustainable Fisheries Management Project (SFMP). Coastal Resources Center, Graduate School of Oceanography, University of Rhode Island.

Cook, R., Acheampong, E., Aggrey-Fynn, J., & Heath, M. (2021). A fleet based surplus production model that accounts for increases in fishing power with application to two West African pelagic stocks. *Fisheries Research, 243*, 106048.

Datta, A. K., & Porter, R. (1971). The Asafo system in historical perspective. *The Journal of African History, 12*(2), 279–297.

Dawson, K. (2018). *Undercurrents of power: Aquatic culture in the African diaspora*. University of Pennsylvania Press.

de Bellefond, N. V. (1670). *A relation of the coasts of Africk called Guinee; with a description of the countreys, manners and customs of the inhabitants; of the productions of the Earth, and the Merchandise and Commodities it affords; with some historical observations upon the coasts. Being collected in a Voyage By the Sieur Villault, Escuyer, Seiur de Bellefond, in the years 1666, and 1667. Written in French, and faithfully Englished*. John Starkey.

De Graft Johnson, J. C. (1932). The Fanti Asafu. *Africa: Journal of the International African Institute, 5*(3), 307–322.

Dickson, K. B. (1965). Evolution of seaports in Ghana: 1800–1928. *Annals of the Association of American Geographers, 55*(1), 98–111.

Dosu, G. (2017). *Perceptions of socio-cultural beliefs and taboos among the Ghanaian fishers and fisheries authorities: A case study of the Jamestown fishing community in the Greater Accra Region of Ghana*. Master's thesis, UiT The Arctic University of Norway.

Edusei K. (1981). *Artistic aspects of Cape Coast Asafo Companies*. Master's thesis, Kwame Nkrumah Nkrumah University of Science and Technology.

Ellis, A. B. (1887). *The Tshi-speaking peoples of the Gold Coast of West Africa: Their religion, manners, customs, laws, language, etc. by A. B. Ellis, Major, 1st West India Regiment*. Chapman and Hall.

Field, M. J. (1935, May 31). *Suggestions for an Achimota Fishery Station*. The National Archives (CO 96/724/18).

Fisheries Act 2002. http://www.fao.org/faolex/results/details/en/c/LEX-FAOC034737

Ford, B. (2011). The archaeology of maritime landscapes: Introduction. In B. Ford (Ed.), *The archaeology of maritime landscapes* (pp. 1–9). Springer.

Gold Coast Leader. (1905, March 4; 1906, December 29; 1908, March 7; 1908, May 23; 1908, June 20; 1908, October 24; 1909, May 1; 1910, December 5; 1911, March 11; 1913, July 14; 1915, April 3; 1920, July 31; 1921, October 1).

Groenewald, T. (2004). A phenomenological research design illustrated. *International Journal of Qualitative Methods, 3*(1), 42–55.

Green, T. (2019). *A fistful of shells: West Africa from the rise of the slave trade to the age of revolution*. Allen Lane.

Gutkind, P. C. W. (1985). Trade and labor in early precolonial African history: The Canoe men of Southern Ghana. In C. Coquery-Vidrovitch & P. E. Lovejoy (Eds.), *The workers of African trade* (pp. 25–50). Sage.

Haakonsen, J. M. (1991). The role of migrating fishermen in West Africa: What we know and what we still need to learn. In J. R. Durand, J. Lemoalle, & J. Weber (Eds.), *La Recherche Face à la Pêche Artisana Symposium* (pp. 709–715). ORSTOM.

Hernaes, P. (1998). Asafo history: An introduction. *Transactions of the Historical Society of Ghana, 2*, 1–5.

Jedwab, R., zu Selhausen, F. M., & Moradi, A. (2021). Christianization without economic development: Evidence from missions in Ghana. *Journal of Economic Behaviour and Organization, 190*, 573–596.

Jones, A. (1983). *German sources for West African history 1599–1669*. Franz Steiner Verlag.

Kalanda-Sabola, M. D., Henry, E. M., Kayambazinthu, E., & Wilson, J. (2007). Use of indigenous knowledge and traditional practices in fisheries management: A case of Chisi Island, Lake Chilwa, Zomba. *Malawi Journal of Science and Technology, 8*(1), 9–29.

Kea, R. A. (1982). *Settlements, trade, and polities in the seventeenth-century gold coast*. John Hopkins University Press.

Khakzad, S., & Griffith, D. (2016). The role of fishing material culture in communities' sense of place as an added-value in management of coastal areas. *Journal of Marine and Island Cultures, 5*(2), 95–117.

Kraan, M. (2009). *Creating space for fishermen's livelihoods: Anlo-Ewe beach seine fishermen's negotiations for livelihood space within multiple governance strucutures in Ghana*. African Studies Centre.

Njaya, F., & Donda, S. (2012). Analysis of power in fisheries co-management: Experiences from Malawi. *Society and Natural Resources, 25*(7), 652–666.

Nti, K. (2011). *Modes of resistance: Colonialism, maritime culture and conflict in Southern Gold Coast, 1860–1932.* Doctoral dissertation, Michigan State University.

Nuttall, M. (2000). Crisis, risk and deskilment in Northeast Scotland's fishing industry. In D. Symes (Ed.), *Fisheries dependent regions* (pp. 106–115). Blackwell Science.

Okafor-Yarwood, I., Kadagi, N. I., Miranda, N. A., Uku, J., Elegbede, I. O., & Adewumi, I. J. (2020). The blue economy–cultural livelihood–ecosystem conservation triangle: The African experience. *Frontiers in Marine Science, 7*, 586.

Parker, J. (2000). *Making the town: Ga state and society in early colonial Accra.* James Currey.

Ransley, J. (2011). Maritime communities and traditions. In A. Catsambis, B. Ford, & D. L. Hamilton (Eds.), *The Oxford handbook of maritime archaeology* (pp. 879–903). Oxford University Press.

Reindorf, C. C. (1895). *History of the Gold Coast and Asante, based on traditions and historical facts, comprising a period of more than three centures from about 1500 to 1860, by Rev. Carl Christian Reindorf, Native Pastor of the Basel Mission, Christiansborg, Gold Coast.* Basel.

Ross, D. H. (2007). "Come and try": Toward a history of Fante military shrines. *African Arts, 40*(3), 12–35.

Russell, A. J. M., Dobson, T., & Wilson, J. G. M. (2018). Fisheries management in Malawi: A patchwork of traditional, modern, and post-modern regimes unfolds. In M. G. Schechter, N. J. Leonard, & W. W. Taylor (Eds.), *International governance of fisheries ecosystems: Learning from the past, finding solutions for the future* (pp. 53–98). American Fisheries Society.

Salm, S. J., & Falola, T. (2002). *Culture and customs of Ghana.* Greenwood Press.

Sam, J. A. (2014). *Drums and drum languages as cultural artefacts of three Asafo companies of Oguaa Traditional area of Ghana.* Master's thesis, Kwame Nkrumah University of Science and Technology.

Sparks, R. J. (2014). *Where the Negroes are masters: An African port in the era of the slave trade.* Harvard University Press.

Thompson, T. (1758). *An account of two missionary voyages by the appointment of the society for the propagation of the gospel in foreign parts: The one to New Jersey in North America, the other from American to the Coast of Guiney. By Thomas Thompson, A.M. Vicar of Reculver in Kent.* Printed for Benj. Dod.

Tilley, A., & Roscher, M. B. (2020). *Information and communication technologies for small-scale fisheries (ICT4SSF).* Food and Agriculture Organization of the United Nations and World Fish.

Walker, B. L. E. (2002). Engendering Ghana's seascape: Fanti fishtraders and marine property in colonial history. *Society and Natural Resources: An International Journal, 15*, 389–407.

20

Lalela uLwandle: An Experiment in Plural Governance Discussions

Kira Erwin, Taryn Pereira, Dylan McGarry, and Neil Coppen

Introduction

Carefully crafted storytelling on how humans make meanings of their worlds in relation to the ocean offers learnings for altering ocean governance frameworks towards both inclusivity and plurality. This chapter explores these learnings through a research and theatre-based project, *Lalela uLwandle*

The research presented in this chapter is prepared under the One Ocean Hub, a collaborative research for sustainable development project funded by UK Research and Innovation (UKRI) through the Global Challenges Research Fund (GCRF) (Grant Ref: NE/S008950/1).

K. Erwin
Durban University of Technology, Kwazulu Natal, Durban, South Africa
e-mail: KiraE@dut.ac.za

T. Pereira · D. McGarry (✉)
Rhodes University, Makanda, South Africa
e-mail: d.mcgarry@ru.ac.za

T. Pereira
e-mail: t.pereira@ru.ac.za

N. Coppen
Durban, South Africa

© The Author(s), under exclusive license to Springer Nature Switzerland AG 2022
R. Boswell et al. (eds.), *The Palgrave Handbook of Blue Heritage*,
https://doi.org/10.1007/978-3-030-99347-4_20

(translated from isiZulu as Listen to the Sea), developed using the Empatheatre methodology[1] under the One Ocean Hub collective. Empatheatre is a research-based, theatre-making praxis in which research, data gathering, analysis and dissemination is collaboratively facilitated across different publics (Coppen, 2019). The *Lalela uLwandle* theatrical script was written and performed through an emergent and iterative process of practice and action-based research that took place over 2018 and 2019 in which co-participants, subsistence and small scale fishers, academics and researchers, marine scientists, traditional healers, religious followers, lifesavers, civil society partners, activists, and marine educators worked to identify and deliberate on matters of concern in relation to the ocean. *Lalela uLwandle* began as a response to a local conflict around a permit application for deep sea gas prospecting by a large oil and gas company off the coast of KwaZulu-Natal in South Africa. This particular local struggle in which the needs of diverse community members were pitted against a large extractive corporation, is one of many examples within a larger national and international move towards the Blue Economy (Bennett et al., 2021). As many countries turn to the seas in the hope of renewed economic growth, issues of environmental justice related to the oceans and climate change are increasing. As outlined in the first section of this chapter, ocean-related policy and governance frameworks interact with competing needs and conflicting desires on how we govern the ocean. In addition, in the blind rush for the dream of GDP growth, policy and governance frameworks may exacerbate historic and contemporary socio-economic exclusions. In South Africa capitalist and environmental conservation endeavours can, and have, caused harm for already marginalised people. These social and environmental justice concerns set an urgent challenge for ocean governance. This challenge lies not only in ensuring inclusion, but in recognising existing power relations between ocean epistemologies in which some ways of knowing, such as science and economics, dominate policy and decision-making forums.

There are many strategies needed, at different scales, to effect changes in governance policy and practice to ensure broad public participation and inclusion. Some interventions tackle the structures and systems of entrenched power head on, through either resistance and protest or critical engagement from within. Other interventions take a step to the side, to create an alternative conversation, with different entry points and framings, that opens up our perception to an expanded set of relations between people and the environment, liberated from the language of management and policy. *Lalela Ulwandle* was an example of the latter, by openly exploring what we may gain by listening closely to a plurality of ocean knowledges. If we are to spark a new imagination for what inclusive and just decision-making feels like then we need to experiment with methods that reject epistemological hierarchies and the problematic view that different knowledge systems are incommensurable. This chapter shares our learnings on how agonistic and plural understandings of the ocean provide an important point of departure for imagining anew just and inclusive ocean governance that works towards planetary well-being.

Ocean Grabbing in Policy and Practice

The oceans are receiving a significant increase in scientific and profit-seeking attention. The latter is discussed in Bennett et al. (2015). For a definition of this concept and the activities that constitute this action. The ocean, a moving body of water, fundamental to all life on earth, is under threat from human activities (Bähr, 2017; Franke et al., 2020; Poloczanska et al., 2018). National states and international bodies juggle competing demands within ocean governance. Extractive ocean industries such as commercial fishing, seabed mining, externalities from land-pollutants, and the hunt for marine genetic resources all make promises of economic growth (Bennett et al., 2021; OECD, 2016). Yet, simultaneously there is mounting scientific evidence advocating for the protection of marine resources as crucial to life on earth, and growing public demands for an end to these extractive industries to mitigate against a climate crisis (Bond, 2019; Hoegh-Guldberg et al., 2019). Many nation states, including South Africa, are developing Marine Spatial Plans (MSPs) in an attempt at governance and monitoring frameworks for their Exclusive Economic Zones[2] and Extended Continental Shelf areas.[3] To work across these demands ambiguous concepts like the Blue Economy are used to "facilitate cooperation between different social worlds" (Schutter et al., 2021, p. 7). These conceptual constructs create policy containers that enable duplicitous legislation to encourage, regulate and prohibit activities in the ocean (ibid., 2021, p. 3).

Amidst these national processes is a growing cry from many different geographies for "a new relationship between humanity and the ocean", one that requires a "a transformative shift from a state-centric approach to a global approach that takes into account the embeddedness of the ocean and associated actors in the wider planetary system" (Rudolph et al., 2020, p. 1). Appeals to collective humanity are necessary for healthy interconnected earth systems, but if they are to be transformative, they must simultaneously grapple with the stubborn power hierarchies and socio-economic fractures across, and within, human geographies and decision-making forums. Given the strong relationship of ocean health to the climate crisis there is a sense of urgency to these national governance frameworks. This urgency too often leads to policy development along a predictable path dependency steeped in capitalist logics and existing power structures. Already in South Africa we see the beginnings of ocean governance replicating weak participatory and inclusion processes (Sowman & Sunde, 2018), as well as entrenching scientific and economic discourses as the only legitimate knowledge systems for decision-making (Boswell & Thornton, 2021). The tensions emerging from this path dependency are clearly manifest in the struggles around subsistence and small-scale fishers. In 2007 fishers on the West Coast of South Africa took the Department of Environment, Forestry and Fisheries to court demanding a more inclusive and responsive policy regime for their sector. The policy instrument that emerged in 2012 had laudable goals for inclusion and indicated a significant move towards legislation for 'community-based' rights for small-scale and subsistence fishers (Sowman & Sunde, 2021). Yet in implementing the policy regulations the state reverted back to predictable economic development aspirations that forced fishers to create co-operatives using a small business model (Sunde & Erwin, 2020). For various reasons these co-operatives have not delivered on the promise of local development, both for the state and for the fishers involved. Slow roll out by the state, broken promises of boats and equipment, capture of these co-operatives for local politics and patronage, and the uneasy fit co-operatives of this kind have with the lived experiences of rural and urban subsistence fishers along the coastline (ibid.). Not only are many fishers excluded from this regulatory regime, but the message these permitting structures send is clear. You are more valued by the state, and your chances of getting a seat at the negotiation table are significantly increased, if you comply with formalising your livelihood into a dominant economic growth model (Ntona & Schröder, 2020).

Fishing for livelihoods along the South African coast is enmeshed in coastal cultural and heritage practices (Sunde, 2014). This is not unique to South Africa (see Gallois & Duda, 2016; Nadel-Klein, 2020; Urquhart & Acott, 2013) but the country's experiences of coloniality and apartheid have shaped these entanglements in specific ways. Many people racialised as black[4] in rural and urban areas were forcibly removed from the coast to make way for industrial and leisure activities for the white elite, violently disrupting their close relationships with the sea (Sunde, 2014). Continuing fishing as a source of pride, and in some cases resistance against this discrimination, further enmeshed fishing into cultural identities for many coastal people (Sunde & Erwin, 2020). Of course, in addition to identifying with a particular culture, religion and class, fishers are also parents and family members whose livelihood and knowledge of the sea support social relations in diverse ways. These socio-cultural livelihoods and logics are often antithesis to obsessions about capitalist growth models for small businesses.

It is not only the "capitalist-industrial visions of ocean space" (Ntona & Schröder, 2020) in the fisheries sector that coastal people are up against. They face similar exclusions through conservation efforts. In South Africa, state-led nature conservation has a damaging legacy of exclusion to answer for. Under apartheid, Marine Protected Areas (MPAs) "are associated with the forced removal of black communities from their lands and their displacement from the waters they traditionally fished" (Sowman & Sunde, 2018, p. 169). Conservation policies "almost exclusively reflected Western scientific values and beliefs, with an emphasis on protecting nature from human impacts" (Cocks et al., 2012). Well-meaning biodiversity protection policies that result in formal exclusions for indigenous and economically marginalised groups are common tensions that arise in marine conservation across national geographies (Crandall et al., 2018). Whilst the South African transition to democracy in 1994 expanded constitutional rights to all people and enabled recognition of these injustices[5] (De Wet & Du Plessis, 2010), restoration processes around land and resource rights for many coastal people have been painfully slow (Sowman & Sunde, 2018, p. 169).

These past exclusions are exacerbated by contemporary neoliberal models for marine conservation that link conservation with fantasies of development (Infield, 2001, p. 800). As Schutter et al. (2021, p. 2) point out the use of international concepts like the blue and green economy,

> share a foundation in ecological modernisation thinking, whereby economic growth and environmental protection can go hand-in-hand through incorporation of environmental issues into markets....[obstructing] the fundamental change required to achieve actual sustainability. (Schutter et al., 2021, p. 2)

Exclusions of local coastal people from protected areas frequently come hand-in-hand with attracting wealthier, mostly people racialised as white South Africans, and international visitors who pay for holidays in these pristine spaces. Rather than valuing traditional and cultural livelihood practices as being an important contributor to conservation efforts, communities who were forcibly removed are now promised (mostly low wage) jobs through growing environmental tourism. It is important to acknowledge there are positive environmental benefits to MPAs in South Africa particularly for the regeneration of fish stocks (Maggs et al., 2013). However, the social impacts of MPAs include weakening of local participatory governance, the loss of tenure rights and access to resources by already marginalised communities, increased food insecurity and reduced household income, as well as negative impacts on culture and identity (Sowman & Sunde, 2018). Too often then contemporary conservation management is experienced by black people as a continuum of apartheid's legacy of forced removals and punitive regulations. Thembela Kepe calls for a recognition of how contemporary conservation efforts, whilst not necessarily conceptualised as racist, as was the case under apartheid, serve as "unintentional acts that highlight race differences, positively or negatively" and "end up raising questions about equality, paternalism and redress in regards to race" (2009, p. 872). In South Africa then these Blue Growth narratives can actively reproduce the "colonization of nature", where dispossession occurs not just through land-grabs but through the rejection of "indigenous knowledge, values and practices in environmental management" (Cock, 2018, p. 140). This failure by the democratic state to adequately address historic injustices related to MPAs, as well as weak contemporary process of participation and inclusion in the promulgation of new MPAs ensures a "growing discontent" around spatial planning for marine protection (Sowman & Sunde, 2018, p. 169).

Given this history of violence and segregation, and continued experiences of exclusion, ocean-related policy forums cannot escape grappling with how unequal power relations impact governance frameworks. As we embark on the exercise of extending existing land-based spatial planning and policy regimes to the oceans, we face a collective challenge. At the very moment in which we need time for critical reflection and debate around why these national and international frameworks have failed to ensure environmental and ecological justice on land, we appear in the face of the climate crisis to have run out of it. How may we work with this urgency *and* insistence on careful participatory processes for governance, alternative social structures and critical reflection? If we are to move beyond the predictable path dependencies of capitalism and siloed governance arrangements in ocean governance, then we need a new imagination for what inclusive and just decision-making feels like.

One method for generating such an imagination is to tackle the dominance of only a few ways of knowing the ocean. The epistemological hegemony of both the natural sciences and economics in ocean-related policy development generates a number of unhelpful rhetorical devices which reproduce fractured and polarised debates. These fault lines, as Jackie Cock calls them, fragment the environmental movement into those perceived as "narrow conservation movement focused on the protection of wild places, plants and animals, and the environmental justice movement, which is organizing around concrete issues in the everyday experience of poor people, especially their exposure to toxic pollution and lack of critical resources" (2018, p. 143). The dominance of scientific and economic epistemologies also creates awkward paradoxes which suggests both the possibility of a win–win scenario through linking marine protection with economic development, and simultaneously de-politicises the numerous trade-offs it sets up; marine conservation vs economic growth; commercial fisheries (jobs) vs small-scale fishers (livelihoods), marine conservation vs local restitution, and international capitalist interest vs. local economies. These repetitive dichotomies in governance forums serve to side-line the growing evidence that involving "local and indigenous communities in planning and decision-making processes enhanced management effectiveness and the achievement of socio-economic and conservation goals" of MPAs (Sowman & Sunde, 2018, p. 169). Rejecting a myopic vision of saving the planet through "greener", or in this case "bluer" economic development that boxes us into the same polarised positions, requires actively working with the relational social and ecological processes between plural ways of knowing the sea. What is required is a concerted push for "a clear recognition of the diversity of values associated with socio-natural well-being as this relates to the ocean" (Ntona & Schröder, 2020). Indeed, if we are to adequately address the scale and scope of climate change then we must open the conversation to diverse ontologies in the production of knowledge.

Beyond understandings of economic and scientific models, cultural, traditional and spiritual knowledge systems equally shape how people act on, and with, ocean life. They also powerfully influence people's responses to state regulations. It is unsurprising that contemporary research on environmentalism indicates "that social narratives are often far more important in leading people to accept or reject climate change than the underlying scientific evidence" (Marshall et al., 2016, p. 5). Diverse epistemologies are not just useful in the abstract notion of inclusivity, but as a form of action "to explore possibilities of pluralism in our responses and politics" (O'Reilly et al., 2020, p. 14). We stand with a growing number of scholars and movements calling for the acceptance of plurality in ocean governance (Bremer & Glavovic, 2013; Corrigan & Hay-Edie, 2013; Flannery et al., 2016; Vierros et al., 2020). As Ntona and Schröder's (2020) caution us,

> specifically, sectorially focused, growth-oriented MSP processes risk forgetting—or, worse, wilfully ignoring—that planning can also serve as an opportunity to develop a socially negotiated, non-economic understanding of oceanic relations, which takes into account the importance of local subsistence and health (physical, mental, spiritual and cultural), as well as local reliance upon the health and resilience of the natural environment.

It is this "socially negotiated" alternative that this chapter explores. It does so through a very local experiment in listening to plural epistemologies along the KwaZulu-Natal coastline of South Africa. The following section outlines how a local contestation around oil and gas prospecting led to this experiment in listening.

An Application to Prospect for Gas of the KwaZulu-Natal Coastline by ENI

On the 28 October 2018 the South Durban Community Environmental Alliance (SDCEA), an environmental justice organisation born from the long fight for clean air and against environmental racism in Durban, called a meeting with the Petroleum Agency of South Africa (PASA). PASA is the independent authority and regulatory body responsible for granting licences to prospect and mine for minerals and other natural resources on land and in the oceans. PASA is legislatively obligated to consider public needs when granting or denying such licences. Besides the representatives from PASA, the meeting consisted of representatives from small towns and communities living up and down the coastline of KwaZulu-Natal. In addition, members of larger environmental NGOs and activist groups were also in attendance. The meeting was called to make it known to the authorities that many people living along the KZN coastline were opposed to the licence application by SASOL[6] and ENI, an international oil & gas company, who wanted to prospect for gas. One of the authors of this chapter, Kira Erwin, was asked by SDCEA to facilitate this meeting. With people representing locations across 600kms of the KZN coast in the hall that day discussions and debates were robust, and translated to and from English to isiZulu to ensure that the discussions were understood by all.

The meeting in many ways represents South Africa's living democracy, where it is not just the government who calls and runs stakeholder consultations on the ocean, but frequently civil society and communities who demand in-person engagements around decision making. People in the room that day came from very different class positions, as well as identified with different racial and cultural identities, and shared different histories in relation to the coast, which shaped their statements of concern and opposition. Whilst some people spoke of scientific concerns for ocean health, possible habitat destruction, and climate change, others spoke of concerns around their livelihoods, risks to their cultural heritage and of traditional knowledge observations around ocean pollution.

Whilst local coastal representatives had a platform to voice their concerns at this gathering there are more than one way for those in power to silence voices other than denying an audience As Arundhati Roy reminds us "there's really no such thing as the 'voiceless'. There are only the deliberately silenced, or the preferably unheard" (2004). One method in which to "unhear" people's concerns is to reduce diverse voices into an artificially constructed homogenous "community". Whilst the PASA officials were sympathetic to some arguments presented, they frequently spoke about "the community" as either "for or against" development, with very little nuanced understanding or acknowledgement of the diverse perspectives, knowledge and concerns vocalised in the room. In doing so the authority officials entrenched a false binary in which they simply had to legally adjudicate between "the community" and the international corporation seeking to mine the seabed. In South Africa, a country with very high unemployment levels and slow economic growth (South African Budget Review, 2021), this binary inherently favours big business' allure of jobs and GDP growth. Here an international capitalist ethos that courts a national growth agenda trumps a myriad of local concerns. What was lost in this insistence to see opposition to sea-bed mining as "one voice", was the complexity of diverse demands for social, economic and environmental justice. It was the plurality of knowledge and experiences in the room that was the real challenge to PASA, a challenge that was "preferably unheard".

If South Africa is to live up to its policy aim of creating an Ocean Economy that contributes to the implementation of the National Development Plan (NDP 2030) through job creation, poverty alleviation and social equity (Findlay, 2018), then we must find alternative methods for listening to plural knowledge systems of the ocean. As will be argued below, different knowledge systems, science and tradition amongst them, do not have to be viewed as incommensurable for the well-being of the oceans. One of the challenges for inclusive ocean governance in South Africa and elsewhere, is not simply a perfunctory opening up of stakeholder forums, but an active approach to creating more just listening and dialogue methods that are comfortable with a plurality of meanings and relationships with the ocean. Methods that make us all more willing to listen to a chorus of voices, where different ways of knowing the sea may hold equal legitimacy within social negotiations on governance. The following section outlines one such local experiment in listening, *Lalela uLwandle*, that was designed in direct response to the above dilemma and the contestation expressed in the meeting with PASA.

The Making and Performance of Lalela uLwandle

Lalela uLwandle was first and foremost a process set out to listen deeply to the stories of coastal people's relationships, histories and concerns for the ocean. We wanted to understand the intergenerational spiritual and historical relationships coastal citizens had with the Indian Ocean. The central questions at the heart of our research was: How does the ocean sustain and nourish KZN coastal dwellers? What are the points of conflict and intersection within and between the range of memories and mythologies, traditional folk (and fairy-tales), religious rites, idioms, scientific understandings, songs and rituals, economic frameworks, points of catharsis, recreation and superstition that the ocean invokes across the people who live along this coastline?

Narrative data that contributed to the script was created through two methods. The early stages of the research consisted of focus groups with marine educators at the local aquarium, subsistence fishers in Durban, and environmental justice partners (notably SDCEA and groundWork). During this stage we identified key texts that offered rich historic readings of people's experiences along this coast line, such as Viroshen Chetty and Neelan Govender's excellent book *The Legends of the Tide*, which records generations of stories around the South Indian Seine Net Fisher folk and industry in Durban, and the PH.D. of Dr. Philile Mbatha exploring the forced removals of coastal KZN communities during apartheid and the impact of coastal mining on traditional healing practices (Mbatha, 2018). In addition, renowned KZN storyteller Gcina Mhlope gave us permission to use and incorporate her beautiful story *Nolwandle: Girl of the Waves* in the play. Gcina's unique South African folk story was useful in establishing a narrative that would open and conclude the play, introducing and echoing so many aspects of our research (loss, learning, medicine, healing) in a magical and profound way. These texts, along with the narrative data, strongly influenced character development in the script. The second stage of the research included interviews with 4 marine scientists (2 social scientists and 2 natural scientists), 2 traditional healers, an environmental activist, a Zulu historian, a lifeguard, an environmental lawyer, a former National government Minister, and a practising Zionist.[7] Whilst the methods in collecting these narratives differ, both the focus groups and individual interviews began with the opening invitation to participants to tell us "your earliest memories of the sea?"

Through these explorations the Empatheatre team worked iteratively to shape the data into a first draft of a theatrical script. In doing so we had to remind ourselves how easily research, when scripted into a play, can feel didactic. We worked to avoid reducing characters to mere mouth pieces conveying critical information. Each character, we decided, needed to be a consummate storyteller, a narrator of their own lives and realities, as well as being able to embody and speak to the many pressing questions and concerns around the sea that had emerged in the research. We have learned over the course of multiple *Empatheatre* processes that the stories of each character needs to have strong story arcs and emotional beats. Admittedly these arcs are often not overtly evident in the research narratives and it's the playwright's delicate responsibility to unearth and shape this whilst remaining connected to the original data. In the end, after much collective discussion, we settled on the development of three characters, each one emerging from different cultural backgrounds and historical contexts existing along the KZN coastline. The characters were: Niren[8] whose family has origins in the Indian indentured labourers brought to South Africa under colonial rule. His stories resonate with the many voices of the KZN fisher-folk as well as environmental activists who have fought years of environmental racism under apartheid and still today; Nolwandle,[9] who plays a marine educator and shares the stories of traditional and religious beliefs of many isiZulu speaking people, as well as the history of forced removals of black communities through the stories of her mother and grandmother; and lastly Faye,[10] a retired marine scientist living in a South Coast town whose science drives her activism and has stories of white privilege that she is confronting in her retirement.

Often Empatheatre processes rely on the use of composite characters, where multiple voices and many strands of research are woven into a single character. Niren, Faye and Nolwandle were all created using this device. This script writing process works in tandem with the research process, for example as the lead playwright Neil Coppen wove these narratives together, the research teams supported this process through fact and member checking with participants, as well as heading back into the field to conduct further interviews if parts of the storyline felt unresolved. The deeper we submerged ourselves in these worlds, the more the story and the incredible and often unexpected parallels between characters revealed itself. We soon discovered that our characters, as disparate as they may have seemed, were connected by the ocean in various profound and unexpected ways, particularly in their devotion to rituals, rites of various kinds and shared respect of the ocean.[11] After a final table reading with key civil society partners who were part of challenging the ENI application described earlier in this chapter, *Lalela uLwandle* went on a week's tour along the KwaZulu-Natal Coast line and ran for 1 week in the city of Durban at the end of 2019. Lalela uLwandle was also in 2020 performed in Port Elizabeth and in Makhanda at Rhodes University.

In total, the play has been performed 21 times, in 8 different towns and cities in KwaZulu Natal and the Eastern Cape, for over 900 audience members. There have been 21 post show discussions and more than 600 feedback forms have been filled out and analysed. Audience members were strategically invited; they were diverse in terms of race and class, and included small scale fishers, mining affected communities, conservation officials, scientists, planners, tourism operators and interested members of the public. The strategic inviting of audience members is a critical aspect of the methodology, playing an important role in terms of the social learning and solidarity potential of the process. Social learning theory reveals that the more diverse an audience the deeper the learning (Wals et al., 2009). Just as critical is the way in which the "post-show discussion" is facilitated to offer the audience a space to share and reflect after seeing the performance. In these discussions, the lead and co-facilitators are chosen from amongst our core team, depending upon the audience present—to ensure that, for example, the discussion is carried out in the language spoken by the majority of the audience, with translation support from co-facilitators on the team. This co-facilitation is very useful if audience members ask questions that another researcher is best suited to address, and makes the space more democratic by inviting other Empatheatre practitioners to share their views. We were often fortunate to have civil society members in the audience who could field answers and to whom we were able to direct the public interested in their work. Audience members themselves can also sometimes present important counter points to their fellow audience members. In this sense the facilitator should "hold the space" rather than too closely direct or control the discussion. At the end of this post-show discussion audiences members could also complete a feedback form on their views of the show and thoughts for ocean governance.

The post-show discussions and feedback forms support collaborative analysis of research findings and data with diverse publics. The audience are involved in identifying implications of the research in relation to their lived experience. The post show discussion can at times open up a tribunal space for audience members to share testimonies of their own experiences in response to those they have seen performed. This opens up the possibilities for the development of new alliances for civic action. In the case of *Lalela Ulwandle,* the KZN tour of the play in October 2019 coincided precisely with the news that the application from SASOL and ENI for licences to undertake exploratory drilling for oil and gas off of the KZN coastline (as per the PASA meeting referenced earlier in this chapter) had been approved by the Department of Mineral Resources, and that there were 30 days to appeal this decision. From the road, whilst touring the play, members of our team were involved in contributing to this appeals process, by drawing in the "expert testimonies" of co-researchers from a range of relevant disciplines; as well as the testimonies of audience members from the *Lalela Ulwandle* audience feedback forms. We worked closely with lawyers assisting with the appeal to gather and compile this input into a substantial letter of appeal; we shared up to date information related to the drilling licences and appeal process with audience members when the issue came up in post-show discussions; and were able to direct audience members who wished to take action towards SDCEA's petition and other organised appeal processes. This appeal process is ongoing at the time of writing, with a collective civil society effort to take the government department to court over the issue. Although not addressed in this chapter the relationships and collective actions catalysed by the *Lalela Ulwandle* process have seeded the development of a growing Coastal Justice Network, responding to the social and environmental justice impacts of the Blue Economy at a national scale.

A Chorus of Voices in Public Story

As we experienced through the *Lalela uLwandle* process storytelling is a powerful way to engage people in a conversation about ways of knowing the ocean and their impact on the world. Sharing research through public storytelling rather than only through publications, research reports and conferences can create a less didactic space for listening and engagement on contentious issues (Erwin, 2020). The stories performed, as well as those shared afterwards by audience members, enabled listening that made allowance for ambiguity, complexity and plurality. This is different from presenting arguments in academic conferences or stakeholder consultations where both the speaker and the audience are already primed to take a stand for or against a position in a critical debate (Polletta et al., 2011, p. 112). As one audience member, himself a scientist, noted "This is a very refreshing take on science and advocacy which is amazing" (DbnP2, 15/10/2019). Whilst critical debate was welcomed and encouraged in the post-show discussion, in our experience it was frequently done in a manner that afforded respect to divergent views. As an audience member in Hluhluwe succinctly wrote "traditions are important even if you don't believe in them you must respect them and the sea" (HluhluweP26, 10/10/2019). In the feedback forms it was interesting to note how appreciative the audience were to be learning about multiple cultural views of the sea, and to realise they shared mutual respect for the ocean, even if the meanings and practices differed. Frequently audience members who identify as Zulu shared in their feedback that they were pleased to watch and learn about these traditional practices. In Zulu cosmology ancestors live under the waters of the deep sea, making the ocean a sacred realm (Hofmeyr, 2020). Feedback comments, such as the ones below, captured feelings of "being recognized":

> *It inspired me because I was not expecting to see something like this but at least it shows that we do have people who are for us and our heritage.* (RichardsBayP10, 12/10/2019)

> *It made me feel excited because now I know history about our ancestors and I also know a lot about the ocean.* (MbazwaneP10, 11/10/2019)

For audience members who follow ancestral beliefs it was affirming that the play presented this knowledge system as equally important to that of the scientist's, or activists'. When responding to how the play made them feel one audience member wrote "moved, often times when speaking of the ocean's conservation it seems to only concern ocean life, but this play profoundly displays how much heritage and people's connection/stories reside within the sea—it affects people just as much" (DurbanDP25, 19/10/2019). Acknowledging the link between heritage, respect and the ocean was a common theme in the feedback data, people wrote explicitly that "they must not put oil in the oceans because it kills the heritage that we believe in" (HluhluweP31, 10/10/19), and that "we must keep it clean so that it will be able to keep the heritage which is what helps us connect with our ancestors" (HluhluweP43, 10/10/19). Audience members who may not have practised ancestral beliefs made a similar link, for example, "I learnt that the sea is important, we must not destroy it because we are harming the lives of the people who believe in it" (HluhluweP26, 10/10/19).

The feedback data and post-show discussions confirmed that "people's emotional ties to nature and their cultural values may offer a stronger incentive for conservation than economic arguments based on the livelihood values of nature, which are generally insufficient to motivate collective action" (Cocks et al., 2012, p. 7). Methodologies of storytelling that make visible emotional connections to nature hold value for marine science education. For example, at the end of a performance with an audience of marine scientists and educators at the local aquarium one of the senior staff said that *Lalela uLwandle* had set them a challenge. She called on her team of scientists and educators to recognise the power of storytelling and imagine ways they could use this in their educational work at the aquarium.

Adding additional voices to the chorus woven into the script, audience members frequently showed pleasure in sharing their own or other peoples' personal connections with the ocean. Talking about our relationship with nature is not a frequent experience for many people. Similarly to Cocks et al. research related to the forest in the Eastern Cape, we too found that both participants and audience members frequently "expressed their pleasure and gratitude for having been given the opportunity to do so" (2012, p. 4). Audience members spoke about the symbolic importance the oceans hold for them and their families, often referring to memories of parents who taught them about the sea, or childhood moments of wonder that have stayed with them into adulthood. At the first performance of the tour an audience member who grew up on an island outside of South Africa wrote down teaching from her mother. She told us,

> I was brought up with respect for the ocean. My mother would say when you enter the sea you are a guest of the fish, be a good guest. (Port ShepstoneP19, 7/10/2019)

This we thought illustrated rather beautifully the theme of respect for ocean life, and in the subsequent performances this maternal proverb was retold by the facilitators as a ritual closing to the post-show discussions. In this way the audience members' symbolic meanings of the oceans started to be shared across performances, and we are very grateful to the participant who shared her mother's wisdom.

The Politics of Storytelling and Working with Translation Protocols

Acknowledging how people value social identities as well as move outside of them through empathetic responses and relational experiences required the research team to be cautious of their own assumptions on how race and culture might be read by the audience. It is important that the storylines in the script reflect research data, but they should not work to confine the audience's readings into expected cultural stereotypes or confirm discriminatory ideologies. In crafting the research narratives into composite characters for the play, it was critical to think through the politics involved in working across different epistemologies in South Africa. Historically and presently, South Africa is deeply divided by racial injustices and economic inequalities. As outlined earlier, experiences of conservation, and relationships of governance around environmental resources, are profoundly shaped by the social constructs of race and class, and the material experiences they engineer at a quotidian scale. Rather than skirt over these inequalities and injustices *Lalela uLwandle* performed these stories of power and injustice. The script explored intergenerational histories illustrating how dominant ideologies and capitalist logics work to reproduce these fractures. It is important to acknowledge the harm caused by racism and class dispossession in our societies. Yet performances of histories and everyday experiences which are so closely mediated and shaped by ideas of race and difference in our country present its own challenge. The gross essentialization of race and other forms of difference in South Africa has engineered another obstacle to working with plural epistemologies for transformative governance. Colonisation and apartheid in South Africa have resulted in the "implicit conflation of identity and knowing", where "the world is constructed around solidarities of knowing that feed into and fuel ethno-racial stereotypes" (Soudien, 2013, p. 151). Telling stories in which epistemologies are presented as rigidly belonging to only one "kind of people" can dangerously entrench essentialist thinking on race, ethnicity and culture. This in turn enables the existing mechanisms of power to repurpose new arguments for ocean well-being into older epistemological hierarchies. Hierarchies that favour science over tradition, and neoliberal economic development over culturally enmeshed livelihoods. Confining ways of knowing into the politics of identity, "who is allowed to believe what", whilst simultaneously presenting a few epistemologies such as science as above ideology, enables hegemonic epistemologies to deny "complicity in the politics of inclusion and exclusion" (Soudien, 2013, p. 151). The *Lalela uLwandle* script then needed to write against essentialist ideas of race, culture and belonging, and against the essentialisation of ocean epistemologies.

There is a growing awareness of and attempts to include indigenous and cultural knowledge on natural resource management into international institutional frameworks on climate change (O'Reilly et al., 2020, p. 17) and ocean governance (Parsons & Taylor, 2021). Included in this is recognition that "long term observations of complex systems" to better understand climate change is not the exclusive domain of science (O'Reilly et al., 2020). Integrating diverse knowledge systems into institutions shaped by hegemonic epistemologies is not an easy process. In South Africa indigenous knowledge is recognised on paper, and in 2019 additional laws were passed to legislate the rights of indigenous peoples. Yet including these knowledge-holders and knowledge systems into environmental decision-making and governance arrangements has proved to be a slow process (Boswell & Thornton, 2021). Ntona and Schröder state that traditional, indigenous and local knowledge, must also be incorporated into planning processes in ways that fully respect their ontological and epistemological underpinnings. Crucially, for this incorporation not to constitute appropriation, information cannot simply be 'extracted' from indigenous, traditional and local knowledge systems and 'transplanted' into decision making. Rather, knowledge-holders themselves must be integrated into the planning process, and their worldviews recognised and meaningfully engaged with. (2020)

Agreeing with the above as a critical starting point we argue, through the learnings from Lalela uLwandle, that more is needed than simply inviting diverse knowledge-holders to the negotiation table. If ways of knowing remain essentialized as belonging to specific "race groups" or "those who are educated", or "those who truly understand the issues at stake" then invitations to diverse knowledge-holders may serve as a checkbox exercise for participatory governance. This leaves untouched the power relations between and within these knowledge systems, and therefore epistemological hierarchies that currently influence policy development are unlikely to transform. For Schutter et al. stakeholder consultation forums "can actually serve to produce apparent consensus that keeps dissent at bay when it only allows for disagreement on specific technology and management choices, not challenging the expansion of a capitalist socio-economic order" (2021, p. 6).

Attentive to reproducing these dangers it was insufficient for *Lalela uLwandle* to simply position different ways of knowing alongside each other (e.g. a character who represents only marine science views, or religion and ancestral beliefs, or culturally entangled livelihoods). Especially since in South Africa audience members might easily read these as mapping onto apartheid constructed racial categories and serve to confirm existing essentialized ideas about people.[12] Performing everyday stories unavoidably means working with some of these familiar frames. Across the performances we have heard 4 voices out of an estimated 900 audience members who verbalised or gave written responses that indicated their desire to read the play through these stereotypes. Three of these four voices did so in a defensive response to believing themselves blamed as white people for the problems today. Far more frequently than these few defensive responses were deeper reflective feedback around race and privilege from the audience. An audience member in Richards Bay shared that she "felt ashamed of the injustices committed by whites in the name of 'conservation'", but that the play "ended on a hopeful notes that there is space for us all" and that "the character's stories are told with conviction that each character becomes a well-loved friend whose story touches your heart" (Richards Bay, 17 12/10/2019). Whilst this play did not aim to focus explicitly on issues of white privilege and conservation, or race and power in conservation, it is worth noting that it provided a productive challenge to these dominant discourses. For example, an audience member a week after seeing the play told one of the authors that the stories had stayed with her for days, and that it really made her think about the damage that apartheid had done to black people in ways she had not thought of before, and how important it was that white people like herself acknowledge this and find a way to work towards a more equal society (personal com. October 2019).

Theatre can move an audience member through a range of emotions. We learnt, however, that to catalyse an alternative conversation *Lalela uLwandle* had to actively work to build translation protocols which opened up the possibilities of freeing ways of knowing from the narrow confines of essentialised identity and belonging. This was done both through the scripting and through creative theatre making in order to weave harmonies and dissonance into a chorus of voices not just between the characters, but within them. Creating didactic essentialised characters that play into expected tropes of environmental conservation within a local context will do little to shift the current power dynamics of ocean governance. Rather characters should embody "agonistic relationships" within their own story arcs and in relation to the stories of the other characters. Agonistic relationships do not look to dismiss or reject difference "but acknowledge and recognise it *as* different whilst still looking for promising, if partial, synergies to serve as the basis for solidaristic relationships that are forged through antiessentialist, relational, and always incomplete identities" (Routledge & Derickson, 2015, p. 392). *Lalela uLwandle* was both a performance on what is lost when knowledge systems are conceived as incommensurable to serve power, and importantly what might be gained when we embody more fluid and plural epistemologies that can mould, expand, broaden and enrich our ways of knowing the ocean. The section below explores some of these weavings in the script and the audiences' responses to these.

Agonistic Performances

All the characters in the play work with plural ways of knowing the ocean. Niren's university shaped environmentalism is woven together with his use of ritual to connect to his grandfather and father and their teachings around fishing. Niren's struggle is not a choice between epistemologies but making peace with choosing a different livelihood from the long heritage of fishing in his family. Faye too performs a blend of marine science and her awe and wonder of the sea, and increasingly with age acknowledges how dreams of her deceased husband shape her environmental activism, in ways similar to how Nolwandle's grandmother worked with dreams to interpret the healing power of marine resources. But perhaps the most powerful performance against the incommensurability of knowledge is embodied in the intergenerational character of Nolwandle. Nolwandle embodies shifting intergenerational knowledge systems that span from her grandmother's traditional ancestral belief to her mother's more contemporary religion, to her own work in scientific education at the local Aquarium. She performs for the audience the strength in working across plural epistemologies. For Nolwandle her grandmother's practices as a *Sangoma* using ocean resources as medicine has something in common with her mother's choice of spiritual belief, Zionism, that uses the "heavenly healing powers of the sea water" for baptisms (Coppen et al., 2019). Throughout the play Nolwandle uses love for the ocean to weave together her family's heritage. Respect for the ocean and its healing power is a direct continuum from her ancestors that continues to flow through her. She tells us,

> Working at the aquarium as a story-teller and educator I am able, like Nolwandle, my mother and grandmother before me, to use the ocean to heal, but it's all of our responsibilities to work to heal her in return.....Every day, at the aquarium I am able to allow children a chance to peek beneath the mysterious blue blanket of the sea and teach them of the consequences of our destructive actions. *(Beat. She chuckles knowingly)* You know what I always tell these children… This heaven they all want a one-way ticket to… it's not in the direction the missionaries promised us it was. No it's in the opposite direction…. it's down there…it's beneath the waves. (Coppen et al., 2019)

As Nolwandle shows, science and spirituality may resonate in order to work towards a healthy ocean for all. In response to the performance of commensurability in the play comments such as those below were very common:

> The way they were all concerned about the same thing, but yet they had different cultures, lives, backgrounds…was really amazing. (HluhluweP10, 10/10/2019)

> It made me realize that even if we have different traditions and cultures the ocean is of significant important to all us. (DbnDayP4, 19/10/2019)

> Respect the oceans and honour other peoples' cultures. (MbazwanaP16, 11/10/2019)

Imagining across divergent cultural and epistemic concepts of the sea reveals "the often hidden economic and political components of cultural representations and more overt patterns of signification in society's constructions of nature" (Jackson, 1995, p. 88). This opened up the possibilities of stepping outside the usual polarised tensions in environmentalism. The agonistic stories scripted into *Lalela uLwandle* offered translation protocols for interpreting exactly how and why environmental and conservation management impacts negatively on the cultural, social and economic well-being of individuals and groups. For many conservation officers, marine scientists and members of the public being a part of a performance that placed "biological conservation efforts into social and historical context" enabled an imagination for how the "articulation of community values into conservation plans…can facilitate local acceptance and participation in management" (Crandall et al., 2018, pp. 9, 11–12). Equally including performed histories of exclusions was a learning experience in acceptance for conservation officials on how current practices and discourses within their field act to exclude. All four authors, as well as the actors, were approached after performances by marine scientists, conservation field guides and environmental impact assessors respectively on how important it was to integrate cultural and living heritage into their fields, and that the play had impacted on how they imagined their disciplinary work going forward. For example, a young trainee conservation field guide in the Hluhluwe area[13] approached one of the authors after the post-show discussions. As part of her field-guide training she was being taught about the medicinal properties of plants in a particular protected area. This she noted was all based on the ecological knowledge of local Zulu communities who were now largely excluded from this government gazetted protected wetland forest. She shared that the irony and unfairness of this had hit her whilst she was watching the performance.

Post-show discussions across venues illustrated the audience's willingness to acknowledge inequalities and injustices, and find connections, synergies and solidarities to imagine mobilizations across differences. Attempting to create alternative listening and learning forums for ocean governance must work with care and attentiveness to the geographies and histories of power shaping local contexts. If done in ways that reject the essentialization of people and knowledge systems then alternative governance forums may rework these "artificial boundaries" that serve existing hierarchies. Acknowledging that we live in plural worlds does not preclude collective action, rather it "constitutes the bedrock of our working together in solidarity, the possibility, through the partial identification of common ground, of a 'performative unity'" (Routledge & Derickson, 2015, p. 392). Working across scales and within the messiness of social structures and relationships required for environmental and ecological justice, demands of us all that we become more literate on plurality for equitable collective action. Creative methods such as *Lalela uLwandle* that perform agonistic possibilities work at an affective level to enact new alliances and situated solidarities (Erwin, 2020; Mouffe, 2013, pp. 96–97). *Lalela uLwandle* offered a productive form of accepting the horror of apartheid in order to work towards dismantling the polarisation and essentialism it has engineered. This enabled recognising how the past and present shape our unequal society in relation to environmental injustices, and the types of respect we need to build to live together in less fragmented and environmentally damaging ways. Whilst affective processes are complex to evaluate, in other words it is difficult for us to quantify who took up the invitation for an alternative imagination, we believe they are a very necessary part of unravelling the harm that humans do to each other and the environment at an ideological level. Working with affective processes that connect and build openings across epistemologies that have been historically, through colonialism, apartheid and capitalism, engineered as incommensurate reminds us that the "script for the future" is not already set (Soudien, 2013, p. 154). There is much to be said on what a critical politics of plurality, working with and beyond hurt and pain, offers for transformative ocean governance.

Conclusion

As the South African government moves quickly ahead with its Blue Economy planning, policy and governance frameworks it remains unclear how it is addressing existing socio-economic inequalities and exclusions, or whether it is just entrenching them. There are, we know, many others who sit across and within the governance fault lines outlined in this chapter who are equally distressed by the relationship between socio-economic inequalities and ecological injustices in South Africa (Cock, 2018, p. 143). We hope this article contributes towards a shifting tide of imagining what more ocean governance might do for environmental and ecological justice, beyond creating policy frameworks whose regulations serve the status quo. How might South Africa engage in an alternative process that works to mitigate against the usual power dynamics in policy forums and governance arrangements? Flannery and McAteer write that in order to ensure progressive Marine Spatial Planning it is important to develop "a political frontier early on" in the process where "pathways for progressive socio-environmental outcomes have been established" (2020). The many voices of dissent and cries of exclusion from coastal people, particularly amongst the most marginalised, suggest that such a "political frontier" has not yet been established in the South African rush for the oceans.

Creating such a frontier for ocean governance requires inclusive forums in which a chorus of voices on how and why the ocean holds social, economic, cultural, spiritual, scientific and ecological importance are heard. However, as this chapter argues, epistemology hierarchies that favour science and economics within decision-making forums produce additional obstacles to creating a political frontier that addresses the wellbeing of the ocean and the most marginalised in society. When ways of knowing are seen as incommensurable, or knowledge frameworks are tied to essentialised identity constructs like race and culture, even an inclusive stakeholder forum may not automatically equate to transformative governance. Shifts in not just who is invited to negotiation tables but the form and the substance of decision making is also required. To enable meaningful and effective participation of diverse knowledge holders it matters who is in the room, but it also matters how the chairs are placed in the room; what language is spoken; whose histories are visible; whose knowledge is listened to, whether issues of power are able to be discussed up front, and whether efforts are made towards agonistic debates through a rejection of essentializing identity and epistemologies.

Empatheatre, through the case study of *Lalela Ulwandle*, offers some insights into how creative methods and public storytelling offer heuristic process for translation between and across ways of knowing the sea. Building on the work already done in calling for plurality in ocean governance, it offers an orientation to empathy to work with plural epistemologies rather than a definitive methodology for transformative governance. It is hoped that sharing our experiences with this particular experimental project helps grow an appetite for action to explore alternative governance methods that unsettles practices and structures of neoliberal capitalism currently dominating our policy and governance regime.

Indeed we would argue we have nothing to lose given the current inequalities within South Africa, and everything to gain by undertaking processes that move out of siloed mentalities by making room for ways of knowing the ocean beyond the confines of the logic of science and capitalism. This makes pragmatic sense for protecting our oceans given the growing evidence that "social factors, rather than physical or ecological factors, ultimately determine the success (or otherwise) of MPAs" (Sowman & Sunde, 2018, p. 169). It also however creates exciting possibilities for connecting and building strategic solidarity across epistemological divides to address pressing issues in climate change. The creative storytelling of *Lalela uLwandle* enabled a shared symbolic language, reference points and aesthetic experiences that offered a strong foundation from which we as researchers and community members can mount a well-informed resistance (and offer alternatives) to ocean harm, and importantly begin to imagine what an ocean governance that works towards planetary well-being looks like.

Notes

1. https://www.empatheatre.com/.
2. The 1982 United Nations Convention of the Law of the Sea enables sovereign states special rights within an Exclusive Economic Zone, that extends 200 nautical miles out to sea from the state's coastland, in relations to the exploration and use of marine resources.
3. Extended Continental Shelf (ECS) Area are portions of the continental shelf beyond the 200 nautical miles from the coast, where states can apply to have sovereign rights to the seabed and subsoil. The extent of a nation's ECS is often undefined or contested.
4. A term used here to include all people who were persecuted and oppressed under the racist apartheid regimes, including people who, under apartheid may still identify as belonging to the constructed racial categories of Black African, Indian and Coloured.
5. NEMPA (National Environmental Management: Protected Areas Act) and the WHCA (World Heritage Convention Act) require 'co-management' (not just consultation with communities), and 'sensitivity to people and their needs' (Physical, psychological, developmental, cultural and social)' respectively.
6. A South African energy and chemical company.
7. The Zionist Churches in Kwa-Zulu Natal combine, to various degrees, traditional isiZulu cosmology with Christianity.
8. Played by Rory Booth.
9. Played by Mpume Mthombeni.
10. Played by Alison Cassells.
11. The podcast of the play Lalela uLwandle can be listened to online here: https://www.empatheatre.com/listen-to-our-lalela-ulwandle-radio-play.
12. See Boswell and Thornton on how "economistic perspectives of ocean management and stereotyping of the Khoisan in South Africa risk producing an exclusionary Blue Economy" (2021, p. 2).
13. An area on the North Coast of KwaZulu-Natal with a long history of land dispossession and conflict surrounding conservation regulations in relation to the World Heritage Site of the iSimangaliso Wetland Park (Hansen, 2013).

References

Bähr, U. (2017). *Ocean Atlas: Facts and figures on the threats to our marine ecosystems.* Heinrich Böll Foundation Schleswig-Holstein, the Heinrich Böll Foundation (national foundation), and the University of Kiel's Future Ocean Cluster of Excellence.

Bennett, N. J., Govan, H., & Satterfield, T. (2015). Ocean grabbing. *Marine Policy, 57*, 61–68.

Bennett, N. J., Blythe, J., White, C. S., & Campero, C. (2021). Blue growth and blue justice: Ten risks and solutions for the ocean economy. *Marine Policy, 125*, 104387.

Bond, P. (2019). Blue Economy threats, contradictions and resistances seen from South Africa. *Journal of Political Ecology, 26*(1), 341–362.

Boswell, R., & Thornton, J. L. (2021). Including the Khoisan for a more inclusive Blue Economy in South Africa. *Journal of the Indian Ocean Region.* https://doi.org/1080/19480881.2021.1935523

Bremer, S., & Glavovic, B. (2013). Mobilizing knowledge for coastal governance: Re-framing the science–policy interface for integrated coastal management. *Coastal Management, 41*(1), 39–56.

Cock, J. (2018). *Writing the ancestral river: A biography of the Kowie.* Wits University Press.

Cocks, M. L., Dold, T., & Vetter, S. (2012). 'God is my forest'—Xhosa cultural values provide untapped opportunities for conservation. *South African Journal of Science, 108*(5), 1–8.

Coppen, N. (2019). *Into Ulwembu: Exploring collaborative methodologies in a research-based theatre production on street-level drug use in KwaZulu-Natal, South Africa.* Master thesis, University of KwaZulu-Natal.

Coppen, N., Walne, H., Mahlope, G., Mthombeni, M., McGarry, D., Pereira, T., & Erwin, K. (2019). *Lalela uLwandle.* Play script.

Corrigan, C., & Hay-Edie, T. (2013). *A toolkit to support conservation by indigenous peoples and local communities: Building capacity and sharing knowledge for indigenous peoples and community conserved territories and areas (iccas).* UNEP-WCMC.

Crandall, S. G., Ohayon, J. L., de Wit, L. A., Hammond, J. E., Melanson, K. L., Moritsch, M. M., & Parker, I. M. (2018). Best practices: Social research methods to inform biological conservation. *Australasian Journal of Environmental Management, 25*(1), 6–23.

De Wet, E., & Du Plessis, A. (2010). The meaning of certain substantive obligations distilled from international human rights instruments for constitutional environmental rights in South Africa. *African Human Rights Law Journal, 10*(2), 345–376.

Erwin, K. (2020). Storytelling as a political act: Towards a politics of complexity and counter-hegemonic narratives. *Critical African Studies, 13*, 1–16.

Findlay, K. (2018). Operation Phakisa and unlocking South Africa's ocean economy. *Journal of the Indian Ocean Region, 14*(2), 248–254.

Flannery, W., & McAteer, B. (2020). Assessing marine spatial planning governmentality. *Maritime Studies, 19*, 269–284.

Flannery, W., Ellis, G., Nursey-Bray, M., van Tatenhove, J. P., Kelly, C., Coffen-Smout, S., Fairgrieve, R., Knol, M., Jentoft, S., Bacon, D., & O'Hagan, A. M. (2016). Exploring the winners and losers of marine environmental governance/Marine spatial planning: Cui bono?/"More than fishy business": Epistemology, integration and conflict in marine spatial planning/Marine spatial planning: Power and scaping/Surely not all planning is evil?/Marine spatial planning: A Canadian perspective/Maritime spatial planning–"ad utilitatem omnium"/Marine spatial planning: "It is better to be on the train than being hit by it"/Reflections from the perspective of recreational anglers and boats for hire/Maritime spatial planning and marine renewable energy. *Planning Theory & Practice, 17*(1), 121–151.

Franke, A., Blenckner, T., Duarte, C. M., Ott, K., Fleming, L. E., Antia, A., Reusch, T. B., Bertram, C., Hein, J., Kronfeld-Goharani, U., & Dierking, J. (2020). Operationalizing ocean health: Toward integrated research on ocean health and recovery to achieve ocean sustainability. *One Earth, 2*(6), 557–565.

Gallois, S., & Duda, R. (2016). Beyond productivity: The socio-cultural role of fishing among the Baka of southeastern Cameroon. *Revue d'ethnoécologie* (10).

Hansen, M. (2013). New geographies of conservation and globalisation: The spatiality of development for conservation in the iSimangaliso Wetland Park, South Africa. *Journal of Contemporary African Studies, 31*(3), 481–502. https://doi.org/10.1080/02589001.2013.807566

Hoegh-Guldberg, O., Northrop, E., & Lubchenco, J. (2019). The ocean is key to achieving climate and societal goals. *Science, 365*(6460), 1372–1374.

Hofmeyr, I. (2020). Imperialism above and below the water line: Making space up (and down) in a Colonial Port City. *Interventions, 22*(8), 1032–1044.

Infield, M. (2001). Cultural values: A forgotten strategy for building community support for protected areas in Africa. *Conservation Biology, 15*(3), 800–802.

Jackson, S. E. (1995). The water is not empty: Cross-cultural issues in conceptualising sea space. *Australian Geographer, 26*(1), 87–96.

Kepe, T. (2009). Shaped by race: Why "race" still matters in the challenges facing biodiversity conservation in Africa. *Local Environment, 14*(9), 871–878.

Maggs, J. Q., Mann, B. Q., & Cowley, P. D. (2013). Contribution of a large no-take zone to the management of vulnerable reef fishes in the South-West Indian Ocean. *Fisheries Research, 144*, 38–47.

Marshall, G., Corner, A., Roberts, O., & Clarke, J. (2016). Faith & climate change: A guide to talking with the five major faiths. *Climate Outreach*. Retrieved April 29, 2022 from https://jliflc.com/resources/faith-climate-change-a-guide-to-talking-with-the-five-major-faiths/

Mbatha, N. (2018). *The influence of plural governance systems on rural coastal livelihoods: The case of Kosi Bay*. Ph.D. thesis, University of Cape Town. https://open.uct.ac.za/handle/11427/29768

Mouffe, C. (2013). *Agonistics: Thinking the world politically*. Verso Books.

Nadel-Klein, J. (2020). *Fishing for heritage: Modernity and loss along the Scottish coast*. Routledge.

Ntona, M., & Schröder, M. (2020). Regulating oceanic imaginaries: The legal construction of space, identities, relations and epistemological hierarchies within marine spatial planning. *Maritime Studies, 19*(3), 241–254.

OECD. (2016). *The ocean economy in 2030*. OECD.

O'Reilly, J., Isenhour, C., McElwee, P., & Orlove, B. (2020). Climate change: Expanding anthropological possibilities. *Annual Review of Anthropology, 49*, 13–29.

Parsons, M., & Taylor, L. (2021). The conversation: Why indigenous knowledge should be an essential part of how we govern the world's oceans. *Geography Bulletin, 53*(3), 15–16.

Polletta, F., Chen, P. C. B., Gardner, B. G., & Motes, A. (2011). The sociology of storytelling. *Annual Review of Sociology, 37*, 109–130.

Poloczanska, E., Mintenbeck, K., Portner, H. O., Roberts, D., & Levin, L. A. (2018). The IPCC special report on the ocean and cryosphere in a changing climate. In *2018 Ocean Sciences Meeting*. AGU.

Routledge, P., & Derickson, K. D. (2015). Situated solidarities and the practice of scholar-activism. *Environment and Planning D: Society and Space, 33*(3), 391–407.

Roy, A. (2004). *Peace and the new corporate liberation theology* [2004 City of Sydney Peace Prize Lecture]. Available at: Peace & The New Corporate Liberation Theology.

Rudolph, T. B., Ruckelshaus, M., Swilling, M., Allison, E. H., Österblom, H., Gelcich, S., & Mbatha, P. (2020). A transition to sustainable ocean governance. *Nature Communications, 11*(1), 1–14.

Schutter, M. S., Hicks, C. C., Phelps, J., & Waterton, C. (2021). The blue economy as a boundary object for hegemony across scales. *Marine Policy, 132*, 104673.

Soudien, C. (2013). Unscripted modernities: Critical questions in working with indigenous knowledges in a time of bounded cultural hegemony. *Education as Change, 17*(1), 149–157.

South African Budget Review. (2021). *National Treasury, Republic of South Africa*. http://www.treasury.gov.za/documents/national%20budget/2021/review/FullBR.pdf

Sowman, M., & Sunde, J. (2018). Social impacts of marine protected areas in South Africa on coastal fishing communities. *Ocean & Coastal Management, 157*, 168–179.

Sowman, M., & Sunde, J. (2021). A just transition? Navigating the process of policy implementation in small-scale fisheries in South Africa. *Marine Policy, 132*, 104683.

Sunde, J. (2014). *Marine protected areas and small-scale fisheries in South Africa: Promoting governance, participation, equity and benefit sharing*. International Collective in Support of Fishworkers.

Sunde, J., & Erwin, K. (2020). Cast out: The systematic exclusion of the KwaZulu Natal Subsistence Fishers from the fishing rights regime in South Africa Policy Research Report. *SDCEA Policy Research Report*.

Urquhart, J., & Acott, T. (2013). *People, place and fish: Exploring the social and cultural meanings of small-scale fisheries through photography and sense of place*.

Vierros, M. K., Harrison, A. L., Sloat, M. R., Crespo, G. O., Moore, J. W., Dunn, D. C., Ota, Y., Cisneros-Montemayor, A. M., Shillinger, G. L., Watson, T. K., & Govan, H. (2020). Considering Indigenous peoples and local communities in governance of the global ocean commons. *Marine Policy, 119*, 104039.

Wals, A. E., van der Hoeven, E. M., & Blanken, H. (2009). *The acoustics of social learning: Designing learning processes that contribute to a more sustainable world*. Wageningen Academic Publishers.

21

'Other' Social Consequences of Marine Protection in Tsitsikamma, South Africa

Jessica Leigh Thornton and Ryan Pillay

Introduction

Although Marine Protected Areas (MPAs) are favoured for the conservation of marine biodiversity and ultimately for securing marine-based livelihoods, social science research is foregrounding the relevance of including coastal communities in oceans management and the challenges that arise from insufficient consideration of human factors in marine protection (Pomeroy et al., 2006, p. 2). Globally, coastal communities are, as a result of climate change and ocean pollution, at increased risk of poverty and poor health. For now,

---

The research presented in this chapter is prepared under the One Ocean Hub, a collaborative research for sustainable development project funded by UK Research and Innovation (UKRI) through the Global Challenges Research Fund (GCRF) (Grant Ref: NE/S008950/1) and NRF UID Grant 129962.

J. L. Thornton
Sociology and Anthropology, Nelson Mandela University,
Port Elizabeth, South Africa

R. Pillay (✉)
Department of Arts, Culture and Heritage, Nelson Mandela University,
Port Elizabeth, South Africa
e-mail: Ryan.pillay@mandela.ac.za

the latter appears indirect, as communities are affected not by the actual pollution of the sea, but the fact that communities are denied access to the sea and coast.

The expansion and network of MPAs in South Africa are highly contested due to the historical injustices of apartheid and ongoing impacts associated with the establishment of marine protected areas (Sowman & Sunde, 2018, p. 170). Alongside this, there is an assumption that because many MPAs are in areas with few direct uses, there are few stakeholders interested in them (Gruby et al., 2017, p. 417). As a result, the social consequences of MPAs have been largely overlooked. As we show in this chapter, restricted access to the sea and coast and to areas now declared as MPAs, is traumatic for local communities. Accessing the sea provides community members with opportunities to improve well-being, community building and connection with the natural landscape. This chapter presents findings of recent research (October 2020) on the social outcomes of the Tsitsikamma MPA in South Africa. We draw on information gathered from in-depth ethnographic research, and oral histories to report on the social outcomes resulting in social hardships and inequities being experienced by the local community and their limited access to the sea; their footprints missing in the sand.

'A Place of Much Water': Tsitsikamma and the Marine Protected Area

Two things became clear during our first drive through the town of Tsitsikamma. First, the village is divided in half. The 'tourist' section encountered during a period of Covid-19 lockdown, appeared quiet, quaint, sleepy and tranquil, while the township housing mostly Coloured and African people seemed to remain (despite Covid-19 restrictions) bustling, sweltering, loud and lively. The first impression of Tsitsikamma is its greenness. It is a place of many forests where alien and indigenous vegetation jostle for space. The impression created is of an idyllic paradise, where there is perfect balance between humans and nature. However, as we moved to the township we saw less and less greenery. The noise of people conversing, the noisy work of small entrepreneurs and of children playing was audible. The children played on the streets and accommodations appeared cramped and not at all idyllic. We found that most people residing in the racially designated township are still poor, even though apartheid was dismantled more than 20 years ago. Those conversed with talked about the lack of food and other essential goods. Their clothing appeared well worn and some people appeared frail and ill, as if they had been denied basic health services. Inhabitants also spoke about land issues. They spoke about the loss of a communal identity largely as a result of an extensive and violent history of colonial invasion that led to the forced relocation and marginalisation of indigenous communities. A major signpost for us, of what was important in Tsitsikamma, was the SANParks sign. The presence of SANParks and the national park signboards prominently marks the area as one of conservation and preservation.

Our findings over the course of fieldwork, however, was that the history of Tsitsikamma is intertwined ecologically, culturally and politically. Nature deeply informs cultural heritage in Tsitsikamma, an issue we are still exploring and seeking to understand. Tsitsikamma is also publicly synonymous with tourism, a recreational and ecologically oriented tourism aligned to its mostly white gentrified suburbs. In brief, in Tsitsikamma we found the classed pursuit of ecological conservation (Boswell et al., 2021, p. 27). Our research also showed however, that the influence of First Nation People in this area, the Khoi-San, plays a significant role in the names and in the practices of both tangible and intangible culture and heritage of the ocean.

Despite its initial appearance as a place of forests, the word Tsitsikamma means 'place of much water'. It is a place named by the indigenous people and recognised in name by the national government. From our initial, outsider perspective, we felt that this bodes well for social inclusion and the integration of indigenous perspectives in ocean management. We soon found out however, that indigenous people have very little, or close to no access to the place of much water. The problem is that the Tsitsikamma MPA still being framed as an endeavour to benefit for ecological conservation and tourism. This means the exclusion of indigenous and local community members.

For this reason, we feel it is important to analyse the social consequences of MPAs. Secondly, while MPAs are beneficial in protecting marine wildlife, their social outcomes remain remarkably under-researched in terms of their magnitude, distribution and variance of social consequence (Fox et al., 2012, pp. 1–10; Jones et al., 2017, pp. 1–7). As policymakers seek to provide input into the governance of MPAs, questions regarding how this affects local stakeholders who depend on the oceans for their livelihood, recreation and overall well-being need to be answered (Sanchirico et al., 2002, p. 15). Presently, the MPA in Tsitsikamma determines human interactions with the sea and coast within the designated area. It indicates who may do what, where, when and how (Pomeroy et al., 2006, p. 7). Our key research question is as follows: what are the immediate and long-term consequences if a reasonable amount of the near-shore habitat is not set aside for communities to use for survival and flourishing? The research is already showing that the consequences thus far are dire. In attempting to provide insight into coastal lived experiences (especially as these relate to the MPA in Tsitsikamma) therefore, this chapter hopes to advance important and urgent dialogues regarding misalignments between community experiences and the goals of the MPA. Quality of life in Tsitsikamma depends on access to the ocean.

The Tsitsikamma MPA expands over 290 km^2 and is the oldest in Africa. For nearly sixty years the MPA has protected marine species, such as linefish species, which are heavily overexploited elsewhere (Marine Protected Areas, 2021). Managed by South African National Parks (SANParks), the MPA is a no-take area, except for 20% of the shoreline where local community members are able to fish if they hold a permit. The MPA is the marine equivalent of the Kruger National Park meaning it is a major tourist attraction. Thousands of local and international tourists visit the area each year, while generations of indigenous inhabitants born to the area do not have equal, let alone some access to the very areas that 'outsiders' have monied access to. By the time that we arrived to do field research in the area in October 2020, it was clear that the local communities are not really benefiting from the arrangement, even though the MPA has effectively protected marine biodiversity and ensured the resurgence of rare fish stocks. In the following section, we offer a brief overview of our method and methodology before offering the interviews collected and our analysis.

Methodology

In analysing the experiences of those encountered, we had to consider literature on quality of life, poverty and exclusion, globally and in Africa (Beegle et al., 2016, pp. 117–120; Plagerson & Mthembu, 2019, pp. 8–12). We considered these in relation to emerging theories of heritage (Siregar, 2018, p. 57), which suggest that the heritage concept be broadened (see chapter one in this book), to include not only celebratory cultural heritages that mark the universal inheritance of humankind (Forrest, 2007, p. 125). These revealed that….in turn, the literature informed our research methods, compelling more empathetic, intersubjective analysis of the situation in Tsitsikamma. The research we conducted was ethnographic in nature, seeking to collect data that offers what Clifford Geertz has called 'thick description' (1973). The purpose of pursuing thick description is so that we relay in as much detail as possible, the rich cultural heritage of communities at Tsitsikamma, and capture the stories and feelings of those interviewed.

Part of our research process also included a self-reflexive component, where we engaged (in discussions beyond the formal research process), with the challenge and issue of who we are as socially embedded researchers in a still racialised, gendered and classed society. In brief, we sought to remain attentive to issues of intersubjectivity in research and ways in which our own personalities and perspectives might influence the research process and its outcomes.

Interviewees were asked the following sample of questions:

1. What is unique about this community in which you live?
2. What does the sea mean to you?
3. Does the sea feature in your community rituals and beliefs? If so, how?
4. What are some of the major changes that have happened here over the past years? Did these changes affect you? How and why?
5. In your view, how does government and/or the local authorities respond to/treat the cultural practices of your community?
6. What would you say has been a major challenge to your way of life in this community?
7. In your view, how does a coastal community differ from any other (i.e. inland) community?

Qualitative research methods were used during the research. These included semi-structured interviews with selected stakeholders, as well as open-ended questions. As researchers we also triaged observation, interviews and conversations to build a comprehensive understanding of coastal existence in the Tsitsikamma area, thereby giving voice to research participants. The approach also allowed participants to fully express their social and sensory experiences of coastal existence.

Adopting the self-reflexive approach, we incorporated substantive pre- and post-interview reflection on the process, nature and outcome of the interviews, as well as critical reflection on the place and role of the researcher in the research process. Using a mirror to see our own reflection placed in the story of local inhabitants in relation to the very access we have, due to the nature of our work, and locating our biases, privilege, emotion and blind spots. This set of self-reflective practices begun a cathartic process of inward looking and immersing ourselves for that day in the ethnographic process. The snowball sampling method was also utilised, whereby key informants were identified prior to fieldwork engagement on site. These initial engagements led to the identification of further recommended interviewees. The process helped us to achieve a more holistic engagement with a diversity of individuals in order to document their experiences, values, rituals and beliefs of the ocean. The interview data was not only analysed for content but also emotion and inflection. The stammer, changing tone, silence and even code switching of languages provides insight not only into 'raw' emotion but also linguistic and cultural conventions in the community. The approach allowed us to 'dig' below the surface to reveal what is often unsaid, especially fears of possible consequences regarding information shared on the MPA. Coupled with this, the role of gender in the responses is noted where women were more able to speak freely on their own. They appeared less silenced by male voices in collective (forum style) spaces where ocean identities and livelihoods were discussed.

To this end, we had to deliberately and very carefully constitute a social and cultural matrix of participants to cover the quadrant of narratives we hoped to surface. Each quadrant would require deliberate sets of questions that are integrated amongst other quadrants to provide an organic conversation, as opposed to a clinical interview with the researcher interviewer being the omnipotent, more powerful knower and the participant or focus group member seen as the less dominant and information giver. Thus, it became imperative to find ways of sharing knowledge gained and seek ways in which to positively impact those we encounter.

Our Cultural Heritage Is the Ocean: Why Can't We Go There?

The voices and contemporary realities of research participants are offered as insight to the social outcome of the MPA to their livelihoods. As the interviews cited here show, the identities and heritages of the people living in Tsitsikamma and its surrounds, are historically, politically and culturally inscribed.

We begin with an interview that was arranged with a conservationist in the Tsitsikamma area. This was a long and informative ethnographic interview and many themes of the research emerged. In the township, our interviews included a fisherman, other residents of the community and traditional healers. These interviews reflect on further consequences of denied or diminished access to the sea, especially the impacts of crime and substance abuse in the local community. Additionally, they offer a composite view of how local people in Tsitsikamma engage with the oceans, how they feel about the oceans and what they have 'lost'. The following shows a brief insight into what people in Tsitsikamma feel they have lost, as a result of the creation of an MPA. In the following interviews we use pseudonyms as a means to protect the anonymity of our participants.

Interview 1: Vernon
I grew up in Tsitsikamma and came back to study. Family born and bred in forestry and planted small agriculture which inspires my work. Late great-grandfather used income as farmer to put grandkids through teacher's training college. I have a great love for nature and nature conservation. [I see] the land and ocean as one thing access, rituals, intangible heritage. [I have] never had the opportunity to go to Marine Protected Area (MPA), I only experienced the ocean outside the scope of Tsitsikamma camping and fishing on the coastline and sleeping in caves. [to me] the story of fishing and access into MPA reminds me of industry. [When I was growing up] Friday afternoons were for teaching children to fish, bonding time with brothers, sons and fathers. Catches were shared with neighbours and to supplement income. Neighbouring communities [beyond Tsitsikamma, places such as Coldstream] were not consulted for MPA. Each settlement in Koukamma area is unique, in its history, economy and general story. At the moment, Tsitsikamma residents feels like it (MPA) is theirs, what they fought for. [The problem in South Africa is that because of history] people don't have access to the ocean in contemporary South Africa, [so] everyone feels entitled to the access and we need to protect the natural environment. The studies we have now (not specified which ones) look at community conditions, registration data to conclude the number of anglers per settlement, male and female. [Today we should] practise smart economy, farming, cattle, bees. Collect medicinal plants while fishing. Shellfish is enough to sustain an individual and they only take what they need, never more because they can't refrigerate it. The harvest is shared with the community. That to me is sustainability, only use what we need. Fast forward to today, there is a slippery slope towards natural resources being exploited to the bone. Environmental education can empower, but there is no follow up, so there is zero confidence amongst the community. Community concern is why should they even bother? Communities still feel marginalised in sustainability and environmental education. Indigenous forests were managed by the ancestors, but residents don't even go there today. That awareness should be a watermark (prioritized). There are small fishery policies where there needs to be permits, yet boats cannot access. This remains a pipedream. Potential for economic activity but little thought goes into capacity [building] to help [people] understand the ocean economy and provide (infrastructural) resources to uplift communities. Tourism [in Tsitsikamma] is water-based, yet I rate Tsitsikamma 1/10, why is this not working? We should ask who is allowed in? It is limited to permit holders. Field guide opportunities and the dolphin trail. Compared to inland working on other reserves there is limited access and exploitation in coastal areas. Boxes are ticked but there is no interest in

allowing access to residents. The river activities add on to ocean economy but even access to this is political and contested. [At the moment] it is difficult …to be able to harvest medicinal plants within MPA. Zoned and proclaimed in 2000 as a "no take" zone. [People are going to the] waterfall trail to collect, harvest and fish but not medicinal plants. Fishing is the one cultural practise that was able to be executed until 2000. The removal of plants was prohibited. In the forests only certain things were allowed. Residents could use the indigenous forests in catchment area to collect medicinal plants. Cut 1kg or less for tea, long leafed honey bush. Residents now have gardens at home with honey bush as there is now less stress on being caught harvesting. Locals [are getting] recruited by sangomas outside Tsitsikamma to harvest.

Interview 2: Joseph (not real name)—Fisherman
[The Ocean and Sea] It's a life to build. Food to eat. Fishing is my life. My first time by the sea was special, I looked at the sea and was not sure what to do. It gives you life as there is lots to do with the sea. What would we do without the sea? Everyone has food to eat for the day. I lived on a farm, but I saw people fishing in Tsitsikamma and learnt how to fish. I am the only person who fishes besides my brother. I learnt from him. [I find that] Kabeljouw is the best to eat [but there is also] Streepies and seekat. We fish on the sly (skelm visvang). I enjoy it but I am scared of being caught. We have to pay fines if we get caught, we wait until coast is clear, it is stressful. (asked about permits) It's not right, the sea is big. Who are they to permit us to fish? It's necessary. It's unfair to ask someone to have a permit to access the sea. Everyone has the right to the sea, how else can we pay for our lives? I am hungry, why do I need a permit? I have no money, but I must pay for a permit. It is a shame.

Interview 3: Oom (Fisherman)
[One can protect the ocean by taking] your rubbish after you fish, that rubbish goes in the sea. Take a black bag with. Put fish back that you won't eat. The medicine from the sea is important for the people. The park is full of medicine. It's our practise to go and take the plants and water from the sea for strength. We live by the sea but there is nothing in our homes that show that? Where are the seashells? There is nothing from the sea in our homes.

Interview 4: Elvis (Fisherman and Herbalist)
We grew up at the sea. On a Friday afternoon we would go to the sea and stay until Sunday afternoon. The sea for us is a goldmine. Our community lives from the sea. If I didn't have food in the house, I could take my rod and go fishing and get food for my family and my neighbours. The whole situation has changed now. We are too scared to go fishing now, they have closed the sea to us, and you can only take 10 fish. What ca you do with 10 fish? There are 13-14 people in the house. That's not enough for the family

or your neighbours. They can't go to the sea; they are old people. The sea has lots of medicine for us to make us healthy. I don't go to the doctor; I am my own doctor. If you want medicine got to the sea. Give me a pot, I'll drink the medicine and I get healthy. But we are forbidden to go to the sea. We can't go to the sea anymore. I don't know how long it will take for us to be able to go back to where we come from. The community is hungry, we have tried to go to SANPark to allow us to go to the sea to fish, but there's just silence. We are a poor community; we don't have the money to pay to go to the sea.

[My first memory of fishing is] when I was 12years old, my father and brothers took me to the sea to fish. They bought me a rod and I found out how nice it was to fish. I like it and the sea became a part of me. I feel free, there is no one to bother me there. I clean my catch and then just look at the waves in the sea, how they roll and smash. There is nothing better than to be between the sea and the rocks. I am free in the water and nature. All I need is coffee and porridge, then my life will be at the sea. That is how I grew up, in the sea. How can we protect the sea?

Interview 5: Mr & Mrs Herman (fisherman and his wife)
The sea is a place of distraction. We can go to relax and fish, to have something to eat. We just want to go to fish and we want to have the freedom to relax. But everything has been taken away, now we have to pay for a permit. We don't know about the decisions they make we just get told that we can't go. It was just taken away; we don't know why or how. A fisherman's heart works differently, it's a part of your souls taken away. The sea means a lot to us. They told us to say no one must go to the sea, they didn't come to talk to us or to explain. The people are the heart of the sea, but we get R1000 fines and court cases. The younger generation drinks too much without being able to go to the sea, there's no relaxing at the sea, so what must they do now? Sit and do nothing? The sea kept us busy. Someone brings us medicine from the sea, and they explain what it is used for. *See salie* is bitter but it makes you healthy, it's for colds and flu.[1] The bitter it is, the better. It has lemon in it to make it taste better and to make us healthy. (The researchers were then offered to try some *see salie*)

Interview 6: Eldrid
You see, all these years we have never had the privilege. The national park has many fishing rules. I'm not sure if it's their own laws or the government laws, but I doubt it's government laws because government gives us the right to go to the ocean, to fish, but SAN Park is standing in our way. They say they don't want people around the coast because of tourism. They're scared we're going to infect (implies affect) the tourism, you see? We're going to infect (implies affect) tourism on the coast, so that's why we can't fish. All the fishermen in this area, Tsitsikamma, would love to fish but we do not have power to get there.

So, if we must beat SAN Park, then we'll get there. There's many fishermen in the area who are interested. You see in the olden days; our ancestors were fishermen and they fished in these places. They fished here before but now, after some time they closed these areas, so that's why we totally can't fish here. There's still many people who are fighting to get the ocean open. We're small scale. We're in a cope (corporative) We're 19 in the corp. that started in March 2018. But in that time, SANS Park is still standing in our way. Yes, we've got the permit. We got our card from the minister. But before we can agree on it, SAN Park can't get to an agreement. Now we don't actually know where the thing(situation) lies. Is it with the minister or is it with the Park? But the minister is speaking the truth, so the thing(situation) actually lies with SAN Park, you see? It seems to me as if they don't have a connect with the Department of Agriculture, Forestry and Fisheries. You see, the people here are a bit scared to go into that area. Because when you go into that area, you'll be arrested, you see? And if you have a permit, you're only allowed to catch certain fish. You're entitled to that only, so it's a waste of time. We're still fighting, you see? We're still in meetings with SAN Park regarding this thing (situation). But then Doves people also need to be here with SAN Park. So that SAN Park's people can understand what it's all about. We're not doing this out of our own. We're receiving instructions from the minister. In this area, yes. If they open the ocean for us then there will be something for all of us to put on the table. If they would just open that ocean for us. But the ocean means a lot to our people, really. And (long pause) that our rights must be taken away is not nice. Really. All the years, the coast was open for fishing. It's so difficult to explain, really. It's very difficult. Things have changed but there are things about apartheid that we got right, you see? There were things we got right in SAN Park. SAN Park was for our people, really. We knew that we could go into SAN Park for free, we could've walked through there without any documents. But in this current day everything gets locked and I don't know why. They say it's freedom so everything should be made more convenient for us. We shouldn't have to struggle to get stuff right. The interview data offered above provide further insight into the broad range of interlocking issues affecting SSF in Tsitsikamma.

According to Vernon, the land and ocean are one and people from the area are socialised as such to cultivate a balanced relationship with the ocean to ensure long-term sustainability and resource stewardship of the oceans and coast. Living in balance with these resources, will allow some people to earn an income from sustainable farming practices and others to enter the world of work beyond fishing-based livelihoods. Vernon further noted that generational socialisation included education and love for the land with respect and dignity towards equitable ecological practices. In this way, being with the sea becomes a conduit for the transmission of intangible cultural values. Insofar as tangible cultural heritage is concerned, being with the sea facilitated the use of diverse material and domestic artifacts (cooking utensils and fishing gear for example), that ultimately encouraged young people to learn about social and cultural practices from an older generation. Tangible 'heritage' artifacts were also exchanged or at least used in families and as the interviews show, Friday afternoons were temporal spaces in which children learned to fish and bond with family.

The interviews show that some people tried to continue these practices beyond the MPA in Tsitsikamma. After the declaration of the MPA, people still tried to maintain sea-based communal practices, such as sharing fish with neighbours to foster a sense of community and belonging. These exchanges helped to build community and solidarity as well as an identity anchored in the forest-sea landscape of Tsitsikamma. Moreover, people were able to feed their families, provide for other family needs and secure livelihoods.

At the start of the presentation of Tsitsikamma, we argued that the place is also a forested environment. Conducting interviews, we found that knowledge attained by indigenous people who are locally accepted as custodians and carers of the forest and ocean are not being passed on to future generations, mainly because such knowledge requires access to the sea and particular sections of coasts denied to the population by the MPA. Overall, however, loss of access to parts of the sea and coast (as well as increased modernisation and mobility within the community since colonial times), means that the transfer of skills and cultural knowledge appears to be declining.

In the interviews, contributors emphasise the importance of only taking what one needs to ensure the perpetuation of a balanced marine ecosystem. The reason for this is that people interviewed saw themselves as co-owners of the sea. They repeatedly spoke about their deep and embodied connection to the sea and coast, while being disconnected from both appears to make those interviewed feel disenfranchised and locked out from a major part of their own identity.

Our observations in the township also revealed that not being with the sea or having the freedom to be with the sea was producing various social ills in the community. Specifically, not only did the sea provide families with a place to learn and bond, it also offered poor communities an opportunity for leisure and relaxation or overall sense of greater well-being. Disconnection from the sea and coast means that people are now confined to the less 'idyllic' setting of the township, where there is overcrowding, poor opportunity for healthy leisure and the spectre of rising crime levels. By only really considering the ecological benefits of the MPA, the government has achieved natural environmental success, however, the impacts on the local communities have not been beneficial.

From the perspective of blue heritages and/or intangible and tangible cultural heritages, we add that the MPA did not merely remove options for healthy leisure and family bonding. The MPA has also resulted in diminished possibilities for the creation of intangible cultural heritage related to the sea and coasts (Mascia et al., 2010, p. 1424). Whereas before the MPA, healers might look to the sea and coasts for medicinal materials, or use the sea to spiritually heal their 'patients', now the healers are confined to selecting and sourcing materials from the forest and further inland. As Michelle Cocks and Valerie Møller (2002, p. 390) demonstrates in her study on the commercialization of indigenous medicines, this has resulted in a veritable market for forest-based indigenous and rare plants, thereby compromising the tangible heritage of the forests. Indeed, and as our interview with one of the healer-diviners in Tsitsikamma revealed, healers now have to travel out of the town to the Eastern Cape towns of Peddie and Makhanda, to find commercially sold herbs and plants. From our discussions we found that while these are the 'correct' plants for medicinal use, they are not ideal, as their potency is diminished, given that they are not sourced from the 'wild'.

The interview with Joseph offers a nostalgic account wrapped in fear and survival, as a young man, he tells a story of how he was introduced to fishing by older siblings and other community members. These practices gave him a sense of belonging and community within the younger generation. With the arrival of the MPA, the dynamic within communities moved from open access and deep feeling of communication to the sea to no access, disconnection and fear. Fishing for *Kabeljouw, Streepies* and *Seekat* for a sustainable livelihood has now become a crime as opposed to a leisurely experience 'with the boys'. For us as researchers, this suggests that even the possibility of developing a healthy, blue heritage with the sea for young adult men, is now denied. If these young men want to continue this practice, they will need transportation to travel further along the coast, away from the MPA. Given the economic circumstances of the communities we encountered, we doubt that this will happen. Plus and as confirmed by the local community members themselves, many young men are turning to crime and substance abuse rather than pursuing a more healthy leisure beyond the community.

Furthermore, we were informed that caring for the ocean by not polluting are age-old practices that communities have been following for generations. It may be fashionable in contemporary society to not litter the ocean. However, conserving the sea and coast forms part of a set of practices expressed in indigenous ways of respect, care and prudence. To this end, homes are not adorned with shells and other remnants of ocean life due to the principle of respect for the ecosystem. A symbiotic relationship is a gift within a caring ecosystem between land, sea, animal and man whereby the ocean flora and water give strength to both the body and the community.

In reflecting on drinking the *see salie* with Mr and Mrs Herman (interview 6), we were at first filled with some apprehension and then we realised that the elixir given was out of pride as a valued heritage. Well wishes in the form of medicine. This reaffirmed that the contributors were more knowing of the ocean, and the true custodians of indigenous knowledge and practice.

The last interview demonstrates the complex intertwining of forest and sea in Tsitsikamma, the mobility of people and natural resources across the Eastern Cape Province and of healing that takes place even on a transnational level (i.e. healing of the English person). The closes access gives further issues of healing (the supernatural) and the ocean, where the different ancestors reside. The people are not just denied access to the oceans and coasts or forests, they are being kept in a state of physical and spiritual illness because of denied access. Although they live in the area, the coastal communities are excluded in decision-making and, further, treated as outsiders compared to the visiting tourists of the National Park and MPA. The contributors refer to these tourists as '*inkomers*'. The irony being that they themselves are on the 'inside' but are pushed 'outside' inland of the coastal fringe. The National Park is a major player in this vicinity and a lot of the interviewees feel like this is limiting their access to their rightful place, their rightful land. There is a deep sense of the people being wronged one way or the other. The people want to have access to the ocean as part of their survival. And it's been taken away from them and if they go there, they ate committing a crime.

Visiting the various villages in the Tsitsikamma area, we looked at change through time and the transfer of generational cultural knowledge since the MPA, as well as impacts on limits to fishing. Previous generations taught their children to fish and to harvest medicinal plants. These are deeply embedded social and cultural practices that showcase the blue heritage or at least, the intangible cultural heritage of the local community. Since the MPA, at least from the perception of the community, there appears to be less knowledge now of the ocean and forest. It is not just that people can no longer make a living from the ocean.

Considering our own participation in the story and self-reflexively analysing our experience, as South Africans we were under the impression that we had heard all the stories of pain, division and limited access. We also thought that we understood the parameters of difference, after all, the South

African apartheid system had made us think about difference in terms of race and gender. Yet after having very intimate, heartfelt discussions with the local community in Tsitsikamma, one begins to understand the true value of social science research. In some respects, it has been painful to hear the stories and to understand the impact of policy around MPA and permits and how it affects people holistically. Social science research in the ocean conservation space has helped us to understand what have they embedded in retelling their story to magnify their voice. The impact of change is felt in all strata of society and everyone needs to hear these stories in order for revised policy, or new policy, to be considered in a way that people feel included.

As a final thought, responses to the questions we were asking suggested that the community felt excluded from the decision-making process regarding access to the ocean, fishing permits and the MPA. Three themes were followed in the conversation:

1. **Issues on MPA and permits**
   Why was the sea taken away from us when vacationers can go anytime? [If] we go to the sea and we get put in jail for trespassing. We must walk 20 km to get to the sea to fish and we must be gone by 18:00. We need a permit to go to the MPA and to the sea, but not everyone can afford this. We must be sneaky when we go fishing.
2. **Medicine and food**
   We go to the sea for food and medicine. The sea means life. It's the air in our lungs. Our medicine. Everyday I was by the sea for medicine and food, now I can't go anymore because I don't have a permit. We can't even just put our feet in the water to cleanse us.
3. **Sustainability**
   People only take just enough for what they need so that there is more for someone else and some for tomorrow. My bag is only small so I only take what fits in my bag. This will feed my family for the day.

Conclusion

When one poses questions to different people, it immediately gets people talking about heart matters, overhead matters, pertaining to the sea and their connections to the sea. It was touching to see how real people got when talking about the sea, their reliance, their co-dependence, their dependence and how different attachments are formed to the ocean, to medicinal plants and this denied access proved to be part of the crux and the real reason to why we toil with doing this research.

It reaffirmed keeping quiet as the researcher and allowing for the participant to speak quite freely within a container of what it is you wish for them to share with you. We allow for multilingualism and layered languages tell a story in a more authentic way for the heart and depth to come through.

Themes that came out were sustainability, the love of the ocean and lack of access. The interviewees are very passionate about the ocean and access to the ocean, as well as for the children and future generations and how that all has impacted on them now. The disrupted access renders communities in limbo in both search of their identity and their past. These challenges are evident in social issues and ills that are rife within coastal communities as so identified by participants in the study. The raw emotion, palpable feelings and heartfelt conversations left us as researchers feeling sad and as marginalised as the participants. This feeling, however, is superficial as we are not members of these communities and will continue without daily existence in another space and time. This continuance is evident behaviour of what communities cite as lack of participation in crafting effective and inclusive MPA protocols and regulations. Communities do not see themselves reflected in the MPA legislation and cannot continue with former practices that they continue to nostalgically hold on to.

Note

1. Traditional medicine, also referred to as *boggo*.

References

Beegle, K., Christiaensen, L., Dabalen, A., & Gaddis, I. (2016). *Poverty in a rising Africa*. International Bank for Reconstruction and Development, The World Bank.

Boswell, R., Pillay, R., Thornton, J. L., Maqabuka, G., Terblanche, T., Xulu, S., & Du Plessis, F. (2021). *Field report: Tangible and intangible cultural heritage*. Tsitsikamma and Algoa Bay Case Study. Report for the UKRI One Ocean Hub (OOH).

Cocks, M., & Møller, V. (2002). Use of indigenous and indigenised medicines to enhance personal well-being: A South African case study. *Social Science and Medicine, 54*(3), 387–397.

Forrest, C. (2007). Cultural heritage as the common heritage of humankind: A critical re-evaluation. *The Comparative and International Law Journal of Southern Africa, 40*(1), 124–151.

Fox, H. E., Mascia, M. B., Basurto, X., Costa, A., Glew, L., Heinemann, D., Karrer, L. B., Lester, S. E., Lombana, A. V., Pomeroy, R. S., Recchia, C. A., Roberts, C. M., Sanchirico, J. N., Pet-Soede, L., & White, A. T. (2012). Reexamining the science of marine protected areas: Linking knowledge to action. *Conservation Letters, 5*, 1–10.

Geertz, C. (1973). *Thick description: Toward an interpretive theory of culture.* 1st ed. Basic Books.

Gruby, R. L., Fairbanks, L., Acton, L., Artis, E., Campbell, L. M., Gray, N. J., Mitchell, L., Jones Zigler, S. B., & Wilson, K. (2017). Conceptualizing social outcomes of large marine protected areas. *Coastal Management, Issue, 45*(6), 416–435.

Jones, N., McGinlay, J., & Dimitrakopoulos, P. G. (2017). Improving social impact assessment of protected areas: A review of the literature and directions for future research. *Environmental Impact Assessment Review, 64*, 1–7.

Marine Protected Areas. (2021). *Tsitsikamma MPA: The heart of MPAS.* Garden Route, Eastern Cape [Online]. Available at: https://www.marineprotectedareas.org.za/tsitsikamma-mpa. Accessed 2 Feb 2021.

Mascia, M., Claus, C., & Naidoo, R. (2010). Impacts of marine protected areas on fishing communities. *Conservation Biology, 24*(5), 1424–1429.

Plagerson, S., & Mthembu, S. (2019). *Poverty, inequality and social exclusion in South Africa: A systematic assessment of key policies, strategies and flagship programmes.* Centre for Social Development in Africa: Johannesburg.

Pomeroy, R. S., Mascia, M. B., & Pollnac, R. B. (2006). *Marine protected areas: The social dimension.* A report for the Food and Agricultural Organization of the United Nations.

Sanchirico, J. N., Cochran, K. A., & Emerson, P. M. (2002). *Marine protected areas: Economic and social implications.* Resources for the Future (pp. 2–26).

Siregar, J. P. (2018). The evolving cultural heritage concept: A theoretical review of cultural heritage in Indonesia. *Journal of Engineering, 8*(7), 57–62.

Sowman, M., & Sunde, J. (2018). Social impacts of marine protected areas in South Africa on coastal fishing communities. *Ocean and Coastal Management, 157*, 168–179.

22

Sustaining the Underwater Cultural Heritage

Elena Perez-Alvaro

Introduction

The sustainable use of the sea is focused on meeting human needs, embracing environmental as well as socio-economic interests. Ideally, the sustainable management of the oceans should include all uses and users, respecting all rights, interests and goals (Pandey et al., 2021). Reflecting on these issues, this chapter introduces the need to consider underwater cultural heritage as an important tool for collaboration and input from all ocean users/actors to co-exist, using a natural/cultural partnership. Underwater cultural heritage can be part of the tangible heritage of a nation, but can also have more abstract 'uses': it can serve as a diplomatic tool, as an identity bearer, as a watery cemetery or as a hidden history container (Perez-Alvaro, 2019).

The 2030 Agenda for Sustainable Development sets out in its Goal 14, Life Below Water (United Nations, 2015), to ensure sustainable development

---

The research presented in this paper is supported by an NRF UID Grant 129662 Ocean Cultures and Heritage. The author, a postdoctoral fellow of the SARCHI Chair Ocean Cultures and Heritage project based in South Africa.

E. Perez-Alvaro (✉)
La Manga Club, Murcia, Spain
e-mail: elenaperezalvaro@gmail.com

© The Author(s), under exclusive license to Springer Nature
Switzerland AG 2022
R. Boswell et al. (eds.), *The Palgrave Handbook of Blue Heritage*,
https://doi.org/10.1007/978-3-030-99347-4_22

of the ocean and marine resources. However, the potential role that underwater cultural heritage can play to help reaching this goal, is not explicit. The United Nations Decade of Ocean Science for Sustainable Development 2021–2030 (United Nations, 2017), on the other hand, briefly mentions underwater cultural heritage (marine cultural heritage) but, there is not a real consideration of the value of this heritage for the Agenda.

The Underwater Cultural Heritage

Underwater archaeology is a flourishing field whose focus is understanding and interpreting the past relationships between humans and the rivers, lakes and oceans that have surrounded us throughout our history. It may include shipwrecks and sunken cities, venerated sites, ancient harbours, plane wrecks or prehistoric landscapes. According to the UNESCO Convention on the Protection of the Underwater Cultural Heritage (UNESCO, 2001), there are historic monuments underwater such as the Lighthouse of Alexandria, and whole cities such as Port Royal in Jamaica, Carthage in North Africa and the submerged temples of Mahabalipuram in India. There are also more than three million vessels lying in the oceans around the world. Consequently, underwater cultural heritage is an invaluable international repository of knowledge about individuals and communities, helping to understand uses and users' issues all around the world.

Underwater cultural heritage is fraught with international issues more than any other branch of archaeology (Maarleveld, 2011). It can be one of the reason for political conflicts since data from marine exploration, including underwater archaeology, affects national defence sensitivity as it may uncover military security information, safety issues, environmental concerns, cultural evidence or food security clues and they are also proof of historical presence, which can bring benefits on territorial claims, as in the case of the use of underwater archaeological remains by China to claim sovereignty and rights over the South China Sea (Perez-Alvaro & Forrest, 2018). In addition, claims on underwater cultural heritage are really contested. One of the main issues is the ownership of the underwater cultural heritage from communities that were colonised, from communities that do not exist today or from states where the territory belongs to a different, new state. For instance, who is the legal owner and claimant of shipwrecks from the former Yugoslavia? On the other hand, underwater cultural heritage is also a proof of intercultural transmission, since, shipwrecks were vehicles of communications and trade and as a consequence they were influential in the mental ideologies of different cultures. In contrast with land archaeological sites that are more related to a national cultural identity, a ship embodies much greater diversity to the history of many countries and cultures and carries out unique cultural values. Consequently, it can provide information about people, past and present, helping to understand the diversity of social existence in all regions of the world and at different times.

One of the commonly agreed principles of preservation of underwater cultural heritage named by the 2001 UNESCO Convention on the Protection of the Underwater Cultural Heritage is the preservation in situ as the first option (UNESCO, 2001, Articles 2.5, 2.10. Rules 1, 7 and 25). The reason for this being mainly because archaeological objects are better preserved under layers of mud and in saline water. The ship, once she has sunk and lies at the seabed, reaches a state of equilibrium with the upper parts destroyed and the buried remains covered. After reaching this state of equilibrium, the wreck will be only disturbed either by human intervention or by geological changes—physical, chemical and biological threats—(Wachsmann, 2011). This is the reason why underwater archaeologists consider preservation in situ as the first option. In order to reach this ideal preservation, monitoring and managing the site is essential. This monitoring can be of benefit for all users of the seas, as this chapter will explore.

In addition, the 2001 UNESCO Convention highlights the importance of public awareness and public enjoyment of the underwater cultural heritage. In its Preface, Articles 2.10, 18.4, 20 and Rule 7 of the Annex, the Convention seems to acknowledge the touristic side of the underwater cultural heritage but only if is not incompatible with the protection and management of underwater cultural heritage. As a consequence, the Convention is already implicitly pointing out the necessity of a sustainable use of this heritage: it tries to balance the desire to conserve and protect versus the benefit to exploit.

There are several arguments and counter-arguments for this sustainable use of the underwater cultural heritage but the reality is that it can help the ocean ecosystems as well as different use and users of the oceans. In fact, they all need to consider the possibility of future common threats, such as climate changes. A joint approach to these threats can bring benefits for all.

An article written by Henderson (2019) explores the role of marine cultural heritage in informing responses to global challenges. The article examines how marine cultural heritage can encourage societies to come together and negotiate different approaches to provide response to the challenges set out in the 2030 Agenda (United Nations, 2015). Henderson focuses on marine cultural heritage as knowledge, exploring some uses of the oceans and how this heritage can serve to achieve effective conservation practices and sustainable economic development. This chapter will add to Henderson's analysis and will acknowledge some important contributions that this kind of heritage can bring to more abstract uses of the oceans. Consequently, the present chapter hopes to surface the materiality and temporality of oceans as a tool for the UN as it seeks to understand how underwater cultural heritage can serve to other uses/users of the oceans. The discussion also highlights some unconsidered use/users such as the oceans as watery graves or as escapes or prisons.

Before creating links between cultural heritage and other uses, the next section will try to summarise each one of the oceans uses/users. This list hopes to include not only the obvious ones, but also others that may not be so evident and to encapsulate the essential information of each one of the uses/users in order to serve as a base for the next sections of the chapter.

Ocean Uses/Users

The oceans are an integrated space for a variety of human uses and is perceived and used differently by its various actors. People and products of all civilisations have been carried by water, and entire continents have been discovered, colonised and defended by sea. Ships have carried crew and passengers of all nationalities; on the *Titanic*, people from more than 28 countries died. Most of those shipwrecks still lie at the bottom of the oceans. Shipwreck sites feature the vessel itself, often a once-magnificent artefact, and often hold cargo, personal items, tools, utensils and human remains. This offers a picture of past human civilisations. As Ransley (2005) explains, the sea can tell us about mobility and connections, adding the human element to the static model of states since it does not respect boundaries, political or geographical.

Oceans provide not only fish and other food resources for humans, but they also provide migration and transport routes, among other things. Dockworkers and artisanal fishers, as well as artists who use the sea as a source of inspiration, are all valid users of the oceans. The oceans contribute to human welfare, including through an economic value estimated to be in the range of US$1.5 trillion per year (OECD, 2020). Humans have been very successful at exploiting the oceans and, recently, new technologies have led to an explosion of new information (Zereik et al., 2018). High-resolution geophysical instruments are fast and cost-effective tools that leave the oceans undisturbed. In the case of archaeological research, these instruments can be used for the non-destructive detection, imaging, research, inspection and monitoring of submerged sites, exposed on or buried. In the case of marine robotics is a field with a large development capable of allowing scientists the means to monitor the ocean and its living and non-living resource at unprecedented temporal and spatial scales. However, this development can also bring detrimental consequences for the oceans. These impacts include oceans' use for tactical military exercises, detrimental use of the oceans for economic advancement or the use of these technologies by treasure hunters to illegally retrieve submerged high-economical values archaeological objects.

The Ocean as a Means of Transportation

Shipping may be the most indispensable use of the sea for society (Ross, 1978). According to the International Chamber of Shipping (2000), the international shipping industry carries around 90% of the world's trade. There are over 50,000 merchant ships trading internationally, and the world fleet is registered in over 150 nations with more than 1,647,500 seafarers. The operation of merchant ships generates an estimated annual income of over half a trillion US Dollars in freight rates. It is not only seafarers who depend on this industry, there are also stevedore crews loading and unloading the cargo, pilots taking the ships to port and land vehicles and crew moving goods inland, among others.

In the past, boats were the main form of transport for people to travel across the oceans, but airplane has become more dominant form of travel for people. However, bulk maritime cargo transport is still cheaper, and global traffic is forecast to triple between 2015 and 2050 (International Transport Forum, 2019) with new commercial shipping routes predicted to open up in Arctic waters as the Arctic melts. For instance, using the Northern Sea Route, which passes along the Russian coast, could reduce voyage distances by 37% (Fernandez, 2013). In the past, the Northwest Passage, connecting the Atlantic and Pacific Oceans, was covered by thick, year-round sea ice, but it is now accessible due to climate change. The Transpolar Sea Route, also connecting the Atlantic and Pacific Oceans, will cut across the centre of the Arctic Ocean, passing close to the North Pole and also reducing distances for global trade, although it is frozen over most of the year. Due to these and other new routes, billions of dollars in transportation costs could be saved each year.

The Ocean as a Natural and Cultural Resource Provider

More than 10,000 offshore structures provide industries and populations around the globe with raw materials, manufactured food and energy since the oceans host a vast variety of geological processes responsible for the concentration of mineral resources, which are currently being exploited. For instance, oil and gas are extracted by drilling into the seafloor on the continental shelves and slopes (Segar & Segar, 1998). The value of this resource in the United States alone is between US$25 and $40 billion per year. In addition, this industry employs submarine pipelines, which is another industry which is becoming very important in the energy industry: the costs of pipelines construction is often higher than that of drilling components.

The oceans also have mineral nodules containing manganese, iron, copper, nickel, phosphates and cobalt and other rare earth elements, such as gold, tin, titanium and diamonds. It is estimated that there may be around 500 billion tonnes of nodules on the seafloor. In addition, about 20% of the world's petroleum products come from the sea (Ross, 1978).

Salt, sand and gravel are also natural resources that come from the oceans. Salt is not only extracted from the oceans by evaporating the water and leaving the residual salts, but it is also mined from large salt beds. Ocean basins are the depositional sites of sediments eroded from the land, forming large residual deposits of sand and gravel.

Marine pharmacology focuses on the substances with active pharmacological properties that are present in marine plant and animal species. The marine organisms provide nutraceuticals for the potential treatment of several human diseases. Many medicines of marine origin have shown promise in the treatment of chronic and unbeatable diseases, such as cancer (Malve, 2016). In addition, marine algae are used for cosmetics because of their high concentration of minerals and trace elements.

Fishing has been one of the most important activities for the world's economy for thousands of years. About 86 million tonnes of fish are caught each year. Fish are caught in a variety of ways, including one-man nets, super-trawlers, dragnetting, longlining and diving. The quantity of fish available is an ever-changing number. Coral reefs also provide vital ecosystem services; they are a source of food, protect coastlines from erosion, provide habitats for important fish species, are a source of new medicines and provide income for local economies.

Another way in which the ocean is used as a natural resource is through the practice of aquaculture. This is the rearing of aquatic organisms, such as plants or animals, under controlled or semi-controlled conditions, which provides food for humans, opportunities for recreational fishing and supports the recovery of endangered species.

The Ocean as a Scenario for Development

There is a common misconception that most international communications are via satellites, when, in fact, over 95% of this traffic is actually routed via submarine fibre-optic cables. Thousands of kilometres of submarine cables lie on or under the seabed carrying telephone calls and internet data. These submarine cables cost around US$100–$500 million to build. In 2013, 283 cables were active with 29 new routes planned (Perez-Alvaro, 2013).

New energies from the oceans are fast developing and can come from several major sources: waves, tides, ocean currents, thermal differences, salinity differences and marine biomass. Since they provide emissions-free energy, they have the potential to be an environmentally friendly form of energy that contributes a significant part of the global electricity demand. However, at the moment, although many advances have been made, some of these technologies are still being driven by small, young businesses. Wind energy is the most advanced in terms of development, with facilities increasing in size and venturing into deeper waters. Wind turbines can be anchored to the ocean floor at depths of more than 40 metres, although floating offshore facilities are also being developed (Bosch et al., 2010). Offshore infrastructure includes wind turbines, offshore electrical platforms, accommodation platforms, met masts, array cables, interconnector cables and export cables.

The Ocean as a Weather Engine: Climate Change

Oceans are important for the Earth's climate and ecology. In fact, the atmosphere interacts with the oceans and exchanges heat, water, gases and particles and produces momentum. In addition, oceans transport heat from the tropics to higher latitudes since heat is stored in the oceans, and then this heat warms the atmosphere. Importantly, it absorbs large amounts of carbon dioxide, making the planet habitable (the ocean absorbed 34 billion metric tons of carbon from the burning of fossil fuels from 1994 to 2007) (Gruber et al., 2019). Oceans also contribute to the thermohaline circulation, as low temperature and high salinity are the main driving forces of convection.

In many senses, the oceans are the source of all life on Earth. Consequently, climate change depends on the oceans, using it. Climate change is warming the oceans, the ice at the poles is melting and sea levels are rising. There are also changes in the chemical composition of oceans, for example in their acidity and salinity. Moreover, current patterns are changing and, as a consequence, ecosystems are becoming increasingly more endangered (Perez-Alvaro, 2015).

The Ocean as a Leisure Supplier

Early in human history, people discovered the significance of coasts as a place of recreation. It was in 414 BC that the Greek philosopher Euripides wrote that the sea washes away every human stain (Bosch et al., 2010). Over the last decade, coastal tourism has been the fastest growing sector of the world's economy. According to the Office for Coastal Management in the United States (NOAA, 2020), ocean tourism and recreation support a wide range of businesses, such as restaurants, hotels, aquariums and marinas as well as boat manufacturers and sporting goods stores. Almost 2.4. million people in the United States alone are employed through ocean-based tourism and recreation, with an economic gain of US$478 billion annually. It has been estimated that 46% of EU citizens spend their holidays on the beach (Bosch et al., 2010). In addition, the hinterland also benefits from coastal tourism, for instance, when beach holidaymakers visit inland towns (Bosch et al., 2010).

Tourism and leisure activities linked to the oceans include the cruise industry, diving activities, wildlife watching, such as shark cage diving, coastal tourism and water sports, such as paddling, surfing, swimming, sailing, yacht racing, coasteering, spearfishing, paddle-boarding, rigid inflatable boats and banana boats or sea kayaking. Furthermore, the use of the seas as a leisure supplier also includes cultural services that are of particular social or spiritual significance for communities, such as the arrival of the Three Wise Men by boat in many different coastal cities in Spain.

The Ocean as Watery Graves

Throughout history, there have been large movements of people for many reasons, such as climate change, conquest, slavery, demographic growth and the formation of new states. This migration has either been forced or voluntary. Most of those movements have crossed continents, and some have crossed the sea. Some of the vessels used for these movements have sunk, and there is still news today of boats sinking with people on board. Many bodies are washed ashore, but most are lost at sea, becoming a watery grave (Perez-Alvaro, 2019). Sea burial is a common practice in Europe and Japan. After cremation, the ashes are scattered in the sea or consigned to the sea in a water-soluble urn. However, this practice is now only allowed in certain sea areas because of pollution (Bosch et al., 2010).

The Ocean as a Maritime Power Projector

Maritime power is the ability of a state to achieve a political goal using the maritime domain (Sanders, 2014). This power includes fighting wars, power projection and maritime diplomacy, preserving maritime resources, ensuring the safe transit of cargoes and people at sea, protecting maritime borders, upholding maritime sovereignty, engaging in maritime security operations, rescuing those in peril, preventing the misuse of the oceans or using the seas as a military exercise area (Sanders, 2014). In fact, the oceans have been used for military purposes since the times of the early Egyptian dynasties (Ross, 1978). Nowadays, nuclear submarines and their detection are probably the most important aspects of the military use of the sea which also include mine countermeasures and minesweeping, radar training, explosive trials, high and low-angle gunnery and air to sea or ground firing.

The sea has recently come to be seen as a new frontier for economic development. Economics, politics, security and strategic interests merge at sea transforming it into a geostrategic space (Otto, 2020). In fact, maritime security is essential to nations. Safeguarding navigation routes, providing oceanographic data to marine industries and protecting rights over marine resources within claimed zones of maritime jurisdiction are just some examples of the importance of the oceans for countries. It also includes ownership of the continental shelf, protecting citizens from ballistic missile attacks or the introduction of non-native marine species to new aquatic habitats, as well as illegal fishing, the smuggling of illicit goods, maritime piracy, migration and cybersecurity.

Land reclamation and the construction of offshore artificial islands is another method for countries to demonstrate their power, although it does not lead to the creation of maritime zones. Coastal land reclamation, which is a process of creating new land from the sea, is increasingly being seen as a solution for urban expansion. In fact, entire islands have been constructed (Yin Chee et al., 2017). Land is changing, and natural habitats are being replaced by urbanisation, which creates both challenges and opportunities.

The Ocean as a Stage for Illegal Activities

The oceans have been referred to by many as the biggest crime scene in the world. Piracy, narcotics smuggling, the illegal dumping of toxic substances, weapons smuggling, illegal fishing or the smuggling of contraband are only some of the illegal activities taking place in the oceans (Stavridis, 2017).

Maritime terrorism is regularly mutating and changing its point of focus. Narcotics and arms smugglers are making some areas of the sea increasingly dangerous (Sloggett, 2013). In addition, since the underwater cultural heritage, especially historic shipwrecks, has a high economic value, it is being increasingly plundered by private treasure-hunting companies. Due to the romantic notion of treasure at the bottom of the oceans, the underwater cultural heritage is being subjected to 'unprofessional' recovery, which can damage the objects and the historic context of the shipwrecks. In addition, the recovered objects are sold at a profit, resulting in money laundering and the financing of terrorists by international criminal enterprises because these transactions appear legal.

The Ocean as a Protected Area: Reserves, Marine Parks and Underwater Museums

Since oceans have been affected by human activities, one of the approaches taken for ocean conservation has been the establishment of marine protected areas in the shape of coastal sanctuaries, ecological reserves, refuges, national marine sanctuaries, marine parks and, in the case of underwater cultural heritage, underwater museums (Hyrenbach et al., 2000). Most of the marine protected areas have been established through legislation. These areas may protect, for instance, threatened species, although it is impossible to establish protected corridors for all threatened species. 7.7% of the ocean is currently protected (Protected Planet, 2021).

The Ocean as a Disposal Area

The unregulated disposal of waste poses human health risks. Industrial and human waste contains large amounts of nutrients. An oversupply of these nutrients can lead to a dramatic increase in the growth of algae and other phytoplankton, which may discolour the water and make it unappealing. In addition, it can deplete the available oxygen in the water, which leads to mass mortalities of fish (Ross, 1978). Previously, the ocean was one of the main areas for the disposal of many industrial and domestic waste products. This included the disposal of waste products from the land and sewage as well as the municipal, industrial and agricultural effluence from rivers. The direct dumping of industrial, municipal and agricultural waste as well as the paint and waste from ships has also been a problem.

After the introduction of many regulations, the vast majority of materials disposed of today are in the form of uncontaminated sediment. In addition, there are designated ocean disposal sites that are controlled by regulating the frequency, quantity and characteristics of the materials being disposed of, establishing disposal controls and monitoring the sites to verify that there are no significant adverse effects (Bosch et al., 2010).

The Ocean as a Research Scenario: Scientific Research and Underwater Archaeology

Marine scientific research comprises physical and chemical oceanography, marine biology and marine geology and geophysics. It includes the study of marine life and ecosystems, the geology of the seafloor, chemical and physical properties and ocean circulation. Underwater archaeology, however, is a relatively young research field. In this case, the study of the material is essential for its interpretation because the position of the objects and the shipwreck provide crucial data. In addition, inside a ship can be found its cargo and contents, fixtures and fittings, such as anchors and cannons, minor structures, such as decks or masts, and major structures, such as the hull and ribs, which are essential for the understanding of the history of humankind.

The Ocean as Both an Escape and a Prison

Further uses of the oceans have arisen with the recent COVID-19 pandemic. This is a situation that has transformed the dynamics of society, with humans been stripped of rights and their uniqueness as individuals as a result of compulsory lockdowns, movement restrictions and social distancing. These circumstances have made the ocean a getaway option. It has been reported that some millionaires have escaped the coronavirus lockdown restrictions by sailing and living on residential ships or luxury superyachts. In addition, living in international waters avoids paying taxes in most of the countries.

In contrast, seafarers have found themselves stranded at sea because of COVID-19. Replacement crews have not been able to fly out and, because ships cannot be left unmanned, crews have had to remain on board. Some crew members have been on the same ship for nearly 16 months (Walker-Vadillo, 2020). They have not been able to stop working while on the vessel or have days off in case an accident occurs.

The Ocean as a Setting: Hotchpotch Uses

Coasts are the areas where land and sea meet, and are not only vital habitats for humans, but are also a back garden for coastal communities, a source of inspiration for art and a landscape to enjoy for those close to the sea. Furthermore, they are parks of marine plants and containers of large collections of animals (somewhere between 500,000 and 10 million marine species live in the ocean). Coasts and oceans also have a religious and spiritual value for many people (Bosch et al., 2010).

The Use of Underwater Cultural Heritage

Underwater cultural heritage can play not only an important role in people and communities, but also in creating jobs and promoting economic growth. This section aims to find some of the uses of underwater cultural heritage as unexploited driver for other users of the oceans.

Since ancient times maritime transportation has been one of the main users of the oceans, many of these transport vessels sinking with all their cargo. Therefore, the bottoms of the sea are the biggest self-constituted museums of the world. These museums offer explanation about the kind of transport vessels, kind of cargo and maritime routes in the past which can benefit the maritime industry today. And although transportation routes can be a factor of risk for the safeguard of that underwater cultural heritage, it is difficult to control its impact and most transportation routes are in the high seas where shipwrecks lie very deep. New sinkings occur and will occur and those will become underwater cultural heritage in the future.

Underwater cultural heritage is also beneficial to the natural and cultural resource users because many shipwrecks have become artificial reefs for fish species, creating new biological maritime ecosystems. Around these shipwrecks there are new fauna and flora which could not co-exist otherwise. This chapter will later introduce the idea of a natural/cultural partnership since underwater cultural heritage can be hardly separated from its natural environment.

New energies and developments in the oceans are a growing but still completely unexplored industry. However, research in this industry can be of benefit to underwater cultural heritage since prior to any construction, research must reveal that no underwater archaeological sites are located in the planned location and new archaeological sites can be discovered. There is still a lot to do regarding policies and procedures in offshore energy extraction and archaeological education of energy farms construction companies and workers are essential for the respect of underwater cultural heritage. In the future, a new issue for ocean users is that offshore oil and gas reserves will trigger one of the largest decommissioning operations undertaken in the marine environment (Sommer et al., 2019). When offshore structures reach the end of their productive life, they must be decommissioned and dismantled, recycled or disposed of. Over the next several decades, > 7500 oil and gas platforms in the waters of 53 countries will become obsolete (Fowler et al., 2018), including platforms that sit on the seabed, floating production systems and other offshore storage and loading units (Simpson, 1998), as well as a fleet of 90,000 ships worldwide. In shallower waters, platforms typically weigh less than 5000 tonnes, although some deep-water steel platforms weigh over 200,000 tonnes. Their decommissioning will bring many issues because, due to their size and weight, the removal of platforms and the recycling or dismantling of large ships is a complex engineering process that will require some of the heaviest lifting operations ever attempted at sea. The most profitable shipbreaking methods are also the most dangerous for both worker safety and the environment (Choi et al., 2016) since many of the structures have been constructed with materials that are now known to be both toxic and highly pollutant. This dismantling/recycling/or preservation in situ of those platforms and ships has much to do with underwater cultural heritage management and both industries can benefit by sharing knowledge and working together.

Oceans as a weather engine will have many repercussions on underwater cultural heritage. Rising sea levels, warmer waters, ocean acidification and changes in currents will almost certainly affect cultural heritage. Warmer waters mean more chemical changes and the proliferation of *teredo navalis*. Ocean currents may cause disturbances to the layer of sediment protecting underwater cultural heritage sites, alteration of the materials and the potential loss of the archaeological record. Although the direct effects of chemical changes (particularly acidification and salinification) are still not well understood, the current rates of metal corrosion and damage to materials may well increase due to their climate change-induced fluctuation. Lastly, the rises in sea levels would reduce the amount of time an air-breathing diver can safely spend under water and hence their productivity. Rises would also mean expansion, which could raise the problem of ocean delimitation. It will also increase the depths of oceans and increase the frequency and magnitude of storms, which may further erode or damage heritage. It may also lead to flooding of cultural heritage that is now on land (Perez-Alvaro, 2019). However, the monitoring of underwater cultural heritages is a practice that can offer information to understand the changes that oceans may be suffering. For instance, the shipwreck *Mazarron II*, a 2600 years-old shipwreck almost perfectly preserved—including cargo—in the coasts of Spain, was discovered and covered by a cage to protect it from human threats in 1995. However, in 2019 the cage was opened and the sand movements had displaced the shipwreck and it was almost touching the cage, with the danger of being damaged. In 2021, the Spanish authorities agreed to recover and extract the shipwreck because the climate changes offered many natural threats to the conservation of the shipwreck in situ. The state of the shipwreck gave clues to the scientist of the sand movements that the coast had been suffered in the last 15 years.

Regarding oceans as leisure supplier, the underwater cultural heritage opens the door to enterprise contribution and people willing to take a financial risk: underwater cultural heritage can be places to visit and learn, giving economic benefits to local communities. While archaeological museums exhibiting objects from land only need to show and explain the object, museums showing archaeological objects from underwater sites have, in addition another goal: to educate about the environmental conditions of the oceans, including the need for clean oceans. In the past, the enjoyment and accessibility of underwater cultural heritage was just kept for the ones who were able to professionally dive.

This sector is a powerful touristic driving force since there are six million active divers around the world and more than 20 million snorkelers, an opportunity to educate them in protecting the natural and cultural heritage of the seas. Their help, their interest and their money can be very important for managing the underwater cultural heritage but also to spread the awareness of the need of a sustainable use of the oceans. The federated associations of divers are the users that are more in direct contact with underwater cultural heritage. By involving them in the protection of the underwater cultural heritage, it is guaranteed that divers will be educated in the same precepts of awareness and protection of the oceans. The reward will be reciprocal. The creation of underwater parks and reserves of underwater cultural heritage will initiate interest on visiting the site diving (or snorkelling in shallow sites which will bring a large amount of tourism).

This option is not only educative for the tourists—the visitors usually receive information beforehand on what they are about to visit—but it also regulates the amount of tourist visiting the place, what makes protection of the sea sustainable with its exploitation. These underwater parks can also be underwater itineraries with replicas, ensuring that the visitor receives a positive experience but that the 'real' heritage is not damaged. These options are experiences visitors cannot acquire with any other form of heritage: underwater trails can change mindsets and engage emotions, where the visitor can experiment the tragedy of the maritime disaster in its habitat with historic shipwrecks but also surrounded of the natural environment.

Also, the location of this heritage in ever-changing environments can bring new meanings to the site, transforming the visitor into interpreter, curator and educator, learning from its own experience and interaction with the site (Scott-Ireton & McKinnon, 2015). In addition, to be so close to a tragic human even can evoke powerful feeling, connecting the visitor to those who lived and perished as a result of its sinking.

Underwater cultural heritage are often graveyards, watery graves. Since most of the earth is covered with water, burial at sea can be seen as an accepted norm for seamen all over the world. And these graves can be tangible or intangible. An example of the first, is for instance, the *USS Arizona* in Hawaii, which is also an example of sustainable tourism on an underwater cultural site, keeping alive the memory of those who died in the shipwreck. Both a touristic attraction and an underwater cemetery, the memorial consists of a platform over the sunken ship. Visitors can see the vessel through the glass platform below their feet. The site preservation options have traditionally been focused on maintaining the integrity of the site as an historic monument and a war grave, while being scientifically studied to examine how corrosion in shipwrecks occur so the shipwreck remains preserved in situ and monitored. The year of the opening, 1962, more than 178,000 people visited the site attracting 50 years later more than 1.5 million visitors per year (Slackman, 2012). The combination of emotions and historical perspective on the shipwreck has been a key to the museum's success on attracting tourism. However, it can happen in the circumstance of a shipwreck where there are not human remains on board, but there were people on board who died. Those shipwrecks could still be considered as 'venerated sites' which are part of a community cultural space and need to be respected.

The use of underwater cultural heritage is also being used as a power projector. Underwater cultural heritage has a power as a diplomatic instrument. Data from marine exploration, including underwater archaeology, affects national defence sensitivity as it may uncover military security information, safety issues, environmental concerns, cultural evidence or food security clues. Therefore, nations want to protect their maritime heritage by safeguarding their data and adapting it to their interests, as in the case of the use of underwater cultural heritage by China to claim sovereignty and rights over the sea (Perez-Alvaro & Forrest, 2018).

The use of the oceans as a stage for illegal activities could have a great impact on the research of underwater cultural heritage, since piracy and smugglers can interfere with underwater archaeologists. However, if the oceans become preserved heritage areas and monitored archaeological sites, they will be better controlled, becoming areas more protected with less possibilities of maritime terrorism. Evaluating the beneficial role of the local population in patrolling these areas requires further consideration.

The delimitation of protected natural areas means that underwater cultural heritage sites can be better protected and vice versa: natural areas can be also be protected if considered part of an underwater archaeological site. Disposal areas, although controversial and of conflicting relations with underwater archaeological research, are currently essential. Designated ocean disposal sites are selected to minimise the risk of the potentially adverse impacts of the disposed material on human health and the marine environment. These designated ocean disposal sites are well investigated beforehand so they do not affect any underwater cultural heritage site. In this phase of research prior designating disposal areas, new underwater archaeological sites can be discovered.

Research brings many beneficial effects to underwater cultural heritage. During the last few years, a large interest has developed among marine scientists and governments in the mineral wealth of the ocean and the new energies that oceans can provide, developing a less polluting energy that could be used by all in the future. In addition, research promotes not only human knowledge but also more practical benefits; for instance, a geophysical survey could be used to analyse potential petroleum reservoirs and mineral deposits, locate groundwater, find archaeological relics or determine the thickness of glaciers. It can also offer information on volcanic processes, coastal erosion, the transmission of light and sound through water, the effects of pollutants or the behaviour of plants and animals in the marine environment. Cooperation between different organisations—biologists, oceanographers and engineers—creates a synergy on actively reporting processes and proposing adaptation measures that could establish solutions for, as previously said, rowing in the same direction.

Digital technologies also promise new improvements in raising awareness and access to underwater cultural heritage and consequently to oceans. State-of-the-art, low-budget digital technology records, analyses and disseminates data from these submerged sites. This option respects in situ preservation, creating virtual reality museums and building 2D and 3D digital models (Varinlioglu, 2016). These advances promote the awareness of oceans through the underwater cultural heritage in an age where visualisation is so important.

With the use of the oceans as settings, oceans become sensorial experiences: places to look, hear, touch, smell and taste, developing the respect for the oceans, one of the most important beneficial interaction for all users. In addition, in the last decades, new options are being developed for making underwater cultural heritage accessible to all the public, since stakeholders have understood that involving society is a good way to preserve the oceans. As a consequence, this heritage can be experimented in more accessible ways, such as walking through museums that take you underwater reaching deeps that most visitors would not be able to experiment otherwise (the unexcavated ship inside the *Maritime Silk Road Museum* in Guangdong, China) or taking the museum out of the building, like the planned above water museum in Alexandria with an underwater area. Shipwrecks encapsulated in museums such as *The Mary Rose* in the UK and the *Vasa Museum* in Sweden are examples of record-breaking number of visitors, bringing numerous benefits to the communities on shore.

Underwater cultural heritage is a kind of heritage which, since it is surrounded by an aura of mystery, can easily attract knowledge and awareness. However, its accessibility has to be weighted with its preservation. Many historic shipwrecks are been moved from beaches to foment other kind of touristic industries, such as water sports or promenades; water pollution from leisure ports can damage underwater cultural heritage and leisure diving can threaten preservation because unaware divers can recollect underwater objects as souvenirs or dive boats can anchor on wreck hulls.

Underwater cultural heritage can also be detrimental to other industries: metal corrosion, pollution or munitions on board of the wrecks can damage local areas. Maybe it is time to adopt the concept of sustainability from environmentalism as the prime motivator for underwater cultural heritage protection and vice versa (Kingsley, 2011). Treating underwater cultural heritage as a completely different element to natural heritage such as fish, reefs or flora prevents the seas being understood as a unique object to protect, as next section will explain.

The Future

The correct long-term use of resources to ensure that they will still be available in the future is one of the pivotal ideas of sustainability (Bosch et al., 2010), which is very closely tied to the responsibility of generations living today for those in the future. However, references to underwater cultural heritage in sustainability of the ocean's arenas are rare. However, as Hall (2016) states, heritage may be employed to further develop clean ecosystems awareness. The present study focuses on the required actions in four steps: recognising the loss, starting the debate, creating a legal framework in which to situate the debate and recognising underwater cultural heritage as a natural resource.

Recognising the loss means that it is essential to understand that although other users of the oceans may destroy the underwater cultural heritage, it also creates new heritage; for instance flooded cities or islands will become underwater cultural heritage in 100 years or sunken transportation vessels may become underwater tourist sites. This is the ethical issue of heritage as a process: understanding that it is inevitable to lose some cultural heritage but that some will be gained. On these premises it is again the time to evaluate the importance of the sites and to undertake actions to preserve the selected ones.

The next step is starting the debate: ocean ecosystems have occupied few concerns in the archaeological and heritage arena. Whatever the causes, the effects may be devastating (Chapman, 2003). This chapter has tried to trigger the debate, which needs to be discussed and disseminated through academic, social and political agendas. Once the topic has been discussed, it is important creating a legal framework: the 2001 UNESCO Convention that guides States on the management of their underwater cultural heritage does not include most of the uses that this chapter has enumerated as a reciprocal threat to underwater cultural heritage. As with any other international instruments, the authors of the Convention would hope to become an example to the states (Carman, 2013). If this Convention does not take most of the other users into consideration neither States will. In this regard, the regulations governing terrestrial matters, which always appear before those governing underwater matters, can help to offer solutions. However, this chapter includes underwater cultural heritage as a part of the oceans because underwater cultural heritage preservation has more similarities to *underwater natural* heritage than to *terrestrial cultural* heritage. Although the methodology—archaeology—and the ethical concerns that both underwater and land heritage face have major equivalences as Henderson (2019) highlights, this chapter considers that, in the aspects of preserving and facing natural and human disasters such as climate change, underwater cultural heritage has to fight the same battles as the natural heritage in the oceans. The changes that will affect them are the same. As a consequence, the same legal and political agendas on climate change affecting the oceans should already include—as coral reefs are included—underwater cultural heritage on their agendas.

This leads to the final point of action: it is essential recognising underwater cultural heritage as a natural resource since the seabed and the sand is covering the archaeological objects. Also, the non-sedentary fishing species live around artificial reefs made by shipwrecks. For this reason, some authors (Rössler, 2006) have tried to link cultural and biological diversity for the better preservation of underwater cultural heritage. In fact, the 1972 Convention Concerning the Protection of the World Cultural and Natural Heritage (UNESCO, 1972) is one instrument designed for the preservation of both cultural and natural heritage. Chapman (2003) also reiterates the close relationship between ecological and archaeological site management and suggests a liaison between archaeologists and other parties interested in the natural environment. Lixinski (Lixinski, 2008, p. 379) claims that the 'nature and culture dichotomy' listed in the 1972 Convention is simply artificial. It is necessary, therefore, to adjust environmental policies on the oceans in an effort to accommodate cultural heritage. As Flatman (2009) highlights, the 'green lobby' is proving successful in ensuring that good consideration is given to the environment. Wildlife biologists make concerted efforts to promote the need for ocean conservation: messages about saving whales and dolphins, protecting coral reefs or promoting catch-and-realise fishing have not only been well received by the public but creates hordes of people defending such principles (Scott-Ireton & McKinnon, 2015). If underwater cultural heritage is treated as one more element in the ocean, its protection would be guaranteed. This could be of mutual benefit: if underwater cultural heritage is preserved, the natural heritage will also gain.

One last consideration is that the study of the underwater cultural heritage is key since human beings interact because of the ocean, live on and around it. For this reason, this heritage should be a primary focus in anthropological research, not only for the study of shipwrecks or submerged cities but for the insights they give about people. The sea tells us about mobility and connections: it does not have boundaries, cultural, political or geographical. Consequently, the oceans' heritage is an invaluable international source of knowledge about individuals and communities since it provides testimony to many periods and aspects of our shared history, such as the slave trade, the gender inequalities or the conflicting relations between countries. Consequently, its study can bring knowledge about patterns in human behaviour, across time and space bringing issues of identity throughout time to the present with the cultural heritage as a base. This perspective of the underwater cultural heritage can bring unexplored benefits to the present use and users of the oceans. It is necessary to keep an open mind to attract new ideas because the sustainable use of the oceans could rely on underwater cultural heritage.

Concluding Remarks

The past is a big business which can be used to serve many masters, such as economic ends, national ends or for gaining knowledge, for instance. Managing an underwater cultural site requires a plan that combines preservation, study and dissemination. Its sustainability is essential for its understanding, appreciation and conservation. Expanded regimes of exploration through the seas, mining of precious and non-precious metals, fishing, marine engineering, the use of the oceans as a cure for climate change through 'ocean nourishment' and the production of marine-zone renewable energy can become either threats or blessings to other users. Underwater cultural heritage could also be a risk or a resource. If it is a risk, its preservation does not need to be guaranteed since can pose a hazard to the environment—pollution—commerce—impeding navigation—or human life—unsafe diving or sailing. If it is a resource, it means that it can be 'used' for research, education and tourism and to reach a sustainable use of the oceans. And although it may be thought that underwater archaeological remains are held to be finite and non-renewable, it also can be argued that the new ships are sunk every day, new shipwrecks are being discovered and new ways to explore the past are investigated, making the past renewable and non-finite.

This chapter has attempted to classify all the uses and users of the oceans, examining their main characteristics. As Steinberg (2001, p. 4) claims

> If each [ocean] actor were to pursue its strategy, the result would be […] the construction of the ocean as unclaimable transport surface, claimable resource space, a set of discrete places and events, and a field for military adventure […].

Consequently, it is difficult to balance the oceans as an open access resource to be exploited by anyone within the limits for sustainability. The ocean shapes and is shaped by social and physical processes, and the uses may conflict. However, the existing conflicts can be controlled—by legislation or just by awareness of other uses/users—and the beneficial interactions between sea actors can surpass the disagreements. Understanding previous generations and where came from, can bring a sense of unity and belonging, creating also a sense of personal and social identity. Underwater cultural heritage can have tangible uses, but it can also have the potential to provide a deeper understanding of the identity and behaviour of human beings and bringing answers for its reappearance, repetitions and repercussions in present-day societies. Underwater cultural heritage is not a different element to natural heritage. A cross-disciplinary and inter-agency collaboration is necessary. It cannot only be the only responsibility of underwater cultural heritage stakeholders: divers, museums, ONG, agencies and communities can also be part of the action. The sustainable use of the oceans is possible but it is necessary to be ready to act with several tools and new ideas.

References

Bosch, T., Colijn, F., Ebinghaus, R., Körtzinger, A., Latif, M., Matthiessen, B., Melzner, F., Oschlies, A., Petersen, S., Proelß, A., Quaas, M., Requate, T., Reusch, T., Rosenstiel, P., Schrottke, K., Sichelschmidt, H., Siebert, U., Soltwedel, R., Sommer, U., ... Weinberger, F. (2010). *World ocean review: Living with the oceans* (pp. 1–6). Maribus. https://worldoceanreview.com/en/

Carman, J. (2013). Legislation in archaeology: Overview and introduction. In C. Smith (Ed.), *Encyclopaedia of global archaeology* (pp. 4469–4485). Springer.

Chapman, H. (2003). Global warming: The implications for sustainable archaeological resource management. *Conservation and Management of Archaeological Sites, 5*(4), 241–245.

Choi, J.-K., Kelley, D., Murphy, S., & Thangamani, D. (2016). Economic and environmental perspectives of end-of-life ship management. *Resources, Conservation and Recycling, 107*, 82–91.

Fernandez, P. A. (2013). Cambio climático y derecho del mar. In J. Juste Ruiz & V. Bou Franch (Eds.), *Derecho del mar y sostenibilidad ambiental en el Mediterráneo* (pp. 271–318). Tirant Lo Blanch.

Flatman, J. (2009). Conserving marine cultural heritage: Threats, risks and future priorities. *Conservation and Management of Archaeological Sites, 11*(1), 5–8.

Fowler, A. M., Jørgensen, A. M., Svendsen, J. C., Macreadie, P. I., Jones, D. O., Boon, A. R., & Dahlgren, T. G. (2018). Environmental benefits of leaving offshore infrastructure in the ocean. *Frontiers in Ecology and the Environment, 16*(10), 571–578.

Gruber, N., Clement, D., Carter, B. R., Feely, R. A., Van Heuven, S., Hoppema, M., Ishii, M., Key, R. M., Kozyr, A., Lauvset, S. K., Lo Monaco, C., Mathis, J. T., Murata, A., Olsen, A., Perez, F. F., Sabine, C. L., Tanhua, T., & Wanninkhof, R. (2019). The oceanic sink for anthropogenic CO2 from 1994 to 2007. *Science, 363*(6432), 1193–1199.

Hall, C. (2016). Heritage, heritage tourism and climate change. *Journal of Heritage Tourism, 11*(1), 1–9.

Henderson, J. (2019). Oceans without history? Marine cultural heritage and the sustainable development agenda. *Sustainability, 11*(18), 5080.

Hyrenbach, K. D., Forney, K. A., & Dayton, P. K. (2000). Marine protected areas and ocean basin management. *Aquatic Conservation: Marine and Freshwater Ecosystems, 10*(6), 437–458.

International Chamber of Shipping. (2000). *Shipping and world trade: Driving prosperity.* Available at: https://www.ics-shipping.org/shipping-facts/shipping-and-world-trade. Last accessed 11 Nov 2020.

International Transport Forum. (2019). *ITF transport outlook.* Available at: https://www.itf-oecd.org/itf-transport-outlook-2019-presentation. Last accessed 21 Dec 2020.

Kingsley, S. (2011). Challenges of maritime archaeology: In too deep. In T. F. King (Ed.), *A companion to cultural resource management* (pp. 223–244). Blackwell Publications.

Lixinski, L. (2008). Book review by F. Francioni and F. Lenzerini: The 1972 World Heritage Convention: A commentary. *European Journal of Legal Studies, 2*(1), 371–386.

Maarleveld, T. J. (2011). Ethics, underwater cultural heritage, and international law. In B. Ford, D. L. Hamilton, & A. Catsambis (Eds.), *The Oxford handbook of maritime archaeology* (pp. 917–941). Oxford University Press.

Malve, H. (2016). Exploring the ocean for new drug developments: Marine pharmacology. *Journal of Pharmacy & Bioallied Sciences, 8*(2), 83–91.

NOAA. Office for Coastal Management. (2020). *Fast facts: Tourism and recreation.* Available at: https://coast.noaa.gov/states/fast-facts/tourism-and-recreation.html. Last accessed 22 Dec 2020.

OECD. (2020). *Better policies for better lives.* Available at: https://www.oecd.org/ocean/OECD-work-in-support-of-a-sustainable-ocean.pdf. Last accessed 13 Aug 2021.

Otto, L. (2020). Introducing maritime security: The sea as a geostrategic space. In L. Otto (Ed.), *Global challenges in maritime security: An introduction* (pp. 1–12). Springer.

Pandey, U. C., Nayak, R., Roka, K., & Jain, T. K. (2021). *SDG14-Life below water: Towards sustainable management of our oceans.* Emerald Group Publishing.

Perez-Alvaro, E. (2013). Unconsidered threats to underwater cultural heritage: Laying submarine cables. *Rosetta Journal, 14,* 54–70.

Perez-Alvaro, E. (2015). Climate change and underwater cultural heritage: Impacts and challenges. *Journal of Cultural Heritage, 21,* 842–848.

Perez-Alvaro, E. (2019). *Underwater cultural heritage: Ethical concepts and practical challenges.* Routledge.

Perez-Alvaro, E., & Forrest, C. (2018). Maritime archaeology and underwater cultural heritage in the disputed South China Sea. *International Journal of Cultural Property, 25*(3), 375–401.

Protected Planet. (2021). *Protected areas (WDPA)*. Available at: https://www.protectedplanet.net/en/thematic-areas/wdpa?tab=WDPA. Last accessed 13 Aug 2021.

Ransley, J. (2005). Boats are for boys: Queering maritime archaeology. *World Archaeology, 37*(4), 621–629.

Ross, D. A. (1978). *Opportunities and uses of the ocean*. Springer.

Rössler, M. (2006). World heritage-linking cultural and biological diversity. In B. T. Hoffman (Ed.), *Art and cultural heritage: Law, policy and practice* (pp. 201–205). Cambridge University Press.

Segar, D. A., & Segar, E. S. (1998). *Introduction to ocean sciences*. Norton Publications.

Sanders, D. (2014). *Maritime power in the Black Sea*. Ashgate Publishing.

Scott-Ireton, D. A., & McKinnon, J. F. (2015). As the sand settles: Education and archaeological tourism on underwater cultural heritage. *Public Archaeology, 14*(3), 157–171.

Simpson, G. (1998). The environmental management aspects of decommissioning offshore structures. In D. G. Gorman & J. Neilson (Eds.), *Decommissioning offshore structures* (pp. 1–22). Springer.

Slackman, M. (2012). *Remembering Pearl Harbor: The story of the USS Arizona Memorial*. Pacific Historic Parks.

Sloggett, D. (2013). *The Anarchic Sea: Maritime security in the 21st century*. Hurst Publishers.

Sommer, B., Fowler, A. M., Macreadie, P. I., Palandro, D. A., Aziz, A. C., & Booth, D. J. (2019). Decommissioning of offshore oil and gas structures—Environmental opportunities and challenges. *Science of the Total Environment, 658*, 973–981.

Stavridis, J. (2017). *Sea power: The history and geopolitics of the world's oceans*. Pinguin Book.

Steinberg, P. E. (2001). *The social construction of the ocean*. Cambridge Studies in International Relations, Cambridge University Press.

UNESCO. (1972, November 16). *Convention concerning the protection of the world cultural and natural heritage*. http://whc.unesco.org/en/conventiontext/

UNESCO. (2001, November 2). *Convention on the protection of the underwater cultural heritage*. Paris. Available at: http://unesdoc.unesco.org/images/0012/001260/126065e.pdf. Last accessed 20 Dec 2020.

United Nations. (2015, September 25–27). *Transforming our world: The 2030 agenda for sustainable development*. Available at: https://sdgs.un.org/2030agenda. Last accessed 9 Aug 2021.

United Nations. (2017, December 5). *United Nations decade of ocean science for sustainable development*. Available at https://oceandecade.org/. Last accessed 9 Aug 2021.

Varinlioglu, G. (2016). *Digital in underwater cultural heritage*. Cambridge Scholars Publishing.

Wachsmann, S. (2011). Deep-submergence Archaeology. In A. Catsambis, B. Ford, & D. L. Hamilton (Eds.), *The Oxford handbook of maritime archaeology* (pp. 202–221). Oxford University Press.

Walker-Vadillo, V. (2020). Stranded at sea: Seafaring in times of Covid-19. *Multidisciplinary Perspectives on the Covid-19 Pandemic*. Helsinki Collegium for Advanced Studies, 35–37.

Yin Chee, S., Ghapar Othman, A., Kwang Sim, Y., Nabilah Mat Adam, A., & Firth, L. B. (2017). Land reclamation and artificial islands: Walking the tightrope between development and conservation. *Global Ecology and Conservation, 12*, 80–95.

Zereik, E., Bibuli, M., Mišković, N., Ridao, P., & Pascoal, A. (2018). Challenges and future trends in marine robotics. *Annual Reviews in Control, 46*, 350–368.

23

Underwater Cultural Heritage, Poverty and Corruption in Mozambique

Anezia Asse

Along Lunga Bay, in northern Mozambique, lies an island, a small strip of land that is also called Mozambique. This island is connected to the mainland by a bridge, and connected to the history of its region by the role its people have played in it (Duarte & Meneses, 1996). Much of the history of the island, like that of Mozambique itself and the wider Western Indian Ocean region, has involved activities that produced rich resources of underwater heritage. The legacies of that history on the island can, therefore, be understood through its people's maritime and underwater heritage and its relationships to their ways of life (their languages, clothes, beauty techniques, songs and food). Cultural blending, enabled by maritime connections, characterizes many aspects of island life: local women, for instance, wear long colourful clothes covering their heads, and adorn themselves with locally manufactured gold or silver jewellery, as well as with a type of face mask called "mussiro" or *m'siro* (ARPAC, 2009; Ministério da Cultura, 2010).

Research from Mozambique's Ministry of Culture (Ministério da Cultura, 2010), meanwhile, indicates that *Nahara* should be considered the most widely spoken of Mozambique island's several languages. This is a creole

A. Asse (✉)
Sociology and Anthropology Department, Nelson Mandela University, Gqeberha, South Africa
e-mail: aneziasse@gmail.com

idiom developed locally from a mixture of African languages and Arabic, brought into contact via sea-borne connections. The Indian Ocean has, since the first millennium AD, shaped patterns of life on the island, supplying food and moulding socio-cultural habits (Chami, 1994). In recent times, around the late 1990s, the Mozambican government decided to leverage the island's cultural heritage, by promoting its tourist potential.

To this end, the Mozambican Ministry of Culture produced its first management plan for the island over a decade ago (Ministério da Cultura, 2010). The plan included all categories of cultural heritage (with exception to underwater heritage) and stated that there had been little tourist interest in the Island until 1997. This scenario began to change when the government of Mozambique implemented new policies allowing national and international private enterprises and entrepreneurs to access the Island of Mozambique. This development led to various significant infrastructural improvements, for example the rehabilitation of the bridge connecting the island to the mainland, the construction of new buildings, and the introduction of banking services. Since then, various entrepreneurs have established international-level hotels and restaurants, increasing local access to employment, and multiplying the annual numbers of national and international tourist arrivals. All this has resulted in the palpable transformation of the community.

One of the prominent companies given access to the island in 1998 by the Mozambican government is Arqueonautas, a multinational commercial large-scale treasure hunting company, operating mostly in the shallow waters of Third World countries. Other significant players in this area include the Mozambican state, UNESCO, Eduardo Mondlane University (EMU) and the Slave Wreck Project, a United States based project working to uncover shipwrecks related to the slave trade, and other research entities. One of these is Arqueonautas: between 1999 and 2014, it explored 25 shipwreck sites in the vicinity of the Island of Mozambique. Such large-scale treasure hunting activities always impact the target area, leading to changes in how the local community understands and uses its underwater heritage.

This chapter presents a case study of the Island of Mozambique. My intention is to understand how local community groups have positioned themselves as recognizable players and agents in the shifting scene of underwater heritage management. To understand this positioning, I conducted semi-structured interviews with 26 people whose work is related to the sea, and especially to the underwater heritage. These included people such as fisherfolk, marine tour guides, informal sellers and boat manufactures. I now turn to a brief background description of the island, which explains what inspired my interest in studying the island's community.

Heritage in Mozambique, and the Island of Mozambique: A Brief Background

A few years ago, to evaluate the underwater heritage exploited by Arqueonautas, Mozambique's National Directorate of Cultural Heritage (NDCH) commissioned a report from the EMU. The report concluded that Arqueonautas' activities were harmful to underwater heritage, stating that treasure hunting activities damaged underwater archaeology sites. The report concluded by underlining the urgent need for proper underwater archaeology work to protect the heritage that remained (see Duarte, 2014). The NDCH accordingly decided to annul Arqueonautas' license. Although this cancellation suggests responsible government protection of the nation's underwater heritage, it must be mentioned that other players continued their extractive practices without meaningful regulation. In addition, new players such as EMU were further commissioned to embark on new projects. This inconsistency in policy is apparent from some pronouncements by fisherfolk (pers. comm., 2019), to the effect that Arqueonautas left because it had exhausted the treasure supplies, rather than because of the government's license cancellation. Thereafter, and in substantiating the government's underhanded complicity in the continued exploitation of the Island of Mozambique's resources, the NDCH gave the new licence to the EMU, allowing it to host the Slave Wrecks Project based at George Washington University. This project aims to (re)trace the wrecks of ships carrying enslaved people on routes from Africa to the new world. Some of this work was located around the island and involved research in the same area explored by Arqueonautas. The EMU team arrived in 2016, together with their partners from George Washington University and Iziko Museums of South Africa and began their field work. The University team comprised one senior underwater archaeologist, four young, and relatively inexperienced, maritime archaeologists (myself included), two established maritime archaeologists, one curator and maritime archaeologist at the Iziko Maritime Museum in Cape Town, one anthropologist from George Washington University, and the head of the USA's National Parks Service.

The work that involved documenting and protecting excavated shipwrecks—those of the *Espadarte* and the *Nossa Senhora da Consolação*—was initially scheduled for two weeks. However, an unforeseen local community demand for participation interrupted the field work for a week. This demand came from members of the Association of Friends of the Island of Mozambique (*Associação dos Amigos da Ilha de Moçambique*), a citizen's pressure group formed by three influential individuals in 2008 to protect both the island's underwater heritage and the local community's interests in relation to that heritage. In making their demand for local participation, the association invoked the 2010 management plan, which designates the local community as the guardians of the cultural heritage of the island. By insisting on joining the work, the community argued that it would be in a better position to prevent further treasure hunting activities. This turn of events highlights the local community's agency in safeguarding communal resources. It also problematises the common, but simplistic, impression that local communities are helpless victims (together with local natural resources) of global liberal tendencies to exploitation (tendencies whose promoters act in collaboration with corrupt local leaders). With this episode, I started to reflect on and interrogate large-scale commercial treasure hunting activities and their impacts on both underwater heritage and on the local community. That community had co-habited with underwater heritage until the arrival of capitalist consumerism and its related economic activities. To pursue these aims, I conducted interviews with different publics including marine tour guides, fisherfolk, boat manufactures and informal sellers. I will first focus on how these diverse groups perceive themselves as players in underwater heritage management.

A Return to the Island of Mozambique

I have visited the Island of Mozambique numerous times, first as part of the Slave Wrecks Project team (2016–2017), then as UNESCO consultant for the underwater management plan (2017), and then, more recently, as a WiSER (Wits Institute for Social and Economic Research) seminar participant for a conference held on the island. Four years after my first visit, I went back to conduct field research, this time as a Master's candidate in Heritage Studies at Wits University. From my prior visits to the island, and interactions with the local community, I had gained some insight into its culture. I had come to know, and work with, some key members of that community, including the core members of the Association of Friends of the Island. These pre-existing networks allowed me to set up interviews and spend time with those people whose livelihoods depend on the sea. Their contributions gave me an intimate insight into their ways of understanding and managing underwater heritage.

In the sceptical eyes of some professionals, the local community is easy-going and even "lazy". When I first heard about this perception, I was studying Heritage Conservation in a classroom at EMU, where my lecturer described them as idle street sellers of artefacts. During my underwater archaeologist stint, the Mozambican' SWP manager saw them as "not hard-working" people, who relied for their living on the easy pickings provided by tourists. I experienced a different side through my lived experiences among them, without wearing a diving suit, and outside a seminar room. This time, sleeping in a room with no air conditioner, and with only a crude window near the street, I could hear the noise of thousands of motorbikes as they travelled the streets from four o'clock in the morning. I saw, also, quotidian wretchedness at close quarters: children wearing big ill-fitting shoes on their way to school, or others who skipped school to dig for molluscs, beads, stones and more on the beach. From time to time some of these children would shadow white tourists, offering to sell them something or begging for money. Youths and elders, meanwhile, were paradoxically dealing with unemployment even as new development projects emerged, and derelict buildings were being restored. Doing first-hand research on the island, I was able to shift from my previous "armchair" view of heritage towards a sensitive, humancentric approach. My transformative experience is relevant, I believe, to Hamilakis' observation (2011, p. 408) that goods experience ongoing transformation from the things and people who live with them. He proposes interviewing as a multidisciplinary approach, one which can help in shifting archaeologists' obsession with artefacts to a view that allows an understanding of other valorizations and the uses of underwater heritage by local communities.

Research Findings: The Interviews

In this section I analyse the interviews that I conducted with various players involved in the underwater heritage management on the Island of Mozambique. These are people who live on and by the sea—marine tour guides, fisherfolk, shipbuilders and informal street sellers. The questions were divided into two sections. The first section mainly focused on introductions, to establish a relationship with them and to gauge their attitudes towards the island and its history. The second section considered how they worked on the sea and with the underwater heritage. In the descriptions that follow, I anonymise my participants, to preserve and protect the identity of my interviewees. The presentation and analysis of the interviews will follow the order in which they were originally conducted. First, I met with five marine tour guides on different occasions by the beach at northwest side (Bairro Museu). Next, I met with informal sellers, before then meeting fisherfolk at the fish market of Bairro Museu (west side). Finally, I met with boat manufactures at Bairro Areal (southeast side). The translations from Portuguese to English are my own.

Marine Tour Guides

I found five self-employed marine tour guides, Adam Uacheque, Chabreque Chambal, Abacar Atumane, Albino Miguel and Assane Atento (all pseudonyms), and one tourist agency, Ilha Blue Island Safaris, that offers sea tour guide services on the island. There are also sea tour services offered by three hotels, the Feitoria Boutique hotel, Rickshaws Pousada and Hotel Omuhipi. These numbers of tourist guides and services may vary according to the season of the year and the rise and fall of tourist demand. At Christmas and years-end, the number of marine tour guides increases, so as to cater for the spiralling interprovincial and international tourists. My research visit fell in summer, just before December. During this time tourists are few, and most of the people who request sea tour services are on the island for work. The self-employed guides and the hoteliers offer the same services. They focus on guiding their customers to strategic spots for snorkelling and submarine wildlife sighting in shallow waters. The professional tourist agencies offer cruises to Ilha da Goa and Ilha de Sete Paus, deep sea sighting where customers may see big turtles and sharks, in addition to kayak outings (see Ilha Blue Island Safaris website). For this research, I focused only on the self-employed sea tour guides who work with underwater heritage. These self-employed tour guides have opinions closer to those of the subaltern, unlike elitist professional institutions, national or multinational.

I first contacted a man who is famous on the island for his work with underwater heritage. He is further prominent for being the most experienced commercial diver who doubles as a sea and underwater tour guide. In his affectionate smile, he told me that he began diving before I was born. I hence assumed that he has close to three decades of professional experience. He came to the island in the early 2000s as a hired diver for the Arqueonautas project, and subsequently chose to make the island his home and start a family. With the end of the Arqueonautas' project, he started his sea guide tourism project. His motivation was to provide people with a view of underwater heritage, drawing on his experience with Arqueonautas and his current work with the EMU. He owns second-hand scuba diving equipment and a boat. His services consist of submarine sighting offerings, snorkelling, diving around underwater shipwrecks and other underwater heritage artefacts.

Here, for ethical considerations to protect his real identity, I will call him Adam Uacheque. Besides his eminence in this community, Uacheque is respected so much in the locality that young people report to him any topical issue concerning underwater heritage or conflicts between other self-employed marine guides. Unsurprisingly, he is a core member of the Association of Friends of the Island (*Associação dos Amigos da Ilha de Moçambique*). In Uacheque, I began to understand that certain people of the island have more power to regulate issues than the larger official political structure itself. In a demonstration of local community agency, this group has organized itself according to different professions, where each profession has a leader. The leader is always someone who has living and working experience outside the island, as this confers a broader perspective. In the case that an external issue is bothering the island, all the associations group to analyse and solve the challenge. In particular, Uacheque has the power to oversee the underwater heritage at local level, and he and his fellows from the friends' association have, together, the higher power to influence the management of Mozambique's underwater heritage at national level. This inverts the typical top–bottom approach to resource management, gesturing towards grass roots management of the same.

Today, Uacheque is in his mid-50s: he started diving when he was a "boy in Maputo city", as he put it (Uacheque, Interview 2019). He decided to work as a commercial diver because "it was cool. I used to work with different people all the time. I love meeting people and traveling. And of course, the world under the water is so cool. It is not like an ordinary job – I feel privileged to wear my working suit". As a diver he joined several projects overseas, which qualified him to join multinational companies such as Arqueonautas (ibid.). This allowed him to get technical skills in the excavation and management of underwater heritage. According to archaeological reports from 2004 to 2015, the divers worked three-month shifts where, depending on the weather conditions, they were obligated to be at sea every day (Mirabal, 2004, 2006, 2010, 2015).

The divers' role in the first phase of the project was to identify archaeological evidence by diving around the abnormalities pin-pointed by a magnetometer. In the second phase, they were required to excavate and lift the artefacts to the surface. Some divers, including Uacheque, joined in the process of selling artefacts by flying overseas, transporting artefacts (Uacheque, Interview 2019). From these processes (attending the discussions sections and seminars hosted by Arqueonautas team), Uacheque learned the history of shipwrecks and of the island. "This is the knowledge I share with my customers" (Uacheque, Interview 2019). Besides that, Adam sees diving as a potential life-changing avenue for young people in his community. He said that, "due to the high level of unemployment here, most young people are making bad decisions in life. I have a co-worker from here. He does his job very well. He is qualified for commercial diving and has been doing a good job outside the island as well" (Uacheque, Interview 2019). He has been preparing another young man to be a commercial diver in the future. Uacheque further shared with me his hopes for the community youth, "I want to show them that there are bright possibilities for their lives, especially the young ones. They need to learn more about their history and find ways to make a living from their heritage" (Uacheque, Interview 2019). He believes in making a living from the underwater heritage, learning from his experience of life, an experience that changed on the island. He told me that the income of a commercial diver is good due to the risk and working condition that it implies.

> Sometimes you can be diving to install cables in a beautiful coral area in summer. But during other days you have to dive in an area that contains hippopotamus in mud waters with zero visibility. Where all you have is trusting the voice commander coming from the guys who control the cameras inside the boat. You turn left or right, picking things up that feels like the bodies of victims of accidents that you are looking for. And you get flustered when you realize that you just plucked somebody's hair from the rest of the body because it has already deteriorated. (Uacheque, Interview 2019)

A part of the reason Adam settled on the island may lie in the pleasure of working in the blue and warm waters. He told me of his aspirations to turn his business into a professional corporation. He arrived on the island just days before my arrival. He had come there from Inhambane province in the south of Mozambique, where he had completed the field training needed for his PADI (Professional Association of Diving Instructors) certificate as a divemaster. He reasoned that, he has been diving for a long time and, he is good in what he does. But that he needs to make his business more professional. He needs to work following international standards, as this will give him more credibility (Uacheque, Interview 2019).

Uacheque's routine, as he described it, and as I observed it during my period (almost two weeks) of field work, is that his workday starts early so as to ensure that he gets into the sea at the right time according to the day's weather pattern.

> If I have scheduled customers. I first check the tides and the wind. It allows me to plan the exact time we should visit certain shipwrecks and for how long.

It is first for safety reasons and second for the quality of the view. High tides make diving much easier in general. And low tides are good for snorkelling. The wind also interferes in diving. So, it is always good to be at sea before the mild wind which brings dust to the sea. (Uacheque, Interview 2019)

After that, he prepares the equipment together with his part-time assistant. He said, "I prepare diving equipment and the motorboat in the afternoon. In the morning, I often snorkel by the bridge (*Pontão*)". When he does not have customers, he spends the day snorkelling, taking pictures of corals and observing the sea [he did not share the reason for this everyday recording], and fishing for family consumption (Uacheque, Interview 2019).

His work has given him formidable popularity on the island. He attracts diverse customers in different ways. Some come to visit the island and, once there, get to know about the snorkelling and diving service through observation or information from the previous tourists or streets talk. Others come to the island already knowing about his services (ibid.).

By upgrading his commercial diving skills to divemaster level, he will, in addition to participating in commercial dives, be able to legally lead any diving excursion. There, he will be keeping an eye on students, demonstrating to them the necessary skills, and refining their skills. This is an area with potential interest for scholars and entrepreneurs. He can also look after a group if the instructor is absent, and he will be responsible for safety equipment and scuba gear (PADI, 2020). His urge to be a divemaster arose when the EMU started its research on the island. They needed someone to look after their equipment, someone who had both a certificate in diving and intimate weather knowledge. Uacheque was part of the community group that interrupted and delayed the start of the university's research (see the descriptive background section above). He, together with his fellow associates from the Association of Friends of the Island, re-appropriated their space and place in the project. While his fellows were still practicing their diving skills, Uacheque proved to be essential in the project. His long-time experience in the area enabled him to wisely schedule dives, equipment choice, and safety issues in certain shipwreck sites. With his divemaster certificate, he aims to further, "bring young people from the community. To teach them how to dive. And my future vision is that next time we will not need people from outside to manage our underwater heritage" (Uacheque, Interview 2019).

Uacheque perceives underwater heritage management as having different levels. And, surprisingly, there are times where he claims to be part of the community of underwater heritage managers, mostly when it is associated with good practices. And when it comes to talk about treasure, or who is doing treasure hunting activities, he excludes himself as a community member, casting himself as the one who came to help the community. However, he has his own way of underwater heritage management, which is different from those who are not from the community. He describes three modes of heritage management: a community management approach, his own personal approach, and an academic research management approach. It is at the first level where his concern is largest, and where his wish to change things is greatest:

> The community does not know its history. They do not care about the conservation of their heritage. They are interested in the money that they can get from it. Here, anyone dives and if there are some artefacts they collect for sale or house decoration. The collection practices start from small kids to old men. They do not have [the] good manners required for underwater management. (Uacheque, Interview 2019)

As for his own personal approach to the management of underwater heritage, he was insistent about certain key features of that approach:

> Underwater heritage must be preserved in situ. I am used to be at the sea almost every day unless I have work to do outside. Sometimes I do exploration diving. I have discovered new artefacts. Before I used to show my discoveries to the university team, but I need to reconsider my attitude because it will be documented as the team's discoveries, and I do not know what they will decide to do with it. At present, I record and register the sites then I share the artefacts and site views with my customers. I think the best thing to do is to take a tour where everyone would enjoy the heritage. This situation is worse during holiday seasons, when the community does its best in diving without equipment with the visitors. It is like everyone is a tour guide. (Uacheque, Interview 2019)

Regarding non-community or academic research approach, Uacheque states that:

> This is the confusing part for me. In the last years of Arqueonautas here, scholars used to call them treasure hunters, and I believed it till I joined EMU team. First, the EMU ignored the community. And then they said that they were here to correct Arqueonautas' destructive work. But in my opinion, this team's way of leading the work is suspect. Arqueonautas had a bigger team where each member had her or his role. I never saw [a key member of the research team] going on diving all the time, he used to stay in the office reading books and doing his research. He was the one who was in charge of telling

people what to do. I remember spending all three months at the sea living in a boat in an area close to Nampula with other divers, marine biologists, photographers, and archaeologists. [A key member of the research team] came once in a while to dive and supervise our work, but we found shipwrecks and valuable artefacts. But with this team which calls themselves underwater archaeologists the work is comical to put it mildly. The archaeologists spend time diving and at the end of the day they are exhausted. I am not surprised that they are yet to find any new shipwrecks. (Uacheque, Interview 2019)

Uacheque insightfully suggests that a financially weak team may cause more damage to the underwater heritage than Arqueonautas did. As he put it, "the lack of diving experience and adequate equipment will destroy heritage by mistake" (Uacheque, Interview 2019). Moreover, it becomes clear that the cancellation of Arqueonautas' operating licence did not do enough to protect underwater heritage resources: in fact, one gets the opposite impression, that the incorporation of new players like EMU worsened matters in relation to underwater heritage protection.

After my interviews with Uacheque were concluded, I interviewed Chabreque Chambal (pseudonym). I met Chambal at his customary spot by Vasco da Gama square, on the northwest side of the island, where he plays games with his friends in between driving his customers to Cabaçeira Grande or Pequena, or from there to the island. He is a young man in his early 20s, but is already a long-time marine tour guide. In the last four years he has been working with underwater heritage resources. I started this interview by asking Chambal to introduce himself and his work.

> Well as you can see Anézia, I am a taxi driver in these waters, just like my late father was. When I was a little boy, I started asking my father if I could follow him to work and he said yes. So, I started helping him shipping customers' luggage and observing how he was driving the rowing boat to the island and then back home to Cabaçeira. I had great fun learning to drive a boat by reading and analysing the wind and tides with him. You must realise that at that time, I was already good at swimming since little children here spent most of their time playing and swimming at the beach. I am so good at swimming that sometimes I cross all the way from Cabaçeira to the island and back swimming. Sometimes I join swimming competitions during festivals while at other times I just compete for fun with my friends. After working with my father, I saved enough money to get my own rowing boat and started working for myself. (Chambal, Interview 2019)

With his excellent swimming skills, I asked Chambal if he knew stories associated with the sea. He laughed and in a contented mood said:

> You know, I work almost every day, excluding our Muslim religious festive days such as Eid Mubarak. We are very religious here. We do not do business during Eid day; we do not believe in spirits at the sea. And all the stories I know are related to accidents involving boats. I remember one day I was working to help a friend of mine in his boat on a windy day and the tide was very high. Violently, the wind blew and turned the boat upside down. We had to swim quickly to the surface to resurface the boat in right position. It was hectic but, as always, we managed to do it. (Chambal, Interview 2019)

Chambal also related other similar episodes involving accidents, which, because of his Muslim religious affiliation, he characterizes as spontaneous mishaps rather than supernatural events. It appears that on most of the occasions when these mishaps took place he was not with customers nor on official business, but was doing other jobs. He also emphasized the importance of teamwork, when he talked about his former assistant.

> As a driver and owner of a rowing boat you need to work with someone younger than yourself to assist in driving the boat. There are two basic requirements in rowing boats; looking after customers in smooth and rough times and driving the boat itself. Up until last year, I had a full-time assistant who was with me from the time I started. But now he has outlived his dependence on me and is determined to manage his own boat business. Coincidentally, I am no longer in need of a permanent assistant, since a motored boat is easier to drive than a manual one. Unless there are petrol shortages on the island, or the customers do not have money to pay for a motored boat, I stick to the latter. In fact, last year I sold my rowing boat to a white lady, but sometimes I borrow it from her if the need arises. (Chambal, Interview 2019)

Chambal works with the community and with tourists:

> I know the whereabouts and needs of my customers. Some are workers on the island, so I transport them to and from here in the morning and evening. Others just come to island to consult the hospital doctor, buy stuff or for recreational fun. But I also guide tourists who want to visit Ilha do Farol and Ilha das Cobras [the tiny, inhabited islands that lie close to the Island of Mozambique]. Ilha dos Farol is the first destination as it is closer to the island and the main touristic attraction there is climbing the lighthouse to enjoy the aerial view of the huge sea waves. In addition, if you are lucky you get to see sharks and big turtles. The next stop is Ilha das Cobras where the first attraction for the tourists is the suggestive serpentine name of the Islet which promises the presence of snakes. But we have never seen even one. I suspect the fearsome name was given to protect the Islet from exploration. There, the tourists go

for camping, or just walking to see the quiet lagoon in the middle of the Islet surrounded by mangroves. (Chambal, Interview 2019)

As for underwater heritage, he said:

> I have to confess to you that I am quite disappointed with it. After Arqueonautas there was a positive expectation on fruitfully working with these objects. I thought that I was going to gain financial rewards from it, in the footsteps of Arqueonautas, but that expectation has remained elusive. I have not gained any special income from it. I took tourists for snorkelling around shipwrecks but there are fewer tourists interested in it. It appears the view from the reefs is better than the view from broken timbers. I also joined the recent project with the University through hiring out my boat and my time to them, but they pay so little that I am looking for another project. (Chambal, Interview 2019)

Fisherfolk

Fishing is one of the oldest professions on the island, with origins that date back to back before the Bantu migration. On my way to the island near the end of the street, before I crossed the bridge, I saw people selling and buying fish at the teeming Jempesse fish market. It was low tide then: when the car slowly negotiated the market and entered the bridge to the island, I looked down and could see several fisherfolk canoeing around the shallow waters. Other fisherfolk were at the beach parking their canoes and boats, or arranging their fishing nets. When the car slowly entered the island on the southeast side towards the northwest where I was staying, I smelled the strong aroma of fresh fish and sea saltwater characteristic of fish markets. Instantly, my appetite for sea food soared.

To understand the fishing practices present at this world underwater heritage site, and the relationship the fisherfolk have with it, I embedded myself in their norms. I first contacted the head of the fisherfolk association (*Associação dos Pescadores da Ilha de Moçambique*) before contacting the rest of the fisherfolk. In an interview (ARPAC, 2009), the former head of the fisherfolk association stated that the principle reason behind the formation of this association was the need to join forces to fight for the common goals of fisherfolk. One of the most important of these goals is the creation of a platform where fisherfolks can safeguard their fishing business. The former head further bemoaned the fact that, although fishing is a major occupation on the island, the fortunes of fisherfolk are more miserable than those of practitioners of other pursuits, who tend to notice positive changes in their lives after years of dedicated labour. But in the fishing business, prospects are dire. This is not helped by the fact that the plenitude of fish in winter results in unsustainable low prices due to the lack of appropriate conservation practices (ARPAC, 2009).

The current head of the association is a man in his mid-60s. For ethical reasons I will call him Caetano Francisco. I met him and described my project to him before requesting for some fisherfolk contacts who could share their experiences with me for this research. Caetano Francisco agreed and promised to organize the interviews during the next coming days. To avoid misunderstandings, Caetano Francisco suggested that the interviews should be conducted collectively rather than on individual basis. I had no option but to agree. Caetano Francisco began the interview by explaining that:

> We have worked with different people before and we are tired of being excluded from the benefits of our contributions. So, my current position is to make sure that my people are not being exploited and that the information they share does not financially enrich other people. Even when it comes to me, I am no longer attending seminars for free. If they want me there, they should pay me first. (Francisco, Interview 2019)

Just like Adam, whom I discussed earlier, Caetano Francisco has made the island his home.

> I was born in Nampula, but I came to this island to live with my relatives and to attend school. In Nampula my parents lived far from schools. After middle school I left the island. I worked outside and returned after a long time. However, even now I still go to work outside but I feel that my skills and contacts are more useful here than anywhere else. As part of my duties I attend meetings at local and national levels. As you know, the government has been working with organizations such as UNESCO Mozambique and usually, after these meetings, decisions that affect the whole community are made. (Francisco, Interview 2019)

During the collective interview I met seven fisherfolk: Rick, Manuel, Tómas, Abdul, Ussene, Jamal and Mohamad. Caetano Francisco further invited Rita da Luz, an entrepreneur working with some fisherfolk. After Caetano Francisco had introduced me and my project to his group, I began the conversation by enquiring why they had decided to work as fisherfolk. Rick, who in his mid-20s is the youngest member of this group, revealed that he had started fishing after failing to find his dream job. "I took a driving licence, I wanted to work as driver, but I could not find any job, so I came here to fish and make some money for my family" (Rick, Interview 2019). In the same vein, Manuel and Tómas who share similar backgrounds inherited the profession from their parents. They knew that their initial job prospects were tied to fishing alongside their parents before they could move on to become fully-fledged professionals.

> We grew up here, seeing our fathers preparing their boats and nets early in the morning and returning in the afternoon. We were always excited by the beach as we saw them returning. We usually jumped into the boats, helping them in parking the boats, carrying the fish and then cleaning the boats. (Manuel & Tómas, Interview 2019)

Rick is still new to this fishing profession and practices it by himself. "I do net fishing by myself. I feel it convenient as I manage my time and still get fish whenever I go fishing" (Rick, Interview 2019). Abdul, Ussene and Jamal all practised artisanal fishing with rowing boats for a long time, before they decided to join Rita da Luz' project, and acquired motors for their boats. Rita proudly described herself as the sole female in this profession in the whole nation.

> I am from Nacala Porto. I am a child of fisherfolk and I always wanted to do something in this area. But I could not find women in this business, so I then tried different businesses. When I got some money, I decided to invest in buying motors to make fishing more profitable for my community. It worked even better than I expected. I now possess many motors and fishing gear around Nacala and I come here to help fisherfolks from this island who seem to have been forgotten by the government. If you go to places like Inhambane you will see that communities there are working with good fishing gear provided by the government. Here I must come to the island and transport fish to sell to other places. (da Luz, Interview 2019)

Abdul, Ussene and Jamal noted positive changes in their fortunes after working with Rita da Luz. They said that it is "cool" to have her here. "Working with rowing boats makes you slow as your navigation depends on the weather. But with a motor you go far into the sea, and you get more from your day" (Abdul, Interview 2019). Caetano Francisco reinforces the viewpoint that the Ilha is a place which provides good fishing during all year even in shallow waters:

> You see what those people are doing now, they are fishing right here (as he points out some people in their canoes fishing in front of us). These practices are new, before we had order in this place but since many white people with

money started to come and run projects for their own benefits like Arqueonautas did, the community also thinks that they can do what others did. (Francisco, Interview 2019)

Manuel added, "do you see the lights in the night, right? Those lights in the middle of the sea. They are Chinese boats fishing all day, which means that in two years' time we will be left with nothing. Can you imagine a fishing community without fish? It will be us" (Manuel, Interview 2019). Caetano then added:

In few days I am attending a meeting in Maputo, I will be meeting the ministers. I really need to talk with someone who has power. This place carries the history of this country. The government uses this historical fact all the time for their benefit, but their actions show that they do not care about the future of Ilha. (Francisco, Interview 2019)

And it was true: I could see the lights that Manuel and others mentioned, distant lights from out at sea, from what looked like industrial fishing vessels. Historically, Ilha has been considered as exclusively a place where the local community practises artisanal fishing using manual rowing and sailing boats (see ARPAC, 2009; Ministério da Cultura, 2010). There were no reports of the kind of industrial fishing associated with China or other countries. The fisherfolk associate the present scarcity of fish with the massive extractive practises of the international players. This impression was expressed by Mohamad and supported by the rest of the group. They told me:

we are not expecting great quantities of fish during summer. We all know that [in] these hot times the fish do not come closer to our hot water, however it is getting lesser and lesser from the amount we used to get in the past. The numerous industrial scale fishing projects are robbing us of our livelihood. (Mohamad, Interview 2019)

Caetano Francisco, therefore, explained what this meant:

You see! We, as a community have organized ourselves into associations to regulate the access to our resources. We will block any researcher who comes to the island for exploration purposes as a way to conserve this place. For that reason, all the associations are working together to deal with people who are not from the community. However, each association is also responsible for choosing members from the community who will be part of responsible research. We are tired of exploitation. Many people come to the island to rob our heritage for their own good, leaving the community in poorer condition.

I think it is only fair that the local community should be the first to benefit from the island's heritage. (Francisco, Interview 2019)

Jamal, similarly, shared his views on fishing practices before the advent of industrial scale interventions through foreign companies:

Fishing was a noble and courageous activity. Fisherfolk were more respected members of the community. But now it has changed, people who are not doing any job get respected. Going fishing is such a dangerous and unrewarding activity. We risk our lives every day. The sea is unpredictable, especially here. Just consider how many ships sunk to the bottom in the past and how many people died. Fortunately, we know what we are doing, and how we do it, and that is why we are still alive. We spent years and years learning before we could go fishing. (Jamal, Interview 2019)

Caetano Francisco shared the details of a certain tradition that is slowly becoming neglected by the young generations. This tradition was also recorded in 2009 by the National Heritage Archive.

We used to conduct a traditional ritual to thank the spirits of the sea for each fishing season. We know that we are not alone and there are other spirits living at sea. They need to be remembered from time to time. This tradition of expressing communal gratitude to the spirits of the sea was a way to reunite fisherfolk from the whole island. We congregated at a religious leader's house where some fisherfolk wives cooked seafood while the fisherfolk offered prayers of thanks, protection, and provision. After the prayers, the fisherfolk gathered food for the spirits. The food was carried on a new canoe prepared specially for this. The leader would choose four fisherfolks to take their boats and transport the special canoe to the sea. There, the food was thrown to the water spirits, and the canoe left there. After the fisherfolks returned from the offering, the party started, and everyone could eat. During this special day no one could go fishing. (Francisco, Interview 2019)

The abandonment of traditional fishing practices and rituals reflects as well on the changing social habits in the community, as Mohamad said:

The mess installed by these projects is reflecting in our kids, they are impatient on learning from us like we did from our parents. They want to emulate the foreigners who run the big projects. They are also looking for treasure. As for me, I grew up seeing many things under the sea, but I knew that I could only pick up what I needed for the house and for nothing else. (Mohamad, Interview 2019)

The last interviewee in this section was a member of the secretariat of the fisherfolk's association. He is an artisan and a former Arqueonautas worker. I met him at his house in the Macute City. This is how he began our conversation:

> I am from Maputo like you, Anézia. I grew up in Mafalala[1] and after visiting this island I fell in love with it and decided to stay. My life is shaped by this place and I have kids. I am an artist and I have worked in many projects. My current passion is working with kids in need. As you can see, poverty is widespread here and kids tend to end up following wrong ways. I help them realise that they can do amazing, clean jobs. (Fernando, Interview 2019)

Fernando told me this while showing me his children's drawings. I appreciated the drawings, which were well drawn, indeed were done with what looked like a professional artist's finesse. Most of those images are representations of the island's landscape. While I was still appreciating his children's artwork, he continued:

> I had a small art school where my pupils used to spend most of their time practicing, but now I do not have money to run a place like that, all that work we do here at my yard. I teach them the history of this country, with a focus on this island, and we do visual representation of our history. (Fernando, Interview 2019)

I saw artistic representations of ships from the fifteenth and sixteenth centuries sunk at the island's deep waters, those excavated and processed by Arqueonautas and others. The details of these artwork were so impressive that I wondered about the process of production of this fine artwork. Fernando explained:

> Look at this piece we have just sold to an American guy. You see this little boy? He is an orphan; he made this piece by himself and I am so proud of him. He is an artist and now I am sure he is confident enough about his life. I learnt a lot from Arqueonautas, I worked as an underwater drawer. I travelled; I went to Germany. I am fluent in German too. Arqueonautas changed my life and, I hope I could do more for the kids as that company did for me. When I joined Arqueonautas I was not that good at drawing, but they had time and resources to teach me. I learnt about diving and I become good at underwater drawing. (Fernando, Interivew 2019)

Like Adam Uacheque, Fernando Mohamad also distanced himself from his community's conservationist tendencies regarding the island's heritage. He told me the following:

I can say that maybe it is sheer innocence or lack of good education, but this community lived for a long time with precious treasures under their eyes without exploring them. Arqueonautas came and showed them how rich they were, all they need to do is to learn and open their eyes. Consider how many people are living miserable lives around this neighbourhood? They need to change their lives through wiser use of their heritage. I am so thankful to some Arqueonautas members who are still helping me with donations of drawing material for my kids. I hope you and more people can do the same next time you come here. (Fernando, Interview 2019)

Informal Street Sellers and Island Heritage

Informal street sellers can be found everywhere around the island, from the Jempesse fish market, the street market at Macute City, to Stone City. Those who are sellers at the latter location offer cooked seafood and craft products to the tourists. For this project, I focused on the sellers from Vasco da Gama square. At this spot each of the sellers has a wooden stall next to one another. The sellers are all different from the previous groups, as they are younger, with their ages ranging from early 20s to mid-20s. They start working after eleven o'clock in the morning till six in the evening. I met four of them (Abacar Ibrahímo, Abdurarremane Malisso, Amade Uandela and José Rocha) and I explained my project to them. They agreed to participate as a group, "since we are all standing close to each other, we can take part at the same time" (Ibrahímo, Malisso, Uandela and Rocha, Interview 2019). Ibrahímo noted that:

I sell beachwear and jewelleries. The first I get from Nampula, while my brother and I make the jewelleries. He sells around Lumbo, and this is my spot. Here I meet customers from overseas. I speak English with them, and they pay me well. They are different from local customers that always negotiate down the prices of everything. (Ibrahímo, Interview 2019)

Uandela observed, "I come here to sell at my best friend's spot, he got married and his white wife gave him a job as a receptionist at the hotel". Malisso is a part-time seller, he is multi-talented and has many jobs. He added. "I am different from many lazy people here. I work in a white lady's house at Cabaçeira Grande. She travels all the time. When she is not here, I have time to make lamp holders, bracelets from seashells and beads" (Malisso, Interview 2019).

The selling of these tourist wares also varies from one touristic season to another, as Ibrahimo submitted. "November and December are undoubtedly the most lucrative season for our business. But there are other holidays in the middle of the year that are as good as well". Malisso added,

> We also collaborate with some touristic related institutions such as hotels and restaurants. We provide them with raw materials if they already have someone who can craft, or we make the products and sell to them. The only challenge is that every day presents new people on the street selling the same things as us. Mozambicans are not creative at all, they like copying each other's business models. (Malisso, Interview 2019)

As Uandela explained, the sellers collect or buy beads and seashells from children and women, and use these items to make their products:

> When we have a lot of jobs, we buy beads and seashells by the beach from children and women, or they come here to sell to us. When we do have time, we go to Ilha das Cobras where there are many seashells to collect for ourselves. During tourist seasons we do not have that time to collect on our own, so we simply buy from others and we create our work here. (Uandela, Interview 2019)

The raw materials are also collected at the nearby towns of Cabaçeira Pequena and Grande. Malisso, who makes baskets from dried leaves, told me, "I live on that side, so I collect white and black seashells and I look for leaves as well. I dry the leaves at home and then take then with me wherever I go making baskets of different sizes and styles" (Malisso, Interview 2019).

The products described above are not the only ones being traded. They later showed me a lamp holder decorated all over with fragmented white and blue porcelain that looked exactly like the famous Chinese porcelain from Ming Dynasty excavated by Arqueonautas (see Chapter 2). I did walk around the beach from the northwest-to-northwest side of the island, and I could clearly find blue, red and black beads and some few fragments of blue and white porcelain. The first are interpreted as some of the principal trading goods that, in the era before European colonialism, were brought from India to the Mozambican coast in exchange for gold and ivory (Duarte & Meneses, 1996). They also sell old coins, some from numerous centuries ago and iron made kitchen wear. The coins from fifteenth to seventeenth century (inscribed on one side of the coin) were recovered from the sea. Of the old kitchenware, some look similar to the ones exhibited at the Museum of the island recovered from the sea. When I asked them where these products come from, the two owners (the rest of the sellers did not have these) submitted that these "are original artefacts used by Vasco da Gama himself. We sell with special prices to tourists who curious to learn our history. So, we give them a slice of this fascinating history" (Informal Sellers, Interview 2019).

Apart from compelling artwork and craftmanship of different products, the sellers are also well skilled in discoursing on the interesting history of the island. During my time with them, I heard, on countless occasions, their accounts of the island's history as they narrated them to tourists.

Manufacturing Boats

The growth of fishing activity around the Island of Mozambique has resulted in increased demand for fishing boats. Therefore, there has been an attendant expansion of the boat construction business. From what I observed, canoes or small boats are in higher demand than rowing boats. While the former is almost exclusively used for fishing, the rowing boats are used for fishing, leisure, tourism and other activities. To see and hear more about this profession at the island I contacted and met José Paruque. He is a young man in his mid-20s. We met at his house in Bairro do Arreal, in Macute City. This time, I needed considerable help from my friend Sulemane Rachid, who became my field work assistant. He walked me around and helped to find the locations of my interviewees. Rachid did the translation from Portuguese to Macua and from Macua to Portuguese, and enabled me to conduct the interview with Paruque.

According to Paruque (Interview 2019), he became a boat builder because:

> My father did the same profession years ago, until he retired. My father taught me all about making boats. The most complicated part is to make it float and comfortable at the same time. But I can do it without much trouble. (Paruque, Interview 2019)

Paruque said that dynamics had changed in the island. He said, that, "the only avenue to enter and explore the sea is via a boat. And I am a boat manufacturer, and everyone needs my services. It is not like I am the only one in this profession, we are many. You will see tomorrow morning, but we are all busy thank God" (Paruque, Interview 2019).

The following morning, I visited their workplace, around the beach close to the main fish market in the west side of the island. This big open space looks smaller thanks to the big tree trunks, unfinished boats in different styles that are parked there. Some of these boats are laid upside down, with their hulls facing upwards, while others are laid on their sides. There are boats there of all sizes, including some so big that they made me wonder why I had not seen such huge boats out at sea. I then reminded myself that when at sea, a part of such a huge boat will be underwater, and that it would be for that reason that they would look much smaller when afloat. Unlike the other groups I worked with, the boat makers did not talk much. Instead, they invited me to join their work, and to see the process of how heavy tree trunks are processed into boat material. Malinda explained:

> We start early in the morning; we have been working more than we used in the previous decades. Now customers ask for fancier styles and bigger sizes. For example, almost all want versatile ones with covers to protect tourists from the sun. The cover must be easily removable to turn the boat into a rowing boat when there are no tourists. We do our best to satisfy our customers. (Malinda, Interview 2019)

Discussion

This chapter has documented some of the attitudes, perceptions and options of the local community of the Island of Mozambique *vis-à-vis* the Arqueonautas company, and other agencies that are working underwater off the island. The interviews conducted have revealed divided opinions, within the community, about the governmental decision to transform the island into a site of competitive touristic attractions. From the marine tour guides' attitudes to the provision of underwater tourism services, we see the advantages of local self-employment. However, it is also apparent that marine tour guides still need training from the government, the EMU and international agencies such as UNESCO, so that they can provide safer and more educational services to their tourist customers. The divers who used to work with Arqueonautas have retained their diving skills and knowledge of the history of shipwrecks around the island: nonetheless, they need further training in order to acquire skills needed for tourism services. The other notable aspect from the former Arqueonautas workers is their interest in developing their community via its cultural heritage.

The informal sellers have become full-time commercial treasure hunters. They have the secret knowledge of the location of artefacts, though it remains unclear how they manage to acquire artefacts for sale during all seasons of the year. Besides the media marketing strategies used by large-scale commercial treasure hunting companies, the sellers on the island have developed face to face marketing. By closely analysing the artifacts offered for sale in this context, I concluded that it was uncertain as to whether all of them were from the underwater sites off the shores of the island (some of these artefacts presented characteristics that differed from the artefacts known to come from the island).

Fisherfolk can be described as the scrutineers of the sea. The success of their work depends on generally correct environmental behaviour while out at sea. Additionally, fisherfolks have organized themselves in order to oppose further large-scale treasure hunting and research activities that do not freely include them as active participants.

The last group is different from the previous groups interviewed. Those who build boats are beneficiaries of both large and small-scale commercial treasure hunting activities. In interviews and other activities with me, they expressed their satisfaction with the rising number of treasure hunting activities, and of academic research, which created opportunities for them.

The division in these opinions arises, in part, from the project's failure to meet the initial high promises it had made. The most unsatisfied group, in this regard, is the fisherfolk who regret that their working place is being made precarious due to a lack of the proper fishing regulation which could prevent overfishing and fish scarcity. The other groups seem to have benefited from these projects, mainly from Arqueonautas' activities.

In order to cope with and regain relative control of their communal destinies in light of foreign driven projects, local communities on the Island of Mozambique have created influential organizational structures to control and prevent further exploration and exploitation of their underwater heritage without their consent. Those who lead these communal structures are individuals with significant and wide experience in public relations acquired through years of working in different places with different institutions. The "grass root" structures thus created are powerful and effective in controlling and policing suspicious actions on the island. These associations are illustrative of communal agency and preparedness in conserving natural resources, filling the huge gap left by indifferent government policies. As the people who bear the brunt of extractive practices the most, it is proper that local communities should police and benefit most from the use of their natural resources.

This case study has revealed different perspectives for underwater heritage management within the community. For Adam Uacheque and Fernando Mohamad, conventional and "modern" heritage management systems provide the most efficient means to protect heritage sites. Their stance is informed by insights gained from their long working experience with Arqueonautas. This system's regime of management argues that underwater heritage sites should the transformed into tourist sites that will benefit the individuals in charge of the programme's implementation. At the same time, fisherfolk are clamouring for a holistic management approach where their knowledge of the sea, their skills in diving and navigation and their treasure hunting should be validated by the government, UNESCO and scholars as local community practices in the underwater management process. Fisherfolk are arguing for a democratic use of the heritage that lies under the sea, and for an equitable distribution of the economic benefits derived from such use. The two last groups—street sellers and boat builders—focus on the economic values involved in underwater heritage.

What also emerges from this research is that it is too simplistic to blame foreign companies like Arqueonautas and Chinese fishing companies for the harm visited on the Island of Mozambique and its people. There are other culprits at work in this case: they include the national government whose corruption, and indifference to the needs of local communities, has condemned the most affected communities to more misery and poverty.

Acknowledgments The author is presently a PhD candidate of the NRF SARCHI Chair in Ocean Cultures and Heritage, UID Grant 129962.

Note

1. Mafalala is a suburb in Maputo province founded by people from Nampula. It is most famous as a place where most of Mozambique's national heroes lived and planned the liberation struggle against colonialism (Gonçalves, 2016).

References

ARPAC. (2009). *Património cultural imaterial Makhuwa Nahara, da Ilha de Moçambique*. Unpublished report, Instituto de Investigação Socio Cultural (ARPAC). Arqueonautas website. http://aww.pt/. Accessed 20 Apr 2020.

Chami, F. (1994). The Tanzanian coast in the first millennium AD: An archaeology of the ironworking, farming communities. Uppsala: Societas. *Studies in African Archaeology, 7*, 287–289.

Duarte, R. (2014). *Resultados da avaliação do project PI/AWW respeitante ao património arqueológico subaquático da Ilha de Mozambique*. Unpublished report, Universidade Eduardo Mondlane.

Duarte, R., & Meneses, P. (1996). The archaeology of Mozambique island. In G. Pwiti & R. Soper (Eds.), *Aspects of African archaeology*. University of Zimbabwe

Publications, *Papers from the 10th Congress of Pan African Association for Pre History and Related Studies* (pp. 555–559).

Gonçalves, N. S. (2016). The urban space of mafalala: Origin, evolution, and characterization. *Journal of Lusophone Studies, 1*(1), 125–138.

Hamilakis, Y. (2011). Archaeological ethnography a multitemporal meeting ground for archaeology and anthropology. *The Annual Review of Anthropology, 40*, 399–414.

Ilha Blue Safaris website. Retrieved from World Wide Web on August 2, 2020. https://ilhablue.com

Ministério da Cultura. (2010). *Plano de gestão e conservação da Ilha de Moçambique, partimónio cultural mundial*. Direcção Nacional do Património Cultural.

Mirabal, A. (2004). *Interim report of the marine archaeological survey performed in the province of Nampula, Mozambique, from September to December 2003*. Unpublished report, Arqueonautas Worldwide Arqueologia Subaquática S.A.

Mirabal, A. (2006). *Intermediate report on underwater archaeological excavations of the Island of Mozambique and Mongicual, from April to November 2005*. Unpublished report, Arqueonautas Worldwide Arqueologia Subaquática S.A.

Mirabal, A. (2010). *Report of the Espadarte (1558) shipwreck (IDM-002), Mozambique*. Unpublished report, Arqueonautas Worldwide Arqueologia Subaquática S.A.

Mirabal, A. (2015). *Inventory of porcelain and gold artefacts from the Espadarte (1558) wreck site (IDM-002)*. Unpublished report, Arqueonautas Worldwide Arqueologia Subaquática S.A.

PADI. (2020). https://blog.padi.com/2019/08/23/the-life-of-a-padi-divemaster-roles-and-responsibilities/. Accessed 2 Aug 2020.

Slave Wreck Project website. Retrieved from World Wide Web on August 14, 2020. https://nmaahc.si.edu/explore/initiatives/slave-wrecks-project-0

Interviews

Aly, A. Technician Museus da Ilha. Island of Mozambique. Personal interview, October 17, 2019.

Andigg, l. Diver and secretary of Association of Fishermen of the Island of Mozambique. Personal interview, October 17, 2019.

Andrade, Fisherfolk and member of fisherfolk association. Personal interview, Nampula October 21, 2020.

Artur, B. Fisherfolk and member of the Associação dos Pescadores da Ilha de Moçambique. Personal interview, Island of Mozambique-Nampula October 21, 2019.

Assane, J. Informal seller. Personal interview, Nampula October 14, 2019.

Cássimo, M. President of Association of Fishermen of the Island of Mozambique. Personal interview, Nampula October 21, 2020.

Catija, M. Entrepreneur. Personal interview, Nampula October 21, 2020.

Eduardo, S. Marine tour guide. Personal interview, Nampula 20, 2019.
Faizal, A. Informal seller. Personal interview, at Vasco da Gama Square, Nampula October 14, 2019.
Jamal, A. Fisherfolk and member of fisherfolk association. Personal interview, Nampula October 21, 2020.
Joaquim, A. Marine tour guide. Personal interview, Nampula 17, 2019.
Joaquim, G. Informal seller. Personal interview, at Vasco da Gama Square, Nampula October 14, 2019.
Mohamad, A. Fisherfolk and member of fisherfolk association. Personal interview, Nampula October 21, 2020.
Mohamad, M. Informal seller. Personal interview, at Vasco da Gama Square, Nampula October 14, 2019.
Paulo, C. Informal seller. Personal interview, at Vasco da Gama Square, Nampula October 14, 2019.
Saide, N. Boat manufactures. Personal interview, Nampula October 21, 2020.

Afterword

George Okello Abungu[1]

This handbook of 'blue heritage', edited by Rosabelle Boswell, David O'Kane and Jeremy Hills, could not have come at a more appropriate time than this, when humans must face and deal with a multitude of devastating and threatening challenges resulting from human actions. The protection and restoration of the Earth's oceans and coasts is an issue at the core of human survival today, not only for the present generation but also for generations to come, who must be protected by the actions of humanity today.

Starting on a high note, this volume begins by citing an example of a Hollywood movie that predicted a planetary doomsday, but this is not a book concerned with sensationalizing the disaster unfolding before us. Climate change, as stated from the beginning is already with humanity: there is no doubt from this book that the subjects covered (namely, the oceans and their coasts or blue heritage and human actions and impacts on them) should be of the uttermost interest and concern to all humanity. This is even made more urgent with the fast, and constantly changing, global scenarios of emerging infectious diseases (and the epidemics and pandemics they bring), and global warming. No place on earth can claim to be immune to the effects of such negative changes, especially when human actions continue to undermine the planet's health. While the writing seems to be on the wall, the numerous preceding chapters of this text have championed many potential answers to

---

[1] Professor George Okello Abungu can be reached at g.abungu@me.com.

the challenges of the threatened catastrophe, as human action or inaction may contribute to the pending disaster, or allow for it to be averted.

Oceanic heritage is at the centre of this dialogue: the ocean is not only a victim, but a potential answer to human developmental needs. However, today human and cultural valuation of the oceans is being superseded by economic and natural resource considerations, and national governments are focussing on narrowly resource-focused discourses and strategies in managing oceans. Yet the Earth's oceans and coasts are the reservoirs of a rich array of tangible and intangible cultural heritages, broader cultural resources that make us what we are. They are also the ultimate biodiversity hotspots.

The book makes it clear that the conservation of marine ecosystems, and the safeguarding of unique coastal cultures and livelihoods, are both key to protecting biological and cultural diversity. This is in addition to the role such conservation must play in sustaining the knowledge necessary for understanding and combating climate crisis. Thus, the focus of environmental conservation and protection must include respect for the seas and their coasts, as well as for all those who have played, and are playing, a part in its appreciation, sustainable use and management. They and their communities are all too often, unfortunately, relegated to the position of silenced and marginalized stakeholders.

They include the common people whom the oceans connect, and who share and cultivate oceanic heritages in diverse ways. Whether within a territory or beyond it, they inhabit these sensitive oceanic and marine areas and, by their actions, they have nurtured, and are nurturing, them. They apply their knowledge of the seas and oceans not to destroy them, but to use their resources with respect. They know that there will soon be a tomorrow with an even greater human population, who will need to inherit and use those same resources. The book is also about these people, and their transnational, shared and valued oceanic heritage, their knowledge, their histories and ethnographies: all these are needed, if we are to understand, treat and sustainably manage blue heritage. It is these people whose voices must now be made audible, and which are made audible in this book, which shows clearly how they and their communities so often share a common vision for the safeguarding of the oceans, seas and coasts, and their associated blue heritages.

The crisis emanating from human actions that have adversely affected and continue to affect the ocean environment should instil a sense of urgency in all of us, in the way we view the ocean, its resources and its sustainable utilization from local to global level. It is not lost on the contributors to this volume that 'the present reality of millions worldwide' is crisis. The fact that the Earth is on the brink of an extinction crisis, for example, evident in the rapid degradation of its oceans and coasts, is clearly illustrated across the book. And all the contributors to this volume, in their different interventions, agree on the need for urgent collective and collaborative actions to safeguard the oceans and coasts and address other challenges associated with the sea in order to tackle the crisis that is looming before us.

As this book demonstrates, the challenges of climate crisis are compounded with other recent challenges, such as the COVID-19 pandemic that has badly affected marginalized communities, if not by deaths directly due to the coronavirus, then by the economic and political disruptions that pandemic brought in its wake, including in those countries classified as Least Developed Countries (LDCs). For many of those countries, the blue economy, the forms of economic life based on the seas and oceans, is critical to all the discussions of power and power relations and their influence on decision making regarding resource utilization in contexts of widespread abject poverty, poor access to livelihoods, and the ever present threats of pollution that are so prevalent on earth today.

The oceans are the 'last frontier' for exploration and exploitation, and the fight for their resources has become real, just as the degradation of these oceans and their coasts has attracted international attention. It is in this regard that the United Nations has called, through Sustainable Development Goal 14, for the sustainable use of the oceans, a demand which has resulted in the realignment of many nation states' growth strategies with that UN SDG. The ocean's potential to advance economic growth and development, to be an avenue for influence and political patronage, as well as its role as an ever fluid and flexible bridge of cultures, has once again put it at the centre of social, cultural and political power-plays. Of course, this comes with obvious ramifications, including pollution and biodiversity loss. And as we see demonstrated in many of the chapters of this book, the 'opening' of the seas and their resources to interests that thrive on exclusion means that the marginalized in society will, again, end up as the losers. This is another reason why this book is so crucial, and comes at the right time for those truly concerned with life in the oceans and what it provides to human survival. Even just to make the point that saving what we have in nature matters means acknowledging that the ways of achieving this are complex, requiring consolidated and multilateral approaches. There is no 'one size fits all' solution for these problems, but there are many knowledge forms which can guide us to more sustainable use of the oceans and coasts.

The advocacy provided throughout the preceding chapters rests on the idea that solutions to our numerous problems call for the concerted efforts of all. These efforts must include engagement with disciplines that were, originally, considered irrelevant to such studies. The interdisciplinary approach that brings both soft and hard sciences together, with where the humanities representing the human element, is just as crucial to the finding of solutions as other contributing approaches may be. This book's rich array of examples from across the globe, the historical richness of the experiences they convey, and the contemporary issues discussed, make it an 'all rounder' document for consultation on blue heritage, blue economy, and oceanic and climate crises.

The oceans are no longer just a bridge between cultures and a means of communication and connection: they are, today, holders of resources that make crucial differences in the economy and wellbeing of nations and humanity, as well a contested political and social space whose survival is threatened, within the survival of its biodiversity being especially threatened. With more than 600 million people (10% of the planet's total population) living within a hundred kilometres of the sea, and given the dependence of humanity on the sea and coasts for food, health, livelihood, identity and trade, the ocean has become the new centre or frontier of economic and political attention and competition. The nationalization of island territories worldwide, the creation of new islands and spheres of influence, and the militarization of the oceans—these have all had their impacts on the oceans, through (for example) regional collaboration, cultural exchange and the valuation of the oceans, all issues tackled in one form or the other in this book.

This book may not be the first publication of its kind, but it is probably the first one to offer globally comparative examples of the human and cultural dimension of the world's oceans and coasts, including through ethnographic and historical analyses of the oceans and blue heritage. Faithful to its title's evocation of blue heritage, it has carefully and beautifully navigated these critical issues through the lens of the human eye, achieving an understanding and appreciation of the sea, and focusing on the human element's connection to the sea and water bodies that collectively form the blue economy. Boswell's observation that 'an important claim of the book is that humans hold diverse perspectives of the sea and coast. Inclusion of diverse, often marginalised perspectives of the sea in development, as well as consideration of marine inspired values and cultural practices deepen democracy and advance environmentally sound practices of ocean management' further captures the humanness of this matter. Who could disagree with these noble aspirations? However, the authors are also not blind to the usual ways of doing things that often hinder acceptance of multiple players and multiple voices. They are aware of the complexities of inclusion, both of people and perspectives.

There can be no doubt, after a reading of the chapters above, that the sea has become a theatre of developmental agendas, one with many players. For the first time this includes the originally marginalized and forgotten communities of the sea. With their deep knowledge of the sea, their interaction with it, and their cultural attachment to it, they now have, probably for the first time, a chance to advocate for their voices to be heard and considered. These voices are not just based on their closeness to the sea but also on their longstanding history of cultural connection, association, interaction, understanding and use of the sea and its coasts. Many of the histories of the peoples on earth have been written on the waters and beaches of the world, whether they travelled those waters as free or enslaved people, or arrived on shores and coastlines as the same. In the light of that history, the principle of inclusion of many voices, rather than their exclusion, has become, rightfully, a main focus of this book.

Through its discussions on climate and other crises, this book brings out deep experiences and diverse challenges across the globe. The climate crisis is very real, especially where global warming's effects on blue heritage is concerned. It is with us and requires our concerted attention, including new approaches that may be different from the usual and 'normal' way of doing things. It proposes that to deal with these mostly manmade crises, we must consider transdisciplinary approaches that recognize the different facets of the issues at hand, and the complex nature of the ocean, its users, uses and emerging potentials at all levels. Regarding this, the book offers critical analysis and case studies of oceanic and coastal heritages and how they have influenced various human developments in Africa, the Pacific, Asia, and Europe.

Policy-informed action vis-à-vis oceanic sustainability is more effective when based on insights derived from transdisciplinary research that does not exclude the human historical and contemporary engagement with the oceans and coasts. This humanization of the solution to the numerous challenges is critical to any attempts made to address the same. Thus, in this case (and possibly for the first time) the dialogue on blue heritage and the blue economy has succinctly recognized the role of the humanities and social sciences in an effort to address the challenges of sustainable oceans management.

Management of the oceans and the seas is often carried on through national policies rooted in western scientific discourse and attitudes of intellectual colonialism vis-à-vis indigenous conceptions (traditional knowledge systems) of the oceans and social identity. The questions that then arises is whether national or global policies and practices should always conflict with the local peoples' practices, needs and expectations? What priorities should national governments in the Global South embed in their oceans management strategies, so as to be inclusive of the expectations and aspirations of coastal, ocean-reliant communities? These are just two of the pertinent questions we need to ask if we are to provide answers to the sustainable conservation and management of the oceans and the coasts, and ensure their protection for the benefit of present and future communities.

With its multitude of examples and case studies, and its critical analyses of the issues, this book has indeed succeeded in its aim of documenting and analysing case studies of 'blue heritage' in the world, and of demonstrating the relevance and meaning of the oceans and coasts to human history, culture and identity from the sixteenth century to the present. A convincing case has been made, throughout this book, for coastal areas as rich historical and cultural archives and sources of knowledge regarding future oceans management. Indigenous holistic perspectives and experiences of the sea have been explicated, and their critical relevance to a better future and more rewarding blue economy has been demonstrated.

To conclude: this is a book that is relevant and useful to diverse audiences: academics in the field of ocean studies, policymakers and practitioners engaged in research on the ocean economy, as well as graduate scholars in the ocean sciences including blue economy studies, oceans sustainability, coastal resilience studies and oceans conservation and heritage studies. It will also be relevant reading for specialists in the natural sciences who have an interest in the Earth's oceans and coasts. Academic associations, professional bodies, international organizations dealing with culture, climate change and oceans will also find this a most valuable source of information. Boswell, O'Kane and Hills have put in a tremendous effort in realizing this all encompassing book on blue heritage, human development and the global climate crisis.

George Okello Abungu is Professor of Archaeology, a Cambridge-trained archaeologist, and a former Director General of the National Museums of Kenya. He is CEO of Okello Abungu Heritage Consultants and a recipient of the Lifetime Achievement in Defense of Art from the Association for Research into Crimes Against Art (ARCA). He is also a Knight of the Order of Arts and Letters (Chevalier de l'Ordre des Arts et des Lettres) of the French Republic for his outstanding contribution to heritage at local and global levels as well as the first African recipient of the World Heritage Fund Award for his contribution to capacity building in the field of heritage in Africa. George

Abungu has researched, published and taught in the disciplines of archaeology, heritage management, and museology, culture and development. He was Kenya's representative to the UNESCO World Heritage Committee, and Vice-President of its bureau (2004–2008). He is founding Associate Professor of the M.A. in Heritage Management at the University of Mauritius and a fellow of the Stellenbosch Institute for Advanced Studies at the University of Stellenbosch, South Africa.

Index

A

African American 111
Artisanal sea-based fisheries 358
Asafo groups 359

B

black hydropoetics 35
Blue Economy (BE) 2, 384, 385, 395, 403
blue heritage 5, 8, 9, 257, 261, 268, 269
Bombay 76, 101
British India 99
built heritage. *See* heritage
Burger, Lynton 141

C

climate change 1, 3, 6, 7, 9, 33, 44–46, 156, 163, 171, 384, 389, 390, 398, 405, 411, 431, 433, 434, 439, 445, 446
coastal Ghana. *See* Asafo groups
colonial Orissa 104

Comunidade 75
Convention on Biological Diversity (CBD) 328
COP26 6
Covid-19 2
critical view of heritage 7
cultural diversity 4, 5
customary law 325–332, 334–336, 338–343, 346, 348–352, 359, 360, 375, 376

D

decolonisation 42

E

ecology 78, 143
Empatheatre 384
enslaved Africans 114
epistemologies 36
    hydro-epistemologies 36

F

Faris, J.H. 162
fishing and tourism 156

G

Ghana 332. *See also* small-scale fishers
Ghana Chieftaincy Act. *See* customary law
Goan Catholic society 75
Gulf of Cambay 93

H

Haida Gwaai 143
heritage 1, 4, 77, 80, 86, 88, 106, 162, 169, 171, 251, 252, 254, 257, 260, 267, 268, 270, 303–305, 307, 313, 315, 318–320, 386, 390, 396, 400, 402, 413, 414, 416, 417, 421–423, 427–430, 436, 438–447
heritage tourism 135
human development 2, 4, 9
human evolution 3, 7, 33, 41
hydrocosmologies 36, 37

I

Indian Ocean 94
India under British Imperial rule 8
Intergovernmental Panel on Climate Change (IPCC) 3
international biodiversity law 332
international human rights law 327
islands
    Malta 25

J

Jason DeCaires Taylor 33

K

Karoo 142, 144, 150, 151. *See also* Southern Africa
Khoisan 37, 141, 144–146, 149–151, 405

L

Lalela uLwandle 8, 383, 391–393, 395, 396, 398–403, 405
Least Developed Countries (LDCs) 2
littoral 33

M

Marine Spatial Planning 403
Mauritius 8, 25, 164
memory and belonging 79
Mozambican. *See* Mozambique
Mozambique 46, 76, 82, 142, 143, 267
Multispecies 34, 35
Mutwa, Credo 149, 155. *See also* Southern Africa

N

new epistemologies. *See* epistemologies
North Carolina 109

O

ocean governance. *See* Marine Spatial Planning
ocean health 5–7, 386, 390
oceanic humanities 7, 32, 34, 165

P

Plural epistemologies 390
Portsmouth 120
Postcolonial theory 36

S

salt 102
salt works 102
SDG 14 4, 169
sea creatures 141
Sea Customs Acts 101
self-reflexive 9, 163, 173, 415
Seychelles 8, 203
Shell Castle 113
Shona beliefs 148
Sierra Leone 303
Sisters of Ocean and Ice 162
slaves 36, 82, 168, 309
small-scale fishers 226, 326, 329, 332, 333, 336–339, 343, 346, 386, 389
Southern Africa 149
spiritual beings 141
Star of Zanzibar. *See* Zanzibar
Sustainable Fisheries Management Project (SFMP) 358
sustainable ocean management 7

T

tenure rights 330
the remittance system 100
Tourism 27

U

UN Decade of Ocean Science for Sustainable Development 4
underwater. *See* underwater cultural heritage
underwater cultural heritage 8, 43, 191, 427–430, 436, 438–447, 451
UNESCO 4, 41, 44, 45, 428, 429, 444, 445

V

vernacular architecture of Goa. *See* vernacular sensibility
vernacular sensibility 77

W

Western India
   British India 102
wet ontologies 32
World Heritage Sites 2
World Marine Heritage Sites in Africa 2

Z

Zanzibar 76